ADVANCES IN MECHANICAL ENGINEERING RESEARCH

VOLUME 3

ADVANCES IN MECHANICAL ENGINEERIG RESEARCH

Additional books in this series can be found on Nova's website under the Series tab.

Additional E-books in this series can be found on Nova's website under the E-books tab.

ADVANCES IN MECHANICAL ENGINEERIG RESEARCH

ADVANCES IN MECHANICAL ENGINEERING RESEARCH

VOLUME 3

DAVID E. MALACH
EDITOR

Nova Science Publishers, Inc.
New York

For permission to use material from this book please contact us:
Telephone 631-231-7269; Fax 631-231-8175
Web Site: http://www.novapublishers.com

LIBRARY OF CONGRESS CATALOGING-IN-PUBLICATION DATA

ISSN: 2159-1989

ISBN: 978-1-61209-243-0

Published by Nova Science Publishers, Inc. + New York

CONTENTS

PREFACE

Mechanical engineering is an engineering discipline that applies the principles of physics and materials science for analysis, design, manufacturing, and maintenance of mechanical systems. This book presents current research in the field of mechanical engineering, including the fatigue and residual stresses during welding of titanium alloys; aircraft engineering and bird strike analysis; modeling to predict ultra-low cycle fatigue fracture of steel bridges; marine engine emissions and the fatigue behavior of composites processed by resin transfer molding processes.

Chapter 1 - Air Transportation growth increases continuously over the years. The increase in air transport activity is accompanied by an increase in the amount of energy used to provide transportation services, and it is assumed to increase environmental impacts. Aircraft pollutant and noise emissions are an important source of pollution and directly, or indirectly, cause harmful effects to human health, ecosystems, and cultural heritage. Any kind of airport brings negative effects to a local environment by means of different factors including noise disturbance, air pollution, and others. Sustainable air transport development necessitates an implementation of measures aimed to reduce environmental impacts around airports.

This chapter presents the results of investigations into noise abatement operational procedures and air pollution control. Obtaining optimal flight procedures for take-off and climbing / approach and landing, aims to minimize noise, pollutants and fuel consumption. The analysis presented compares standard operational procedures performed nowadays which have been optimized using Integrated Noise Model. Noise and pollutants abatement procedures (NPAPs) are developed and considered as a necessary measure of a balanced approach for annoyance control in particular around airports. The presented results fit with environmental challenges of ACARE vision and of ICAO recommendations. Although noise abatement procedures may have quantifiable environmental benefits, effective implementation may be difficult: procedures must be developed, tested, and evaluated for benefits and ATC impacts; approved and accepted by the airport and the ANSP (Air Navigation Service Provider), and adopted by the airlines and other airport users.

Evaluation of an airport's impact on surrounding environment in terms of noise and air pollution could be achieved by defining the environmental capacity of an airport. Environmental capacity means such service conditions of airport operations and development that ensure fulfillments of normative conditions of noise and air pollution at the control points around the airport.

In addition, there are many methods to assess pollutant and noise emissions used by various countries using different and separately methodology. This causes a wide variation in results. Some lack of information, and the use of certain methods will require justification and reliability that must be demonstrated and proven. In relation to this issue and because there are no methods for the combination for assessing Aircraft Pollutant and Noise emissions, this contribution identifies, improves, develops, and combines the methodology of pollutant and noise emission integrated with the ICAO concept of the Balanced Approach for short and long term predictions. This analysis should help Airport Authorities, Air Traffic Control, and political decision makers to decide, to manage, and to get reliable information of the impacts of aircraft pollutant and noise emissions in short-term and long-term prediction.

Chapter 2 - Titanium alloys are very suited for highly specialized applications such as naval constructions where resistance to corrosion is the most important issue to be considered in materials selection, aerospace structures where the primary goal is to achieve high specific strength, and orthopedic prosthetic treatments where biocompatibility between implants and human tissues is mandatory.

Welding of titanium alloys is still considered a fairly uncommon process because industrial applications of these materials cover only the above mentioned niches where the unique properties of titanium and its alloys represent the best option in terms of minimizing the lifetime cost of the structure yet meeting stringent requirements on structural integrity and safety. However, whilst high quality welded components without inclusions and distortions and with a mild notch at the weld toe can be obtained through appropriate selection of welding process parameters, it should be considered that the weld cord is anyhow a geometric discontinuity in the structure, which modifies stress distribution. Furthermore, residual stresses are an unavoidable consequence of the thermal welding cycle.

Fatigue strength of welded joints, the most important information needed in aerospace applications of titanium alloys, is obviously lower than its counterpart for the base material. In spite of this, data available in literature are still not sufficient to make design procedures general and reliable. For this reason, in order to investigate the mechanical behavior of titanium welded joints and to build a technical database useful to designers, this chapter presents some fatigue strength curves recently obtained for butt joints made of Titanium grade 2 and grade 5. The joints tested in this research are welded by using different techniques. Experimental results of fatigue tests are discussed in terms of both nominal stress amplitude σ_a and local strain amplitude ε_a.

Fatigue curves determined experimentally are then compared with fatigue design curves reported in a very recent AWS standard that was specifically released for titanium welded joints. The role played by weld seam geometry is studied with respect to the fatigue strength reduction factor.

In order to complete the experimental analysis, residual stresses are measured by means of the hole drilling method and then correlated with mechanical response of welded components and microstructure modifications.

In addition to experimental tests, very detailed finite element analyses are carried out. Numerical results are in good agreement with the experimental evidence and confirm the importance of considering properly the interactions between weld seam geometry, defects eventually included in the weld and changes in microstructure induced by welding thermal cycles.

Chapter 3 - Bird strike is a major threat to aircraft structures, as a collision with a bird during flight can lead to serious structural damage. Computational methods have been used for more than 30 years for the bird-proof design of such structures, being an efficient tool compared to the expensive physical certification tests with real birds. At the velocities of interest, the bird behaves as a soft body and flows in a fluid-like manner over the target structure, with the high deformations of the spreading material being a major challenge for computational simulations based on the finite element method. This chapter gives an overview on the development, characteristics and applications of different soft body impactor modeling methods by an extensive literature survey. Advantages and disadvantages of the most established techniques, which are the Lagrangian, Eulerian or meshless particle modeling methods, are highlighted and further topics like the appropriate choice of impactor geometry or material model are discussed.

Chapter 4 - The concept of ultra-low cycle fatigue (ULCF) was recently originated with some of sudden failures of existing bridges, which were characterized by large scale cyclic yielding due to occasional loadings such as earthquakes, typhoons. Generally, experimental approaches are popular for ULCF fracture prediction. As for the authors view, only one theoretical study has been published in year 2007, and the observed failure mechanism is based on void growth process. The fracture is calculated to occur when cyclic void growth index (VGI_{cyclic}) exceeds its critical value. The VGI_{cyclic} demand is calculated based on complex integrations of a function, which depends on triaxiality and incremental plastic strain. However, it is required to modify commonly available FEM programs to cater this integration and finally it hindered the usage of general propose FEM packages as it is to estimate ULCF fracture. As a result, found applications of this fracture criterion are very less.

Therefore, this chapter presents a simplified model to assess the real ULCF fracture of steel bridges using available general-purpose FEM packages. However, this approach is limited to the situation where triaxiality remains relatively constant during its loading history. As highlighted in previous studies, in many realistic situations, this statement can be applicable. The fracture mechanism of this model is also similar as previous model. But the fracture criterion is totally different from the previous approach such that the ULCF fracture is calculated to occur when significant plastic strain exceeds its critical value. As for alternative component of simplified ULCF fracture criterion, a simple non-linear hardening model was also proposed to employ in elasto-plastic analysis for the situations, where parameters of available mix hardening models are difficult to determine. The behavior of the proposed hardening model was verified with experimental behavior of few materials. The verification of the simplified fracture criterion was done by comparing the results with previous criterion-based estimations of some bridge components and hence importance of this model was clearly illustrated. Finally, chapter tends to conclude that the proposed model gives reasonably accurate prediction to ULCF fracture of steel structures where triaxiality remains relatively constant during the loading history.

Chapter 5 - Emissions originating from maritime transport in European waters continue to increase at such a rate that the foreseeable evolution may exceed the levels previously forecast. MARPOL, in ANNEX VI, establishes certain conditions for marine engines constructed from the year 2000 with the goal of reducing emissions of NO_x. Although the manufacturers of marine diesel engines make considerable efforts to design and manufacture low emission engines, it is very possible that a maladjustment of the injection system may cause an increase of emissions that has not been seen when the engine is subjected to bench trials. However, no reference at all is made to the limits that are likely to be exceeded when the engine is actually in operation with poor maintenance conditions.

This chapter describes the engine parameters which affect the in-cylinder formation of emissions. It also describes how maladjustment in some parameters like the injection system, changes in the injection pressure, the state of conservation of the nozzles and changes in the scavenging air pressure and temperature influence the emissions. With this objective, some studies have been described on particular engines with both normal or maladjustment conditions of all of parameters mentioned above.

Chapter 6 - Structural polymer composites have been widely applied in the aeronautical field. However, composite processing, which uses unlocked molds, should be avoided in view of the tight requirements in service and also due to the possibility of environmental contamination. To produce composite aeronautical structural components with low cost, the aircraft industry has shown interest in resin transfer molding process (RTM) as an adequate option to substitute for conventional process with the advantages of faster gel and cure times, besides the low percentage of voids and high fiber volume percentage, which are the essential parameters to design aircraft structures. Since the low viscosity resin is injected into the closed mold, in this process, the edge effect can promote incomplete wetting of the fiber reinforcement, dry spot formation and other defects in the final composite. Knowledge of material behavior is essential to design structures as aircraft landing gear, for example. Compared with isotropic materials, polymeric fibrous composites submitted to cyclic loading present a degradation phenomenon of mechanical properties as a consequence of residual stress redistribution into the structure. It was established that fatigue mechanisms associated to fibrous composites occur in four stages: nucleation of local damage, stable propagation of crack due to the cyclic load, local crack propagation which is dependent of fibers orientation, the matrix ductility and the interfacial adhesion and propagation of last loading cycle, which is analogous to the tensile test fail. Because of the many mechanisms involved during the degradation of the composites, scatter in fatigue is higher and special care need to be taken when approach the S-N curve. In this chapter, the review of data presented in the literature with focus in the fatigue behavior of polymeric composites reinforced with carbon fiber processed by RTM are compared with some experimental data obtained during three years of study. As a specific aim this chapter proposes a new methodology for fatigue behavior on composites, as this field lack of reliable predictive methods, and the main drawbacks of composites applied in aircraft structures.

Chapter 7 - This chapter presents the model of the components of a hybrid propulsion system and their integration of a light urban vehicle. This concept of vehicle originated from a specific requirement to design a vehicle with a concrete application: to collect used batteries from urban bus-shelters. The propulsion system is a series hybrid configuration with an internal combustion engine as the main charge source for the power batteries feeding an electric motor. The system is also fitted with a regenerative brake and solar panels to recharge the batteries.

The models of the components try to describe the internal processes that take place in them, so they are modelled in detail. More specifically, a complete internal combustion engine is developed considering the following submodels: Air intake model, Fuel injection model, Air-fuel mixture model, Combustion model, Vehicle dynamics model, Exhaust Gas

Recirculation (EGR) model and Vapour Canister Purge (VCP) model. The model of the batteries gives information on the state of charge (SOC) at any time and calculates the changes in stored energy as a result of: charge and discharge cycles, self-discharge and variations in temperature. Finally, control strategies are proposed.

The model can be used with two main purposes:

- During the design phase, it can be used for components selection according to the initial specifications.
- It can also be used to evaluate the vehicle performance under different operating conditions.

Chapter 8 - The combustion by compression of a lean and homogeneous (or premixed) air-fuel mixture has emerged in the last few years as an effective alternative to achieve simultaneous reductions in nitrogen oxides (NOx) and particulate matter (PM) in internal combustion engines. The main subject of this chapter focuses its attention on the analysis of different strategies that promote the premixed burn phase in a heavy duty diesel engine, typical of those used in on-road transportation, with the aim of reducing NOx and PM emissions simultaneously.

The methodology used in this study is based on a parametric experimental study of these strategies. Using both, cylinder pressure measurements and operating conditions of the engine, the combustion diagnosis is carried out. This makes it possible to obtain relevant parameters for the combustion analysis such as pressure evolution, mean gas temperatures, adiabatic flame temperature, premixed burnt mass fraction, fifty percent of heat release, etc. When all these parameters are obtained systematically, the way in which the operating conditions affect the combustion process and the relationship between these parameters and the performance of the engine (measured in terms of efficiency and emissions) can be properly analysed.

Even though a completely premixed burn could not be achieved, the studied strategies have proven to be effective ways of achieving a simultaneous reduction in NOx and PM. Nevertheless, there was also an increment in the HC and CO emissions and higher fuel consumption.

Chapter 9 - Aircraft maintenance technical orders are guidelines for aircraft system maintenance and repair. Maintenance technicians must follow instructions of technical procedures step by step. For aviation safety the writing quality of technical orders must be enhanced to provide aircraft maintenance technicians with a reliable management system. This study adopts the Six Sigma Process improvement procedure known as DMAIC. The readability index is defined as readability of statements written in maintenance technical orders, while the importance index is defined as the significance of aviation safety obtained through composing maintenance technical orders. These indices evaluate quality and vigilance in writing maintenance technical orders. The DMAIC simultaneously rates and analyzes maintenance of technical orders with a high satisfactory value on the importance index, but low on readability. It proposes a way to enhance writing quality while devising a management mechanism for maintenance technicians in order to ensure quality technical procedures to enhance aviation safety.

Chapter 10 - Infrastructures to provide access to custom integrated hardware manufacturing facilities are important because they allow Students and Researchers to access professional facilities at a reasonable cost, and they allow Companies to access small volume production, otherwise difficult to obtain directly from manufacturers. This paper is reviewing the developments at CMP to offer various types of MEMS manufacturing to Students, Researchers and Companies since 1994. CMP has been the first service of its type to introduce MEMS fabrication. Today CMP is offering bulk micromachining on CMOS and GaAs, and various MEMS processes. CAD tools provided by CMP are also reviewed.

In: Advances in Mechanical Engineering Research, Volume 3 ISBN: 978-61209-243-0
Editor: David E. Malach

Chapter 1

OPTIMIZATION STRATEGIES REDUCING AIRCRAFT NOISE, FUEL CONSUMPTION AND AIR POLLUTION: LOW-NOISE AND LOW-POLLUTANT FLIGHT PROCEDURES

S. Khardi with the collaboration of J. Kurniawan and O. Konovalova
INRETS – LTE. Laboratoire Transports et Environnement. 25 avenue François Mitterrand, case 24, 69675 BRON Cedex – France

Air Transportation growth increases continuously over the years. The increase in air transport activity is accompanied by an increase in the amount of energy used to provide transportation services, and it is assumed to increase environmental impacts. Aircraft pollutant and noise emissions are an important source of pollution and directly, or indirectly, cause harmful effects to human health, ecosystems, and cultural heritage. Any kind of airport brings negative effects to a local environment by means of different factors including noise disturbance, air pollution, and others. Sustainable air transport development necessitates an implementation of measures aimed to reduce environmental impacts around airports.

This chapter presents the results of investigations into noise abatement operational procedures and air pollution control. Obtaining optimal flight procedures for take-off and climbing / approach and landing, aims to minimize noise, pollutants and fuel consumption. The analysis presented compares standard operational procedures performed nowadays which have been optimized using Integrated Noise Model. Noise and pollutants abatement procedures (NPAPs) are developed and considered as a necessary measure of a balanced approach for annoyance control in particular around airports. The presented results fit with environmental challenges of ACARE vision and of ICAO recommendations. Although noise abatement procedures may have quantifiable environmental benefits, effective implementation may be difficult: procedures must be developed, tested, and evaluated for benefits and ATC impacts; approved and accepted by the airport and the ANSP (Air Navigation Service Provider), and adopted by the airlines and other airport users.

Evaluation of an airport's impact on surrounding environment in terms of noise and air pollution could be achieved by defining the environmental capacity of an airport. Environmental capacity means such service conditions of airport operations and development that ensure fulfillments of normative conditions of noise and air pollution at the control points around the airport.

In addition, there are many methods to assess pollutant and noise emissions used by various countries using different and separately methodology. This causes a wide variation in results. Some lack of information, and the use of certain methods will require justification and reliability that must be demonstrated and proven. In relation to this issue and because there are no methods for the combination for assessing Aircraft Pollutant and Noise emissions, this contribution identifies, improves, develops, and combines the methodology of pollutant and noise emission integrated with the ICAO concept of the Balanced Approach for short and long term predictions. This analysis should help Airport Authorities, Air Traffic Control, and political decision makers to decide, to manage, and to get reliable information of the impacts of aircraft pollutant and noise emissions in short-term and long-term prediction.

Keywords*:* Aircraft pollutant and noise emission, balanced approach for Aircraft noise management, noise abatement operational procedures, air pollution, airports.

1. Aviation and the Environment

Transportation growth increased continuously over the years. However, the growth has not uniform in the various transport modes and sectors, and varies from country to country. The general increase in transport activity has been accompanied by an increase in the amount of energy used to provide transportation services. The increase in air transport activity and energy consumption is assumed to increase environmental impacts. This also contributes significantly to greenhouse gas emissions. The continuing growth in air traffic and increasing public awareness have made environmental considerations one of the most critical aspects of commercial aviation today. It is generally accepted that significant improvements to the environmental acceptability of aircraft will be needed if the long-term growth of air transport is to be sustained. This is an open issue. The Intergovernmental Panel on Climate Change has projected that, under an expected 5% annual increase in passenger traffic, the growth in aviation-related nuisances will outpace improvements that can be expected through evolutionary changes in engine and airframe design.

Any kind of airport brings negative effect to a local environment by means of different factors including noise disturbance, air pollution and others. Evaluation of an airport impact on surrounding environment in terms of noise and air pollution could be realized by defining environmental capacity of an airport. Environmental capacity means such service conditions of airport operations and development that ensure fulfillments of normative conditions of noise and air pollution at the control points around airport. Sustainable air transport development necessitates the implementation of measures aimed at reducing the aircraft noise nuisance at airports with particular noise problems. Among environment concerns, excessive aircraft noise and its control has become a major objective of airport authorities. The capacity of airports, particularly in Europe, is eliminated by noise impact, and it can be increased only with the implementation of effective transport management system, which reduces this impact on the local community.

During recent decades, aircraft noise levels have been successively reduced up to 20 dB. Nevertheless, the large numbers of people who live in communities near airports are affected by aircraft noise which has increased tremendously in scope. Decisions have been made to enable the choice of possible solutions of aircraft noise control around airports (Bies, 2003A; Antoine, 2004; Burton, 2004; Crocker, 2007). Several procedures have been used in the worldwide aircraft operation such as low-noise during the take-off and landing flight procedures, optimal route distributions, flight route optimization around airports, etc. (Clarke et al., 2006; ECAC, 1997; Directive 2002). Nevertheless, the noise in the vicinity of airports, in particular under the take-off and landing flight paths remains high disrupting the quality of life of local residents. Complaints are increasing despite the withdrawal of the noisiest aircraft, the fleet renewal, and the vote of international resolutions recommending that airports, faced with the problem of noise, introduce restrictions of operations.

The aircraft manufacturers foresee a demand for aircraft to cope with the increased traffic and fleet renewal in the coming years. The proportion of heavy-lift is progressing towards almost half the fleet. This growth will differ by two essential characteristics: 1. mass transport anticipating the scarcity of take-off slots; 2. transport playing on the increasing frequency and flexibility of operations continuing despite the anticipated shortage of oil. All the experts agree that around 2020, taking into account the known oilfields and the potential extraction, the production of oil will reach a maximum level and then decrease especially with the growing economic power of China and India. Whatever the efforts to conserve energy and promote renewable energy, air transport will continue to grow, even with very expensive oil. This problem can only be solved within the framework of a global vision for sustainable development involving new technology engines and fuselages, breakthrough technologies, the design of new procedures and flight paths (Zaporozhest and Tokarev, 1998; Zaporozhest and Khardi, 2004), airspace management, new regulation rules and certification (ECAC, 1997). Sustainable air transport development necessitates the implementation of measures aimed at reducing the aircraft noise nuisance at airports with particular noise problems.

Noise: While considerable progress has been made to reduce the noise signature of airliners, the public's perception of noise continues to grow, as illustrated by the ever-increasing number of public complaints. This can be attributed to increasing air traffic as well as further encroachment by airport-neighboring communities. As a result, noise has become a major constraint to air traffic, with 60% of all airports considering it a major problem and the nation's fifth largest airports viewing it as their biggest issue (NASA, 2003). The construction of new runways and airports raises massive issues due to public fears of increased air traffic and the associated louder, or more frequent, noise. In response to these public concerns, airports have adopted operational restrictions on top of the International Civil Aviation Organization (ICAO, 1993, 2002, 2004, 2006, 2007-2009) certification guidelines. A survey of the world's airports reveals a two-fold increase in the number of noise related restrictions in the past ten years (FAA, 1983, 1997, 2005, 2009). These include curfews, fines, operating restrictions, and quotas.

The historical trend in aircraft noise has shown a reduction of approximately 20 dB since the 1960s (Smith, 1960) largely due to the adoption of high bypass turbofans and more effective lining materials. Reductions since the mid-eighties have not been as dramatic (Figure 1). The point seems to have been reached where future improvements through technological advances will be possible only by significantly trading off operating costs for environmental performance. As shown on the notional graph in Figure 1.3, the outlook is that further reductions in the environmental impact of commercial aircraft will exact increasingly severe penalties in operating costs.

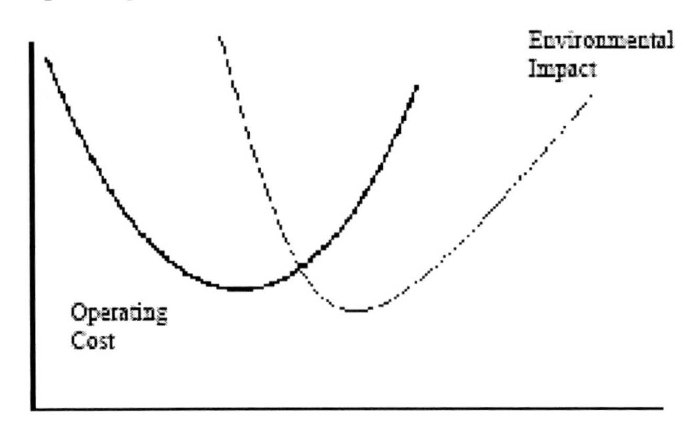

Figure 1. Technological advances reduce the environmental impact of aircraft but only at rising operating costs.

The challenge is to determine the designs offering the optimal trade-off between operating and environmental performances.

The ICAO Assembly has endorsed the concept of a 'Balanced Approach' that aims to address noise issues by working simultaneously on four parameters: aircraft noise at the source, flight and operating procedures, operating restrictions at airports, and land-use planning and management (Figure 2).

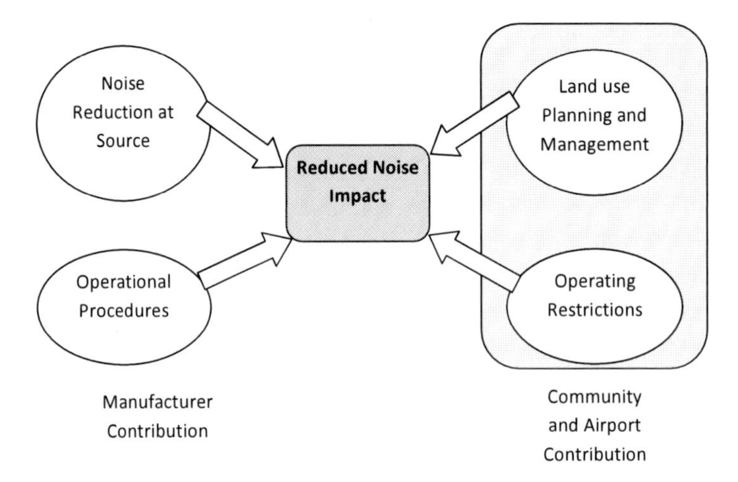

Figure 2. The ICAO Balanced Approach: successfully reducing the noise impact of commercial aircraft on communities must include contributions from the manufacturers, airports, and communities.

ICAO Balanced Approach concept consists of short term and long term measures. Short term measures must facilitate the solution at local level (noise abatement procedures and mitigation of aircraft operations). Long term measures must force the solutions at the regional level (reduction of noise and pollutant emissions at source, phase-out the solution of non certificated airplanes, noise and pollutant emissions charge and land use planning and management). In any case huge relationship between them must be provided, because of reaching more efficiency and advantages.

Pollutant emissions: The release of exhaust gasses in the atmosphere is the second major environmental issue associated with commercial airliners. The world fleet releases approximately 13% of CO_2 emissions from all transportation sources, or 2% of all anthropogenic sources (IPCC, 1999). The expected doubling of the fleet in the next twenty years (Boeing) will certainly exacerbate the issue: the contribution of aviation is expected to increase by factor of 1.6 to 10, depending on the fuel use scenario (Peeters et al., 2005). Conscious of this problem, engine manufacturers have developed low-emission combustors, and made them available as options. These combustors have been adopted by airlines operating in European airports with strict emissions controls, in Sweden and Switzerland, for example. The following figure, presents progress made with some individual pollutants than with others. Aircraft emissions have also declined over time when considering the emissions from transporting one passenger one mile.

Current emission regulations have focused on local air quality in the vicinity of airports and the research will also focus on this area. Emissions released during cruise in the upper atmosphere are recognized as an important issue with potentially severe long-term environmental consequences, and ICAO is actively seeking support for regulating them as well.

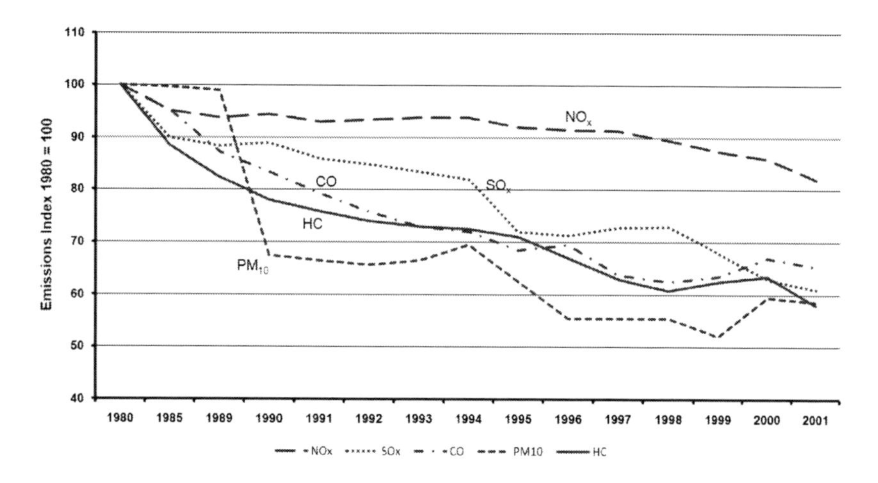

Figure 3. Local air quality pollutants have declined steadily over the past several years. NO_x has been the most challenging pollutant to constrain and progress has lagged that of other pollutants (FAA 2005).

Operations of aircraft are usually divided into two main parts (EEA 2000):

- The Landing Take-off (LTO) cycle which includes all activities near the airport that take place below the altitude of 3000 ft. This includes taxi-in and out, take-off, climb-out and approach-landing (ICAO, 1993).
- Cruise which is defined as all activities that take place at altitude above 3000 ft. No upper limit altitude is given. Cruise includes climb to cruise altitude, cruise, and descent from cruise altitudes.

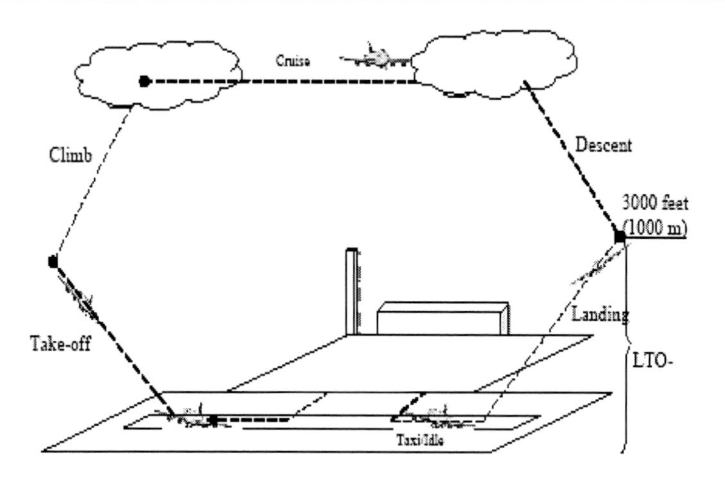

Figure 4. Landing Take-off (LTO) Cycle.

2. Identification of Methods of Assessments and Measurements of Pollutants

Emissions from aircraft originate from fuel burned in aircraft engines. Aircraft jet engines produce carbon dioxide (CO_2), water vapor (H_2O), Nitrogen Oxides (NO_x), Carbon Monoxide (CO), Oxides of sulfur (SO_x), unburned or partially combusted hydrocarbons (also known as volatile organic compounds (VOC)), particulates and other trace compounds. A small subset of the VOCs and particulates are considered hazardous air pollutants (HAPs). Aircraft engine emissions are roughly composed of about 70 percent CO_2, a little less than 30 percent H_2O, and less than 1 percent each of NO_x, CO, SO_x, VOC, particulates, and other trace components including HAPs. Aircraft emissions, depending on whether they occur near the ground or at altitude, are primarily considered local air quality pollutants or greenhouse gases, respectively.

Water in the aircraft exhaust at altitude may have a greenhouse effect, and occasionally this water produces contrails, which also may have a greenhouse effect. About 10 percent of aircraft emissions of all types, except hydrocarbons and CO, are produced during airport ground level operations and during landing and takeoff. The bulk of aircraft emissions (90 percent) occur at higher altitudes. For hydrocarbons and CO, the split is closer to 30 percent ground level emissions and 70 percent at higher altitudes. Aircraft are not the only source of aviation emissions. Airport access and ground support vehicles produce similar emissions. Such vehicles include traffic to and from the airport, ground equipment that services aircraft, and shuttle buses and vans serving passengers. Other emissions sources at the airport include auxiliary power units providing electricity and air conditioning to aircraft parked at airport terminal gates, stationary airport power sources, and construction equipment operating on the airport. Emission from Combustion Processes CO_2 – Carbon dioxide is the product of complete combustion of hydrocarbon fuels like gasoline, jet fuel, and diesel. Carbon in fuel combines with oxygen in the air to produce CO_2. H_2O – Water vapor is the other product of complete combustion as hydrogen in the fuel combines with oxygen in the air to produce H_2O. NO_x – Nitrogen oxides are produced when air passes through high temperature/high pressure combustion and nitrogen and oxygen present in the air combine to form NO_x. HC – Hydrocarbons are emitted due to incomplete fuel combustion. They are also referred to as volatile organic compounds (VOCs). Many VOCs are also hazardous air pollutants. CO – Carbon monoxide is formed due to the incomplete combustion of the carbon in the fuel. SO_x – Sulfur oxides are produced when small quantities of sulfur, present in essentially all hydrocarbon fuels, combine with oxygen from the air during combustion.

Particulates – Small particles that form as a result of incomplete combustion, and are small enough to be inhaled, are referred to as particulates. Particulates can be solid or liquid. Ozone – O_3 is not emitted directly into the air but is formed by the reaction of VOCs and NO_x in the presence of heat and sunlight. Ozone forms readily in the atmosphere and is the primary constituent of smog. For this reason it is an important consideration in the environmental impact of aviation. Compared to other sources, aviation emissions are a relatively small contributor to air quality concerns both with regard to local air quality and greenhouse gas emissions. While small, however, aviation emissions cannot be ignored. Emissions will be dependent on the fuel type, aircraft type, engine type, engine load and flying altitude. Two types fuel are used. Gasoline is used in small piston engines aircraft only. Most aircraft run on kerosene and the bulk of fuel used for aviation is kerosene. In general, there exist two types of engines; reciprocating piston engines and gas turbines (Olivier, 1991) and European Environmental Agency - EEA (2000).

The International Civil Aviation Organization (ICAO) is responsible for a worldwide planning, implementation, and coordination of civil aviation. ICAO sets emission standards for jet engines. These are the basis of FAA's aircraft engine performance certification standards, established through EPA regulations. Currently ICAO has covered three approaches to quantifying aircraft engine emissions: two in detail and one in overview:

a) Simple Approach is the least complicated approach, requires the minimum amount of data, and provides the highest level of uncertainty often resulting in an over estimate of aircraft emissions. This approach considers the emission of NO_x, CO, HC, SO_2, CO_2.
b) Advanced Approach reflects an increased level of refinement regarding aircraft types, EI calculations and TIM. This approach considers the emission of NO_x, CO, HC, SO_2.
c) Sophisticated Approach is provided in overview, will be further developed in an update of this guidance (e.g. CAEP/8) and is expected to best reflect actual aircraft emissions.

a) Simple Approach Model:

Emission of species X = ∑([Number of LTO cycles) × (Emissions Factor) *(in kg) All Aircraft of Aircraft Y for Species X.*

This equation does not account for specific engine types, operational modes or Time in Modes (TIM) as it assumes that the conditions under study are the same or similar to the default data being used.

b) Advanced Approach: $E_{ij} = \sum (TIM_{jk} \times 60) \times (FF_{jk}/1000) \times (EI_{jk}) \times (NE_j)$

- E_{ij} = Total emissions of pollutant i (e.g. NO_x, CO, HC), in grams, produced by aircraft type j for one LTO cycle.
- EI_{jk} = The emission index for pollutant i (e.g. NO_x, CO, or HC), in grams per pollutant per kilogram of fuel (g/kg of fuel), in mode k (e.g. takeoff, climb out, idle and approach) for each engine used on aircraft type j.
- FF_{jk} = Fuel flow for mode k (e.g. takeoff, climb out, idle and approach), in kilograms per second (kg/s), for each engine used on aircraft type j.
- TIM_{jk} = Time-in-mode for mode k (e.g. idle, approach, climb out, and takeoff), in minutes, for aircraft type j.
- NE_j = Number of engines used on aircraft type j.

ICAO does not have emission certification standards for SO_x. However, SO_x emissions are a function of the quantity of sulphur in the fuel: $E_k = \sum (TIM_k \times 60) \times (ER_k) \times (NE_k)$

- E_k = Total emissions of SOx, in grams, produced by aircraft type k for one LTO cycle.
- NE_k = Number of engines used on aircraft type k.
- $ER_k = 1 * (FF_k)$
- ER_k= emission rate of total SO_x in grams of SO_x emitted per second per operational mode for aircraft k.
- FF_k = the reported fuel flow by mode in kg/s per operational mode for each engine used on the aircraft type k.

c) Sophisticated Approach: $E_{ij} = (TIM_{jk} \times 60) \times (FF_{jk}/1000) \times (EI_{jk}) \times (NE_j)$

- E_{ij} = Total emissions of pollutant i (e.g. NO_x, CO, or HC), in grams, produced by aircraft type j for one LTO cycle.
- EI_{jk} = Performance based emission index for pollutant i (e.g. NO_x, CO, or HC), in grams per pollutant per kilogram of fuel (g/kg of fuel), in mode k (e.g. takeoff, climb out, idle and approach) for each engine used on aircraft type j.
- FF_{jk} = Fuel flow for mode k (e.g. takeoff, climb out, idle and approach), in kilograms per second (kg/s), for each engine used on aircraft type j.
- TIM_{jk} = Time-in-mode based on aircraft operational performance for mode k (e.g. idle, approach, climb out, and takeoff), in minutes, for aircraft type j.
- NEj = Number of engines used on aircraft type j.

This approach requires the use of propriety data or models that are not normally available in the public domain.

2.1. MEET (Methodologies for Estimating Air Pollutant Emissions from Transport)

Three main classes of air transport can be distinguished when analyzing its operational and emission related characteristics:

- ✓ flights performed under Instrument Flight Rules (IFR),
- ✓ military operational air traffic,
- ✓ flights performed under Visual Flight Rules (VFR).

There are some minor overlaps between the classes. However, each category has its own typical data set available for traffic characteristics and engine emissions. Accuracy of data input is different for the three categories, but is their contribution to total air transport emissions. About sixty to eighty percent of emissions originate from IFR flights. Normally IFR flights are operated as flights controlled by Air Traffic Services (ATS) within controlled airspace only and generally flights with civil aircraft. Only those military flights which belong to general air traffic (GAT) are included in IFR data. Jets, turbofans and turboprops generally represent this sort of traffic, which typically involves big commercial aircraft flying long distances using fixed operational routines. Flights performed under VFR generally are not operated as controlled flights so neither a Flight Plan nor detailed information on the route flown is available. However, VFR flights represent less than 5 % of fuel consumption and pollution caused by air traffic. For IFR flights, emission indices (EI), i.e. the mass of pollutant produced per mass of fuel used, are provided for eight typical operational conditions, which combined to cover most of an aircraft's operation during a flight. For ground operations while aircraft is standing (refueling, use of auxiliary power units, engine start) just few data exist for three aircraft/engine combinations. The standard operating conditions are:

- ✓ taxi out / off
- ✓ climb / cruise
- ✓ descent / landing / taxi in.

Emission factors are based on engine certification data in the ICAO Engine Exhaust Emission Databank. It contains data sets of thrust (engine performance), fuel flow and emissions of components CO, NOX and VOC which apply to four different power settings, Mach number 0 and altitude 0 m.

Aircraft emission factors (Approach and available data within MEET): Within the MEET-Project, Kalivoda and Kudrna (1997) review the existing air traffic emission models and present an emission simulation model. The model uses artificial flight profiles and emission indices for the thirty most frequent used aircraft types in Europe.

The fundamental methodology to estimate the air pollutant emissions for the flight of an aircraft is based on the duration of specific operational states (engine start, taxi-out, take-off, climb, cruise, descent, landing, taxi-in and ground operations) and the corresponding specific emission factors. Using typical flight profiles (for each aircraft type separately), with the common cruising altitude of the aircraft and the flight distance being the basic parameters, the total fuel consumption can be estimated.

Emission factor calculation: The methodology adopted for the calculation of air traffic emissions differs for IFR (instrument flight rules), VFR (visual flight rules) and military air traffic. VFR flights usually are those of small aircrafts. They represent approx. 5% of the total fuel consumption caused by air traffic (Kalivoda and Kudrna, 1997). Generally, no detailed information on the flight route and height is stored in the air control centers for VFR flights, so that only a tentative estimation of the air pollutant emissions related to this category of flights is available (Gardner et al., 1997; Kalivoda et al., 1997). However, emissions from VFR flights generally are emitted at lower altitudes than IFR flights, often even within the planetary boundary layer.

The basic formula for one flight (for a specific aircraft/engine combination, i. e., different engine types for the same machine are treated as different aircrafts) is:

$$E = \sum_{j=1}^{8} T_j FC_j EI_j$$

This formula is a compilation and reformulation of the basic approach by Kalivoda and Kudrna (1997), where

- E [g] total emission of air pollutant
- j [–] index running over the 8 operational stages, i. e., taxi out, take off, climbing, cruise, descent, landing, taxi in, ground operations.
- Tj [s] duration of operational stage j
- FCj [kgfuel/s] fuel consumption during operational stage j
- $EI j$ [gpoll./kgfuel] emission index of pollutant for operational stage j

Emissions Index: Emission index for IFR flight for each aircraft/engine are given in Emission Index sheet (EIS). Calculation has to be carrying out for 9 operational states (OS). Input for the calculation using the aircraft emission index sheet for a complete mission from airport to airport has to be:

- aircraft type,
- total distance between the two airports (e. g. distance as the crow flies, actual route, actual rout including holding patterns) and
- cruise altitude.
- Additional information which is useful:
- average duration of taxi out [s]
- average duration of taxi in [s].

2.2. Emission Data Sheets for IFR Step by Step

Ten steps are used and described below.

Step 1 (distances): Distance flown *(Dr)* and cruise altitude *(CRALT)* are the two basic input parameters to determine the geometric shape of a flight profile on a mission. The total distance of a city-pair is the sum of the distances flown during the three operational states climb, cruise and descent. The average climb and descent rate has been defined so climb and descent distance are just a function of cruise altitude *(CRALT):*

$$Y = b_0 + b_1 * x + b_2 * x^2 + b_3 * x^3$$

where x means the actual *CRALT in feet* and the parameters $b_0 \ldots b_3$ come from the EIS. This leads to $D_{cl} = f_{j,cl}\,(CRALT)$ and $D_{dsc} = f_{j,dsc}\,(CRALT)$.

Distance of cruise is defined by: $D_{cr} = D_r - (D_{cl} + D_{dsc})$

D_{cl} = Distance of climb [km]
D_{dsc} = Distance of descent [km]
D_r = Distance between city pair (=route r) [km]
$f_{j,cl}$ = function of CRALT for climb
$f_{j,dsc}$ = function of CRALT for descent
CRALT = cruise altitude [ft]
j = Aircraft/engine category
r = Route from airport to airport
cl = climb
dsc = descent

Step 2 (taxi out - OS2). The EIS provides both:

- a specific fuel consumption *(SFC)* and default duration *(DUR)* for taxi out
- a total fuel consumption *(FC).*

If the average taxi out time at an airport differs from the default duration in the EIS, the fuel consumption of taxi out is: $FC_{j,txo} = SFC_{j,txo} * DUR_{txo,act}$

$DUR_{txo,act}$ = actual taxi out time
$SFC_{j,txo}$ = Specific fuel consumption of aircraft category j for taxi out [kg]
$FC_{j,txo}$ = Fuel consumption of aircraft category j for taxi out [kg]
j = Aircraft/engine category
txo = taxi out

If no data on taxi out time is available, the default fuel consumption of EIS can be used.

Emission E of a pollutant p is fuel consumption multiplied by the specific emission factor:

$$E_{j,txo,p} = FC_{j,txo} * SE_{j,txo,p}$$

$SE_{j,txo,p}$ = Specific emission of pollutant p, of aircraft/engine combination j for taxi out [g/kg$_{FUEL}$]
$FC_{j,txo}$ = Fuel consumption of aircraft category j for taxi out [kg]
$E_{j\,txo,p}$ = Emission for pollutant p for taxi out [kg]

j = Aircraft/engine category
p = Pollutant
txo = taxi out

Step 3 (take off - OS3). The EIS provides both:

- a specific fuel consumption *(SFC)* and default duration *(DUR)* for takeoff and
- a total fuel consumption *(FC)*.

If the average take off time for an aircraft type differs from the default duration in the EIS, the fuel consumption for takeoff becomes: $FC_{j,tff}=SFC_{j,tff}\,DUR_{tff,act}$

$DUR_{tff,act}$ = actual take off time
$SFC_{j,tff}$ = Specific fuel consumption of aircraft category j for takeoff [kg]
$FC_{j,tff}$ = Fuel consumption of aircraft category j for takeoff [kg]
j = Aircraft/engine category
tff = take off

Emission E of a pollutant p is fuel consumption times specific emission factor

$$E_{j,tff,p} = FC_{j,tff} * SE_{j,tff,p}$$

$SE_{j,tff,p}$ = Specific emission of pollutant p, of aircraft/engine combination j for takeoff [g/kg_{FUEL}]
$FC_{j,tff}$ = Fuel consumption of aircraft category j for takeoff [kg]
$E_{j,tff,p}$ = Emission index for pollutant p for takeoff [kg]
j = Aircraft/engine category
p = Pollutant
tff = take off

Step 4 (climb - OS4): Both fuel consumption and specific emission factors for climb are a function of the cruise altitude. The functions $f(x)=Y$ for fuel consumption and specific emission factor of NO_X are:

$$Y = b_0 + b_1*x + b_2*x^2 + b_3*x^3$$

where variable x is *CRALT in feet* and $b_0...b_3$ are the coefficients from the EIS. There is a different function $f(x)=Y$ for the specific emission factor of HC and CO: $Y = a_0 + a_1*(1/x)$ where variable x is *CRALT in flight level, which is 100 feet* and a_0, a_1 are the coefficients from the EIS. This leads to:

$FC_{j,cl} = f_{j,cl,fc}\,(CRALT)$ and $SE_{j,cl,p} = f_{j,cl,p}\,(CRALT)$

and emission E of a pollutant p is fuel consumption times specific emission factor:

$$E_{j,cl,p} = FC_{j,cl} * SE_{j,cl,p}$$

$FC_{j,cl}$ = Fuel consumption of aircraft/engine combination *j* for climb [kg]

$SE_{j,cl,p}$ = Specific emission of pollutant *p*, of aircraft/engine combination *j* for climb [g/kg_{FUEL}]

$E_{j,cl,p}$ = Emission index for pollutant *p* of aircraft/engine combination *j* for climb [kg]

$f_{j,cl,fc}$ = function of aircraft/engine combination *j*, of fuel consumption for climb

$f_{j,cl,p}$ = function of aircraft/engine combination *j*, of pollutant for climb

CRALT = cruise altitude [ft,100 ft]

j = Aircraft/engine category

fc = fuel consumption

p = Pollutant

cl = climb

Step 5 (cruise - OS5): Fuel consumption for cruise is the product of cruise distance and specific fuel consumption as a function *f(x)=Y* of cruise altitude (*CRALT=x*):

$$Y = b_0 + b_1 * x + b_2 * x^2 + b_3 * x^3$$

where the variable x is CRALT in flight level, which is 100 feet and b0...b3 are the coefficients from the EIS.

$$FC_{j,cr} = D_{cr} * f_{j,cr,fc}\ (CRALT)$$

Due to the temperature dependency of NOx the specific emission factor for NO_x is also a function f(x)=Y of cruise altitude (CRALT=x):

$$Y = b_0 + b_1 * x + b_2 * x^2 + b_3 * x^3$$

where variable x is CRALT in flight level, which is 100 feet and $b_0...b_3$ are the coefficients from the EIS:

$$SE_{j,cr,p=NOx} = f_{j,cr,p=NOx}\ (CRALT)$$

Specific emission factors of HC and CO are almost independent from cruise altitude and a constant:

$$SE_{j,cr,(\ p=CO,HC)} = const.$$

Emission E of a pollutant p is fuel consumption times specific emission factor:

$$E_{j,cr,p} = FC_{j,cr} * SE_{j,cr,p}$$

$SE_{j,cr,p}$ = Specific emission of pollutant p, of aircraft/engine combination j for cruise [g/kg_{FUEL}]

$FC_{j,cr}$ = Fuel consumption of aircraft category j of cruise [kg]

$EI_{j,cr,p}$ = Emission index for pollutant p of aircraft/engine combination j for cruise [kg]

D_{cr} = Distance of cruise (=route r) [km]

$f_{j,cr,fc}$ = function of aircraft/engine combination j, of fuel consumption for cruise
$f_{j,cr,p}$ = function of aircraft/engine combination j, of pollutant for cruise
$CRALT$ = cruise altitude [100 ft]
j = Aircraft/engine category fc = fuel consumption p = Pollutant
cr = cruise

Step 6 (descent - OS6): Both fuel consumption and specific emission factors are a function of cruise altitude using a polynomial function $f(x)=Y$: $Y = b_0 + b_1 * x + b_2 * x^2 + b_3 * x^3$
where variable x is $CRALT$ *in feet* and $b_0 ... b_3$ are the coefficients from the EIS. There is a different function $f(x)=Y$ for the specific emission factor of HC and CO: $Y = c_0 + c_1 * \ln(x)$ where variable x is $CRALT$ *in feet* and c_0, c_1 are the coefficients from the EIS. This leads to:

$$FC_{j,dsc} = f_{j,dsc,fc}(CRALT) \ \& \ SE_{j,dsc,p} = f_{j,dsc,p}(CRALT)$$

Emission E of a pollutant p is fuel consumption times specific emission factor:

$$E_{j,dsc,p} = FC_{j,dsc} * SE_{j,dsc,p}$$

$FC_{j,dsc}$ = Fuel consumption of aircraft/engine combination j for descent [kg]
$SE_{j,dsc,p}$ = Specific emission of pollutant p, of aircraft/engine combination j for descent [g/kg$_{FUEL}$]
$E_{j,dsc,p}$ = Emission index for pollutant p of aircraft/engine combination j for descent [kg]
$f_{j,dsc,fc}$ = function of aircraft/engine combination j, of fuel consumption for descent
$f_{j,dsc,p}$ = function of aircraft/engine combination j, of pollutant for descent
$CRALT$ = cruise altitude [ft] j = Aircraft/engine category
fc = fuel consumption p = Pollutant dsc = descent

Step 7 (landing - OS7). The EIs provide both:

- a specific fuel consumption *(SFC)* and default duration *(DUR)* for landing, and
- a total fuel consumption *(FC)*.

If the average landing time (thrust reverse) at an aerodrome differs from the default duration in the EIS, the fuel consumption of landing is: $FC_{j,ld} = SFC_{j,ld} * DUR_{ld,act}$

$DUR_{ld,act}$ = actual landing time
$SFC_{j,ld}$ = Specific fuel consumption of aircraft category j for landing [kg]
$FC_{j,ld}$ = Fuel consumption of aircraft category j for landing [kg]
j = Aircraft/engine category
ld = landing

Emission E of a pollutant p is fuel consumption times specific emission factor:

$$E_{j,ld,p} = FC_{j,ld} * SE_{j,ld,p}$$

$SE_{j,ld,p}$ = Specific emission of pollutant *p*, of aircraft/engine combination *j* of landing [g/kg$_{FUEL}$]
$FC_{j,ld}$ = Fuel consumption of aircraft category *j* of landing [kg]
$E_{j,ld,p}$ = Emission for pollutant *p* of aircraft category *j* of landing [kg]
j = Aircraft/engine category *p* = Pollutant *ld* = landing

Step 8 (taxi in - OS8). The EIS provide both:

- a specific fuel consumption *(SFC)* and default duration *(DUR)* for taxi in, and
- a total fuel consumption *(FC)*.

If the average taxi in time at an aerodrome differs from the default duration in the EIS, the fuel consumption of taxi in is:

$$FC_{j,txi} = SFC_{j,txi} * DUR_{txi,act}$$

$DUR_{txi,act}$ = actual taxi in time
$SFC_{j,txi}$ = Specific fuel consumption of aircraft category *j* for taxi in [kg]
$FC_{j,txi}$ = Fuel consumption of aircraft category *j* for taxi in [kg]
j = Aircraft/engine category
txi = taxi in

Emission *E* of a pollutant *p* is fuel consumption times specific emission factor

$$E_{j,txi,p} = FC_{j,txi} * SE_{j,txi,p}$$

$SE_{j,txi,p}$ = Specific emission of pollutant *p*, of aircraft/engine combination *j* of taxi in [g/kg$_{FUEL}$]
$FC_{j,txi}$ = Fuel consumption of aircraft category *j* of taxi in [kg]
$E_{j,txi,p}$ = Emission for pollutant *p* of aircraft category *j* of taxi in [kg]
j = Aircraft/engine category *p* = Pollutant *txi* = taxi in

Step 9 (engine start - (OS1) and ground operations - (OS10)): No representative and sound database is available for the emissions from engine start and aircraft related ground operations like refueling and APU operation.

Step 10 (summing up totals): Total fuel consumption of a mission is the sum of fuel consumption from all single operation states: $TFC_{j,Dr} = \sum_i FC_{j,i}$

Total emission of NO_x, CO, HC is the sum of emissions of NO_x, CO, HC from all single operation states: $p= NO_x, CO, H\,TE_{j,p} = \sum_i E_{j,i,p}$

Total emission of CO_2, H_2O, SO_2 is the product of total fuel consumption and the specific emission factors of the component.

$p= CO_2, H_2O, SO_2$ $TE_{j,p} = TFC_{j,Dr} * SE_p$
$Te_{j,p}$ = Total emission of pollutant p of aircraft/engine combination j[kg]
$TFC_{j,Dr}$ = Total fuel consumption of aircraft category j, of total distance Dr [kg]

Se_p = Specific emission of pollutant p, [g/kg$_{FUEL}$]
$FC_{j,i}$ = Fuel consumption of aircraft category j for operational state i
$E_{j,i,p}$ = Emission for pollutant p of aircraft/engine combination j [kg] and operational state
Dr = Distance between city pair (=route r) [km]
j = Aircraft/engine category i = operational state p = Pollutant
r = Route from airport to airport

2.3. Airport Local Air Quality Study "Alaqs"

Typical air quality assessment; consist of emission inventory and spatial allocation, dispersion modeling, determination of background concentration and, visualization of the results. ALAQS methodology (ALAQS, 2009) consists of developing Pan-European emission inventory methodology with spatial information and future application of dispersion modeling to this inventory with use of GIS technologies. Hence the toolset and database to support ALAQS of European airports are:

- Pan-European ALAQS central databank for emission factors of different sources: All related emissions factors for different pollution sources are defined and aggregated from different sources and harmonized in Access database. This will provide the opportunity to change or compare different emissions factors used for the same type of sources.
- Scalable approach for developing emission inventory and dispersion modeling.
- ALAQS-AV GIS application. Optional interface to GIS software will allow:

1) The capture of the data i.e. spatial location and attributes
2) Testing and implementing different methods to calculate emissions by means of Emission Toolbox
3) Option to use emission factors from different sources
4) Preparing emission sources for dispersion modeling
5) Importing/exporting data to and from different models
6) Importing and analysis of dispersion results
7) Calculation of emission for gates, taxiways and runways based on LTO cycles.
8) Improved data capture interface

ALAQS Aircrafts Emission methodology: Flight operations encompass the entire landing and take-off (LTO) cycle as defined by the ICAO. Emissions of each aircraft type are computed by knowing the emission factors for the aircraft's specific engines at each power setting or mode of operation and the time spent in each mode.

In ALAQS-AV methodology for a specific scenario, aircraft movements table is prepared for this specific period. For each movement: date, time, aircraft type, arrival/departure flag, gate (stand) and runway are specified. ALAQS-AV toolset uses the movements table to calculate hourly emissions at gates, taxiways, queues and runways.

Aircraft exhaust emissions are calculated for the following operating modes:

- Engine Start / Taxi in and taxi out (TX, 7% thrust)
- Queuing (TX, 7% thrust) / Approach (AP, 30% thrust)
- Landing roll (AP, 30% thrust) / Takeoff roll (TO, 100% thrust) / Climb-out (CL, 85% thrust)

Except for engine start emissions - aircraft engine emissions during a particular operating mode of the Landing and Take-Off (LTO) cycle are given by the product of the time-in-mode, the fuel flow rate and the emission index for the appropriate engine thrust setting engaged. Data is extracted from the system database. (i.e. aircraft-engine combination, number of engines etc..)

$$ACe = FF_{mode} \times EF_{mode} \times T \times N$$

Ace = Aircraft total engine emissions
FF_{mode} = Fuel flow rate (kg/s) per engine in mode
EF_{mode} = Emission factor (kg/kg) per engine in mode
T = Time-in-mode (s)
N = Number of engines

✈ APU / GPU: APU (Auxiliary Power Units) are on-board generators in larger aircrafts that provide electrical power while its engines are shut down (For example when the aircraft is parked away from the terminal building), GPU (Ground Power Units) are mobile diesel powered units that provide power to an Aircraft. Those are operated on mostly gates. Therefore emission calculation for APU/GPU is integrated to the gate emissions calculations.

Start emissions are the Volatile Organic Compounds (VOC) emitted when the aircraft engines are started before departure. The amount of emission released by GSE, APU and GPU are a function of aircraft size and type of stand. However, VOC emissions released at engine start are a function of the aircraft class only. For APU and GPU an emission rate (grams per minute) is specified. The total GPU and APU emissions are obtained by multiplying the running time in minutes specified for arrival or departure with the emission rate. For engine start emissions a total amount of HC (VOC) emissions (grams per aircraft) is specified.

Calculation Method:

1) For each hour get all aircraft movements and regroup movements by gate.
2) For each gate:

- Get aircraft group from aircraft system table.
- Get gate scenario from project gate shape file.
- Get scenario emission parameters for aircraft group from project gate scenario table.
- Calculate GSE, APU, GPU and Start emissions.
- Add emissions to gate totals.

3) Store each gate total in hourly emissions table.

- Taxiway Emissions: Taxiway emissions are released by aircraft engines while an aircraft travels from gate to runway and vice versa. Emissions are calculated on the assumption that all engines are idling on a 7% (ICAO settings) power thrust setting (mode TX). For taxiway emissions, a more refined approach is implemented. Instead of allocating fix idle times to all aircraft (such as in ICAO database), for each aircraft movement and gate scenario taxi routes are defined. Based on taxi speed, total taxiing times are calculated. Then from aircraft –engine emission table emission factors for idle phase is extracted to calculate taxiing emissions.

In ALAQS-AV taxi routes between gates and runways are selected automatically from the user-defined taxi routes based on the end roll position. The end roll position is the sum of the arrival profile's landing roll and the touchdown offset specified for the runway. The selection of taxi route is done on four criteria: Gate, Runway, Arrival/Departure flag (A or D) and End Roll Position. For a gate-runway combination the end roll position is compared to the routes' exit positions. The route corresponding to the exit position situated immediately ahead of the end roll position is selected. Engine Emission Indices or obtained from the Engine Table based on Engine ID and Aircraft Mode (taxi mode = TX). Taxi Times are obtained from the Taxiway Shape file and emissions. For each taxiway, calculated hourly emissions totals are stored in the Hourly Emission table.

Queue time per aircraft at selected hour:

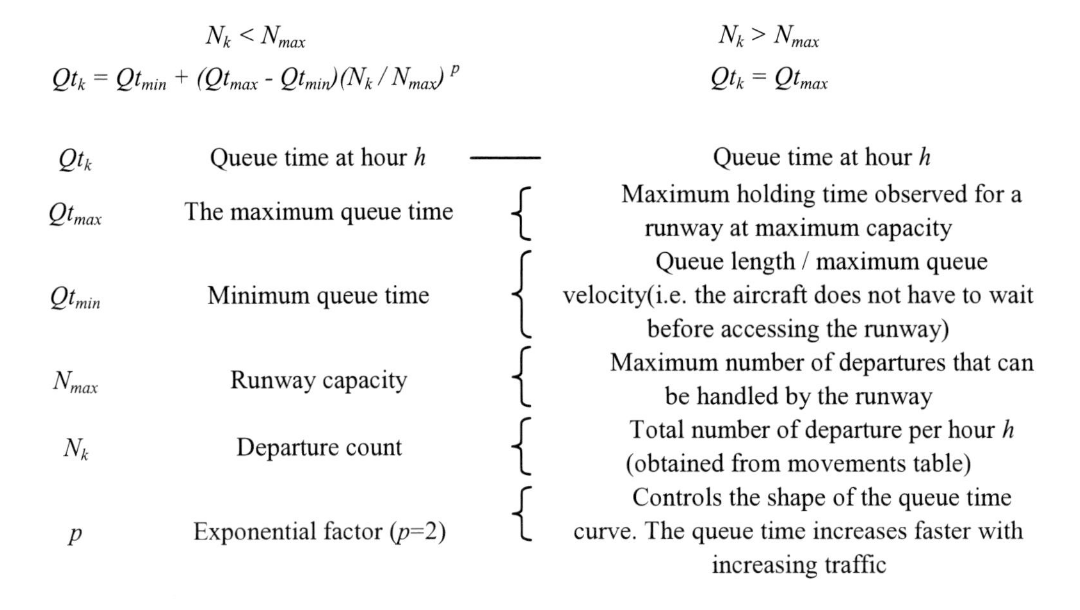

$N_k < N_{max}$

$$Qt_k = Qt_{min} + (Qt_{max} - Qt_{min})(N_k / N_{max})^p$$

$N_k > N_{max}$

$$Qt_k = Qt_{max}$$

Qt_k	Queue time at hour h	Queue time at hour h
Qt_{max}	The maximum queue time	Maximum holding time observed for a runway at maximum capacity
Qt_{min}	Minimum queue time	Queue length / maximum queue velocity(i.e. the aircraft does not have to wait before accessing the runway)
N_{max}	Runway capacity	Maximum number of departures that can be handled by the runway
N_k	Departure count	Total number of departure per hour h (obtained from movements table)
p	Exponential factor (p=2)	Controls the shape of the queue time curve. The queue time increases faster with increasing traffic

- Queue Emissions: Flights incur at some percentage of the delay on the ground during the departure process between their schedule departure from the gate and take-off. This cause queuing at the runways and has to be calculated for their emissions. Queue emissions are emissions released by aircraft on the last taxiway section before entering the runway for take-off. In ALAQS-AV the amount of time an aircraft spends in the queue is a function of the maximum queue velocity, the maximum queue time and the number of departures that can be handled by the runway.

Those queuing parameters are extracted in ALAQS-AV tool. Engine emissions are calculated on the assumption that all engine are idling on a 7% power thrust setting. Emission Indices are obtained from the engine table based on Engine ID and Aircraft Mode (taxi mode = TX).

Calculation Method:

1) For each hour get all aircraft departures.
2) For each hour process each departure one at a time.

    a) Count the number of departures and calculate the queue time.
    b) From movements table get Aircraft Type and Runway.
    c) From aircraft table get Engine ID and Engine Count.
    d) From engine table get Engine Emission Indices.
    e) For each queue calculate emissions and add to queue totals.

3) Store each queue total in hourly emissions table.

- Runway Emissions: Runway emissions are emissions released by aircraft on or above the runway during takeoff roll, climb-out, approach and landing roll. Conventionally runway emissions for inventory purpose are calculated up to an elevation of 3000 ft (914.4m) above the runway. However, emissions released above an elevation of 400m above the runway have little impact on the air quality at ground level.

Calculation Method:

- For each hour get Aircraft Type, Runway and Arrival/Departure flag from movements table.
- From aircrafts table get for each aircraft Arrival Profile ID, Departure Profile ID, Engine ID and Engine Count.
- Based on Runway and Profile ID get profile segment data (Time-in-mode and Mode) from runway space table.
- From engine table get Engine Emission Indices based on Engine ID and Mode (takeoff [TO], climb-out [CL] or approach [AP]).
- For each segment calculate emissions and add runway space block totals.
- Store each block total in hourly emissions table (hr_emis).

Runway emissions include also runway roll emissions (takeoff roll and landing roll) and emissions released in the vertical plane above the runway (climb-out and approach). Additionally there is a pre-processor which intersects profile trajectories with runway space blocks.

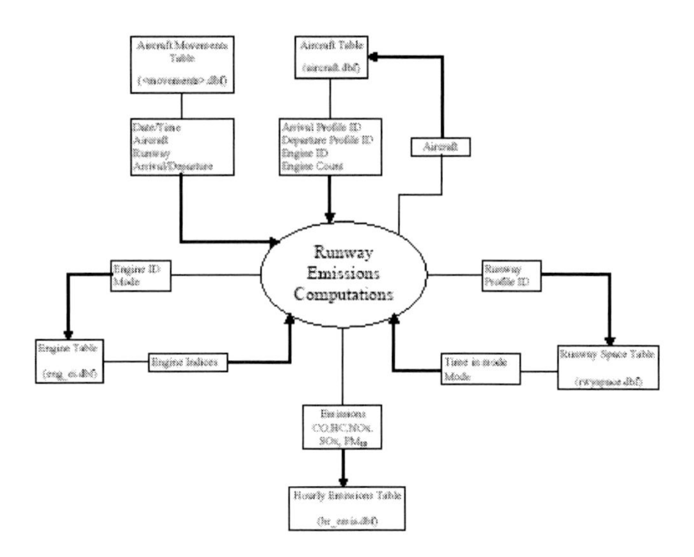

Figure 5. Runway Emission Calculation Process (ALAQS).

2.4. Sourdine II

The objective of Sourdine II project (Sourdine, 2005) was to propose and evaluate new, innovative procedures for reducing the impact of emission and aircraft noise on the ground, otherwise known as Noise Abatement Procedures (NAPs). For the emission assessments of the SII NAPs, a specific tool called TBEC (Thrust Based Emission Calculator) has been developed. This tool calculates aircraft emission levels associated to a given INM-like flight profile, on the basis of the ICAO Engine Exhaust Emissions Data Bank. However, this tool remains a prototype with several limitations. Further investigations need to be carried out in order to refine and validate its modeling principles. Emission results produced with such a tool should therefore be taken with caution and analyzed in a relative way (i.e. relative variations of emission levels between the SII procedures and a baseline/reference procedure).

Thrust-Based Emission Calculator: The Thrust-Based Emission Calculator (TBEC) is a Microsoft Access application which has been specially developed for Sourdine II in order to calculate aircraft emissions resulting from the different SII procedures. It uses the ICAO Engine Exhaust Emissions Data Bank, which provides, for a large series of engine types, fuel flow (kg/s) and emission indices (g/kg of fuel) at four specific engine power settings (from idle to full take-off power). The overall principle of TBEC consists of calculating (by interpolations) emission levels, based on the actual thrust along the vertical *fixed-point* profiles associated to the SII procedures. To calculate emission levels of different pollutants, it is necessary to have fuel flow information along the flight profiles. It was originally planned to approximate these by interpolations on input thrust values, as the ICAO databank provides fuel flow data associated to specific power settings. However, the International Civil Aviation Organization Committee on Aviation Environmental Protection (CAEP)'s Modeling Working Group (WG2) considered that estimating fuel flow based on thrust was unsatisfactory without having a greater knowledge of individual aircraft/engine performance parameters, data that is not yet readily available. Consequently, realistic fuel flow data have been supplied by Airbus for all the studied SII procedures (along with the baseline procedures) and for the eight Airbus aircraft. These fuel flow data have been incorporated, as an additional parameter. Based on the fuel flow and thrust values along the flight profiles, TBEC calculates total fuel burn (a straight forward process), and emission levels of different pollutants. Calculation of arrival emissions stops at touchdown since the fuel-flow data available stop at that point. Reverse thrust emissions are not taken into account. These would vary as a function of the landing speed of the aircraft, which is very slightly higher in the Sourdine II procedures than the baseline due to the different landing configurations used.

TBEC inputs: TBEC calculates fuel burn and emission levels for the fixed-point profiles of the SII flight profile database, which include the additional fuel flow parameter (Airbus aircraft only). These input fixed-point profiles provide altitude (ft), speed (kts), corrected net thrust (lbs) and fuel flow (kg/s) as a function of the ground distance (ft) from brake release (for departures), or to touchdown (for arrivals).

TBEC outputs: For a given procedure (i.e. a flight profile), TBEC calculates total fuel burn and total emissions (in kgs) of the following components:

- Hydrocarbons (HC);
- Carbon Monoxide (CO);
- Oxides of Nitrogen (NO_x);
- Sulphur Dioxide (SO_2);
- Carbon Dioxide (CO_2);
- Water (H_2O);
- Volatile Organic Compounds (VOC);
- Total Organic Gases (TOG).

VOC are Acetaldehyde, Acrolein, POM16PAH, POM7PAH, Styrene. TOG are Formaldehyde, Propianaldehyde, Toluene, Xylene, 1-3Butadiene, Benzene, and Ethylbenzene. The calculation of total fuel burn is a straight forward process: it is obtained by the time integration of the input fuel flow data along the profile. HC, CO and NO_x are obtained by linear interpolations in the ICAO databank, using as input data the corrected net thrust and the fuel flow on the successive segments of the profile. CO_2, SO_2 and H_2O emissions are proportional to fuel burn (or fuel flow), and are obtained using emission coefficients (kg/kg fuel flow, or g/kg fuel flow for SO_2). The VOC and TOG emissions are obtained in a similar way from the calculated emissions of HC. All these emission coefficients are independent of the engine type.

Calculation principle: The flight profile is defined by a series of small segments, each segment being defined by two consecutive points of the fixed-point profile. The overall calculation principle consists of estimating the fuel burn and emission levels produced by each segment, and summing them (over the flight profile) to obtain the total fuel burn and emissions of each pollutant.

Fuel burn: The fuel burn on a segment FB_{seg} is calculated as follows: $FB_{seg} = \Delta T_{seg} * FF_{seg}$

Where

- ΔT_{seg} is the duration (in seconds) of the flight segment. ΔT_{seg} is calculated using the distance between the two end-points of the segment, divided by the average speed of the aircraft on the segment;
- FF_{seg} is the average fuel flow on the segment (kg/s), calculated using the input fuel flow values at the two end-points of the segment.

HC, CO and NO_x: The ICAO Engine Exhaust Emissions Data Bank provides emission indices (g/kg fuel flow) at four different power setting levels, namely: Take-Off, Climb-Out, Approach, and Idle. These four power states correspond to a percentage of F_{oo}, the maximum engine thrust available for take-off under normal operating conditions at ISA sea level static conditions. By definition, the four tabulated power settings correspond respectively to 100%, 85%, 30% and 7% of F_{oo} (Similar to ALAQS). The emissions of HC, CO and NOx on a segment are calculated through a linear interpolation between the above tabulated emission data. The different steps of the process are described below. The Emission Indices $EI(P_i)$ of each pollutant provided by the ICAO data bank at the four power settings are converted into segment-specific emission flow $EF_{seg}(P_i)$ as follows:

$$EF_{seg}(P_i) = EI(P_i) * FF_{seg}$$

- $EF_{seg}(P_i)$ is the emission flow for the segment associated to power setting P_i (in g/s)
- P_i is one of the tabulated engine power settings for which emission indices are provided in the data bank (7%, 30%, 85% or 100%)
- $EI(P_i)$ is the emission indices associated to power setting P_i (in g/kg of fuel)
- FF_{seg} is the average fuel flow on the segment (in kg/s), calculated using the input fuel flow values at the two end-points of the segment.

The segment-specific power setting parameter P_{seg}, at which the emission levels will be interpolated, is approximated as follows:

$$P_{seg} = \frac{CNT_{seg}}{MaxStaticThrust} \times 100$$

- P_{seg} is the segment-specific power setting (%)
- CNT_{seg} is the average corrected net thrust (lb) on the segment, calculated using the input *CNT* values at the two end-points of the segment
- *MaxStaticThrust* is the engine-specific maximum sea level static thrust, available in the INM (INM) database (lb).

The following describes the interpolation process to estimate the emission level of a given pollutant on the segment (Figure 6).

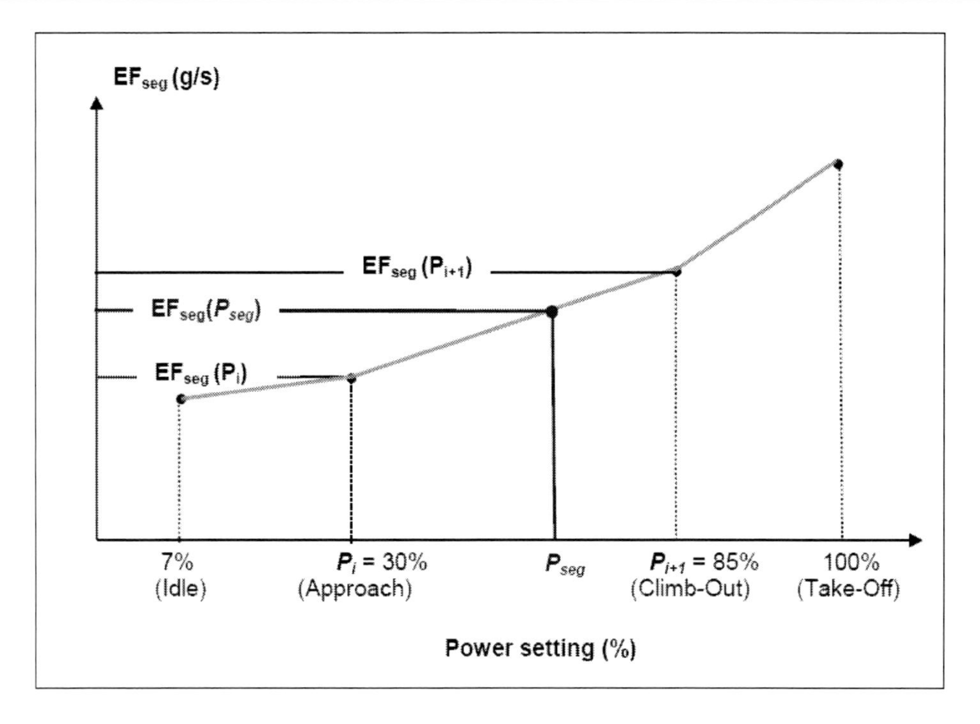

Figure 6. Emission flow interpolation.

The emission level of a given pollutant on the segment EL_{seg} is expressed as:

$$EL_{seg} = \Delta T_{seg} \times \left[EL_{seg}(P_i) + \frac{P_{seg} - P_i}{P_{i+1} - P_i}\left(EF_{seg}(P_{i+1}) - EF_{seg}(P_i)\right)\right]$$

- EL_{seg} is the emission level of the pollutant produced on the segment (g);
- ΔT_{seg} is the duration (in seconds) of the flight segment. ΔT_{seg} is calculated using the distance between the two end-points of the segment, divided by the average speed of the aircraft on the segment;
- P_{seg} is the segment-specific power setting (%);
- P_i and P_{i+1} are the two tabulated power setting values bounding P_{seg} (%);
- $EF_{seg}(P_i)$ and $EF_{seg}(P_{i+1})$ are the emission flow values (g/s) associated to P_i and P_{i+1}.

CO_2 , SO_2 , H_2O: CO_2 , SO_2 and H_2O emission levels are directly proportional to the calculated fuel burn and are estimated using the following emission coefficients:

Components	Emission coefficients
CO_2	3.149 (kg/kg fuel)
SO_2	0.84 (g/kg fuel)
H_2O	1.23 (kg/kg fuel)

Limitations / Validity: The first limitation of TBEC is that it does not take into account the variation of the emission indices with altitude due to temperature and pressure changes. Indeed, the ICAO databank provides emission indices for ISA conditions; these are, however, assumed to be valid for altitudes below 3,000 ft. Implementing the Boeing Method 2 (*BM2*), described in [AEM], which TBEC does not do for the moment, would allow the modeling of the effects of non-ISA temperature and pressure conditions at the airport. Another limitation is due to the assumption that emission indices vary linearly with the thrust level, which is obviously not the case in real life. Implementing the *BM2* would enable the modeling of non-linear variations between the four thrust settings in the ICAO databank (Take-Off, Climb-Out, Approach and Idle). The method used to calculate the power setting parameter required to perform the interpolations might be questionable. Using the N1 parameter instead could be more appropriate. Further investigation of this point is required.

TBEC should be, therefore, considered as a prototype, which can only be used to derive general trends between different procedural scenarios, rather than to assess the exact amount of gaseous.

2.5. Environmental Protection Agency (Epa)

EPA's recommended emissions calculation methodology for a given airport in any given year can be summarized in six steps (EPA, 1999), e-CFR Data):

1) Determine the mixing height to be used to define a landing and takeoff (LTO) cycle.
2) Determine airport activity in terms of the number of LTOs.
3) Define the fleet make-up at the airport.
4) Select emission factors.
5) Estimate time-in-mode (TIM).
6) Calculate emissions based on the airport activity, TIM, and aircraft emission factors.

Steps two through five are repeated for each type of aircraft using a given airport. This methodology is essentially the same as that used in the FAA *Aircraft Engine Emissions Database* (FAEED) model (USDOT, 1995).

Time in Mode Calculations: The duration of the approach and climb out modes depends largely on the mixing height selected. EPA guidance provides approach and climb out times for a default mixing height of 3000 feet, and a procedure for adjusting these times for different mixing heights. The adjustments are calculated using the following equations:

Climb out: $TIM_{adj} = TIM_{dflt}\left[\frac{MixingHeig\ ht-500}{3000-500}\right]$

Approach: $TIM_{adj} = TIM_{dflt}\left[\frac{MixingHeig\ ht}{3000}\right]$

TIM_{adj} is the adjusted time-in-mode for approach or climb out, and TIM_{dflt} is the default time-in-mode. Mixing height is by default given in feet. The equation for climb out assumes that 500 feet is the demarcation between the takeoff and climb out modes. Expressed in metric units, the approach and climb out adjustment equations are as follows:

Climb out: $TIM_{adj} = TIM_{dflt} \left[\frac{MixingHeig\ ht-152}{915-152}\right]$

Approach: $TIM_{adj} = TIM_{dflt} \left[\frac{MixingHeig\ ht}{915}\right]$

Default mixing height is 915 meters, with the demarcation between approach and climb out modes at 152 meters. Consistent with EPA guidance (EPA, 1992), a four-minute default approach time was assumed for this study.

Emissions Calculation: The weighted-average emission factor represents the average emission factor per LTO cycle for all engine models used on a particular type of aircraft. The weighted-average emission factor per 1000 pounds of fuel is calculated as follows:

$$\overline{EF}_{ijk} = \sum_{m=i}^{NM_j} (X_{mj} . EF_{imk})$$

EF_{imk} = the emission factor for pollutant i, in pounds of pollutant per 1000 pounds of fuel (or kilograms pollutant per 1000 kilograms fuel), for engine model m and operating mode k;

X_{mj} = the fraction of aircraft type j with engine model m; and

NM_j = the total number of engine models associated with aircraft type j.

Note that, for a given aircraft type j, the sum of X_{mj} for all engine models associated with aircraft j is 1. Total emissions per LTO cycle for a given aircraft types are calculated using the following equation:

$$E_{ij} = TIM_{jk} * \frac{FF_{jk}}{1000} * EF_{ijk} * NE_j$$

TIM_{jk} = time in mode k (min) for aircraft type j;

FF_{jk} = fuel flow for mode k (lbs/min or kg/min) for each engine used on aircraft type j;

EF_{ijk} = weighted-average emission factor for pollutant i, in pounds of pollutant per 1000 pounds of fuel (kilograms pollutant per 1000 kilograms fuel), for aircraft type j in operating mode k; and NE_j = number of engines on aircraft type j.

Once the preceding calculations are performed for each aircraft type, total emissions for that aircraft type are computed by multiplying the emissions for one LTO cycle by the number of LTO cycles at a given location:

$$E_i = (E_{ij} * LTO_j)$$

E_{ij} = the total emissions for pollutant i from aircraft type j;

LTO_j = the number of LTOs for aircraft type j.

Total emissions for each aircraft type are then summed to yield total commercial exhaust emissions for the facility as shown below:

$$ET_i = \sum_{j=i}^{N} \left(E_{ij} * LTO_j \right)$$

ET_i = the total emissions for pollutant i from all aircraft types;
E_{ij} = the emissions of pollutant i from aircraft type j;
LTO_j = the number of LTOs for aircraft type j; and
N = the total number of aircraft types.

2.6. Emissions and Dispersion Modeling System (EDMS)

The EDMS (Emissions and Dispersion Modeling System) has been considered a preferred model for airport air quality analysis. The modeling system EDMS was developed in the mid-1980s as a complex source microcomputer model designed to assess the air quality impacts of proposed airport development projects. EDMS is a combined emissions and dispersion model for assessing air quality at the airports. The model is used to produce an inventory of emissions generated by sources on and around the airport or air base, and to calculate pollutant concentrations in these environments, which consist of:

- Aircraft / Auxiliary power units
- Ground support equipment / Ground access vehicles / Stationary sources

EDMS is one of the few air quality assessment tools specifically engineered for the aviation community. It includes:

- Emissions and dispersion calculations
- The latest aircraft engine emission factors from the International Civil Aviation Organization (ICAO) Engine Exhaust Emissions Data Bank
- Vehicle emission factors from the latest version of the Environmental Protection Agency's (EPA) MOBILE6 model
- EPA-validated dispersion algorithms

In the early 1970s, the FAA and the USAF recognized the need to analyze and document air quality conditions at and around airports and air bases. Each agency independently developed computer programs to address this need. The USAF developed the Air Quality Assessment Model and the FAA developed the Airport Vicinity Air Pollution Model (AVAP). These models were used to perform limited air quality assessments in the late 1970s. Recognizing the inefficiency of maintaining two non-EPA approved models, the agencies agreed to cooperate in developing a single system that would have regulatory, operational and economic benefits. The result was the EDMS development effort jointly supported by both agencies and leading to a model listed among the EPA's preferred guideline models. Emissions modeling in the FAA began with the early Simplex A modeling efforts using the HP-97 calculator. The Simplex A algorithms included calculations for aircraft takeoff plume dispersion. In the 1980s, the model was moved to the Apple II computer and the Simplex A algorithm was expanded to include dispersion calculations for roadways, parking lots, and power plant sources. The revised and enhanced Simplex A model became known as the Graphical Input Microcomputer Model (GIMM). GIMM was ported to a PC and further enhanced by improvements in processing speed and refinement of the emissions inventory calculations. This enhanced version of GIMM became known as EDMS. In 1997 EDMS was reengineered for Microsoft Windows and included the algorithms from the Environmental Protection Agency (EPA) dispersion models PAL2 and CALINE3. With the release of version 3.0 in 1997, EDMS became the FAA-preferred model for air quality assessment at the airport and air bases. In 2001 EDMS 4.0 was released which marked the transition to EPA's next generation dispersion model AERMOD as the main dispersion engine behind EDMS, and the introduction of aircraft performance data to allow EDMS to estimate the contribution to concentrations from aircraft up to 1,000 feet above the ground. In 2004, the FAA re-engineered EDMS to take advantage of new data & algorithm developments and released the software as EDMS Version 4.2. This version of EDMS allowed users to select the version of EPA's MOBILE model (5a, 5b, or 6.2) to use for on-road vehicle emissions estimation. An interface to EPA's AERMAP terrain processing module was also provided for the first time in this release. AERMOD version 02222 was bundled with the EDMS software and was the most current version of AERMOD available as of September 30, 2004. Incremental releases of EDMS 4.3 in 2005, EDMS 4.4 in 2006, and EDMS 4.5 also in 2006 provided updates to the system data, and updates of EPA models. In particular, EDMS 4.4 contained an upgrade of AERMOD and AERMET to version 04300, which was the first version of AERMOD promulgated by the EPA. Also in 2004, the FAA embarked on development of its next generation of airport analysis tools, known as the Aviation Environmental Design Tool (AEDT). The development of this toolset is a 6-year effort that will result in the ability to model noise and emissions interdependencies. AEDT is being developed in phases and leverages the investment made in EDMS and the Integrated Noise Model (INM). The first phase of development is complete, which represents a 2 year effort, and harmonizes the underlying system data from both of those models as well as the aircraft performance calculation methods. EDMS 5 has been given a new architecture and includes over 150,000 new lines of code to support additional enhancements to its capabilities and the evolution toward AEDT. A study can now contain multiple scenarios, multiple airports and span multiple years, with emissions or dispersion being run for all at once. The First Order Approximation (FOA) version 3.0 / 3.0a has been incorporated for estimating PM emissions from jet aircraft. Aircraft fleet data have been harmonized with INM, and a common dynamic flight performance module exists in both tools as well, that accounts for aircraft weight and meteorological conditions. EDMS 5 represents the state of the art for airport emissions modeling and an important step toward the development of AEDT.

Functional Flow – Emissions: Overall, the fundamental usage of EDMS is to first perform an emissions inventory, after which the user can chose to continue to model the dispersion of the emitted pollutants calculated. As shown in Figure 7, to perform an emissions inventory the user would follow the following steps:

1) Set up the study by adding scenarios and airports, and choose the modeling options to use.
2) Define all emissions sources, including operational usage.
3) Define the airport layout if sequence modeling was selected.
4) Select a weather option: annual average or hourly (requires running AERMET).
5) Select Update Emissions Inventory

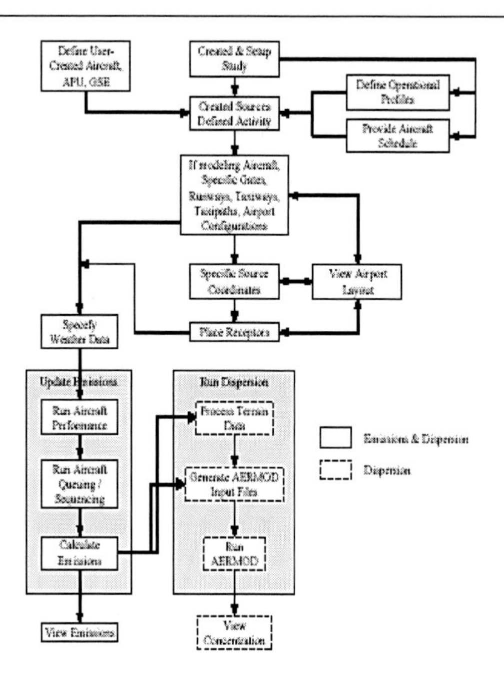

Figure 7. Functional Flow.

The simplest way to generate an emissions inventory and obtain a course estimate of the total annual emissions is to perform the first two steps, and use the ICAO/EPA default times in mode along with the default operational profiles, and the annual average weather from the EDMS airports database. Doing so would only consider the total number of operations for the entire year without regard to when those operations occurred. If a more precise modeling of the aircraft taxi times using the Sequencing module is desired (required if dispersion will be performed), then the user must define the airport gates, taxiways, runways, taxi paths (how the taxiways and runways are used) and configurations (weather dependent runway usage). The resulting emissions values can be viewed by selecting *Emissions Inventory* on the *View* menu. These results can be printed by selecting *Print* under the *File* menu while viewing the emissions inventory.

Functional Flow – Dispersion: To run a dispersion analysis, the user must first generate an emissions inventory while dispersion is enabled. Because of this, the methodology used for generating the emissions inventory is also the same one used to calculate the emissions for dispersion purposes. This inventory will take many times longer than the same one without dispersion enabled, because EDMS must generate (.HRE) files which contain all of the emissions broken into hourly bins by source and the (.SRC) files that define all the sources. Also, enabling dispersion forces the selection of *Performance Based* times in mode, *Sequence Modeling* and hourly meteorological data. In addition, to run dispersion the user must define receptors, which are points at which the concentration of pollutants will be computed. Next, the user can optionally run AERMAP, which will adjust the elevations of all emissions sources and receptors to the terrain data supplied. This will override any user-defined elevations that had been entered. Next, the user must specify the AERMOD run options and generate the AERMOD input (.INP) files. And finally, the user runs AERMOD within the EDMS GUI to generate the concentrations at the receptors. The resulting concentrations can be viewed by selecting *Concentrations* under the *View* menu. These results can be printed by selecting *Print* under the *File* menu while viewing the concentrations.

Features and Limitations: EDMS incorporates both EPA-approved emissions inventory methodologies and dispersion models to ensure that analyses performed with the application conform to EPA guidelines. Since EDMS is primarily used in the process of complying with EPA air quality requirements (e.g. through an environmental impact statement) it is imperative that the application uses the most current data available. For this reason, it is the FAA's intention for the database to contain a comprehensive list of aircraft engines, ground support equipment, aerospace ground equipment, auxiliary power units, vehicular, and stationary source emission factor data. However, there may be cases where the database does not contain a specific data element (e.g. a newly available emission factor). In these cases, EDMS tries to make allowances for the user to enter their own data and will perform parameter validation where possible. The pollutants currently included in the emissions inventory are CO_2, CO, THC, NMHC, VOC, TOG, NO_x, SO_x, PM-2.5, PM-10, and 394 organic gases. CO_2 is calculated only for aircraft, and THC is calculated only for aircraft and APUs. Aircraft PM emissions are only available for aircraft with ICAO certified engines. From the 394 organic gases, 45 of them are considered to be Hazardous Air Pollutants (HAPs), while the other 349 are non-toxic compounds. EDMS performs dispersion analysis by generating input to EPA's AERMOD dispersion model, and provides an interface to the complex terrain module of AERMOD. To use this function, the user can run AERMAP (the AERMOD terrain pre-processor) as a part of EDMS. The pollutants currently included in EDMS for dispersion analysis are CO, THC, NMHC, VOC, TOG, NO_x, SO_x, PM-2.5 and PM-10.

Emissions Calculations: An emissions inventory is a summary of the total annual emissions of the modeled pollutants for the sources defined in a study. Depending on the purpose of the study, the emissions inventory may be an end in itself or an intermediate step towards performing a dispersion analysis.

Emission Inventory Pollutants: EDMS calculates emissions of the following pollutants:

a) CO_2 (carbon dioxide) for aircraft only,
b) CO (carbon monoxide),
c) THC (total hydrocarbons) for aircraft and APUs only,
d) NMHC (non-methane hydrocarbons),
e) VOC (volatile organic compounds),
f) TOG (total organic compounds),
g) NO_x (nitrogen oxides),
h) SO_x (sulfur oxides),
i) PM-10 (particulate matter, 10 microns)
j) PM-2.5 (particulate matter, 2.5 microns), and
k) 394 organic gases

- 45 Hazardous Air Pollutants (HAPs)
- 349 non-toxic compounds

Aircraft PM emissions are only estimated for aircraft with ICAO certified engines using the FOA3 methodology.

EDMS input required: EDMS generates input files for use with the AERMOD dispersion model. Since EDMS is a model specifically developed for use at airports, the inputs relate directly to the placement of aircraft and other source activity and movement on the airport. Data input includes:

1) Creation and specification of

- runways,
- queues,
- taxiways,
- buildings
- gates.

2) Emission data of

- aircraft activity,
- ground support equipment,
- on-road vehicles,
- parking facilities,
- stationary sources, and
- training fires.

Aircraft activity is expressed in landing-takeoff (LTO) cycles. Each LTO consists of taxiing, queuing, takeoff, climb out, approach, and landing. There is no provision to specify arrival numbers and departure numbers independently of each other. EDMS contains default aircraft-specific data relating to the emissions:

1) Times in mode (TIMs) are the durations per LTO cycle that an aircraft spends in each of the four modes of aircraft operation. Takeoff, climb out, approach, and landing mode TIMs are aircraft-specific in EDMS.
2) Ground support equipments (GSE) are made to be based on aircraft type with an operating time associated with each aircraft LTO cycle.
3) Aircrafts are assigned to an unlimited number of taxiways and runways.
4) Engine emission factors are aircraft-specific.

Emissions Inventory: An emissions inventory is a summary of the total pollutants generated by all active sources in the study. Using EDMS to perform an emissions inventory requires the user to identify the emission sources, the annual activity for each of these sources and, in the case of user-defined sources, the emission factors. EDMS then calculates the total annual pollutant emissions for each of the identified sources and presents it in both a summarized report and a detailed report.

Dispersion Calculations: EDMS generates input files for use with EPA's AERMOD dispersion model, its meteorological preprocessor, AERMET, and its terrain preprocessor, AERMAP. AERMOD is a steady-state plume model that assumes a Gaussian concentration distribution in both the horizontal and vertical directions in the stable boundary layer. In the convective boundary layer, dispersion is Gaussian in the horizontal direction, with the vertical direction being modeled by a bi-Gaussian probability density function. Detailed information about AERMOD is available from user guides and additional information contained on the EPA's internet site (www.epa.gov/scram001/dispersion_prefrec.htm#aermod).

Dispersed Pollutants: EDMS can calculate hourly emissions, and generate AERMOD input files for the following pollutants:

a) CO (carbon monoxide),
b) THC (total hydrocarbons) for aircraft and APUs only,
c) NMHC (non-methane hydrocarbons),
d) VOC (volatile organic compounds),
e) TOG (total organic compounds),
f) NO_x (nitrogen oxides),
g) SO_x (sulfur oxides),
h) PM-10 (particulate matter, 10 microns), and
i) PM-2.5 (particulate matter, 2.5 microns).

Inputs Required: The amount of data required to perform a dispersion analysis is significantly greater than the data necessary for just an emissions inventory. All of the inputs necessary for the emissions inventory are also necessary for dispersion modeling. In addition, some modeling options that are optional for just emissions inventory are required when dispersion is enabled, including:

1) Accurate operational profiles or a schedule,
2) Aircraft performance modeling,
3) Aircraft delay & sequencing modeling,
4) Hourly weather data, and
5) Placement of receptors.

An emissions inventory must first be generated before dispersion can be performed, since the set of emissions that are dispersed is the same as that produced from the annual inventory. The dispersion algorithms use the selected operational profiles or aircraft schedule to vary the source activity based upon time. It is important that accurate profiles be developed to represent the variation of individual source activity as this can affect the outcome of dispersion significantly. Two similar parameters found in all of the emissions source screens are the values for *Yearly* and *Peak Quarter Hour* activity. The dispersion pre-processing routines use the *Peak Quarter Hour* value in the computation of an emission rate. If the *Yearly* activity were the only known variable then the user would use operational profiles to derive the *Peak Quarter Hour* value. Upon entering the value for *Yearly* activity and choosing the appropriate *Quarter Hourly*, *Daily*, and *Monthly* operational profiles the program will automatically compute the corresponding *Peak Quarter Hour* value. Even if the *Peak Quarter Hour* value is known, and entered directly, accurate operational profiles will still have to be defined and selected for each source in the study. AERMOD itself uses hourly time bins. The use of quarter hours is only to provide better fidelity from the aircraft sequence modeling.

Since EDMS is a model specifically developed for use at airports and air bases, there are several screens that relate directly to the placement of aircraft and other source activity and movement on the airport. Data input includes the creation and specification of runways, taxiways, buildings, and gates. These inputs are converted into a collection of appropriate sources for modeling dispersion in AERMOD.

Dispersion Modeling Calculation: The intent of dispersion modeling is to assess the air pollutant concentrations at or near the airport or air base resulting from identified emissions sources. These pollutant concentrations are calculated to determine whether emissions from the site result in unacceptably high air pollution levels downwind by comparison with the National Ambient Air Quality Standards (NAAQS) or other relevant air quality standards. To perform dispersion modeling, EDMS requires the coordinates (in meters or feet relative to the user-specified origin) of each emissions source, the specification of an emissions rate (derived from emission factors) and its variation through time. For some sources, the release height, temperature and gas velocity are also required. The identification of spatial points in the coordinate system for concentration estimation (receptors), and the availability of weather data for individual hours are also required. The basic Gaussian equation, a mathematical approximation that simulates the steady-state dispersion of pollutants from a continuous point source is given below (the Gaussian approximation):

$$C(x,y,z,H) = \frac{Q}{2\pi\sigma_y\sigma_z\mu} exp\left[-\frac{1}{2}\left(\frac{y}{\sigma_y}\right)^2\right]\left\{exp\left[-\frac{1}{2}\left(\frac{z-H}{\sigma_z}\right)^2\right] + exp\left(-\frac{1}{2}\left(\frac{z+H}{\sigma_z}\right)^2\right)\right\}$$

C = point concentration at receptor, in mg/m3

(x,y,z) = ground level coordinates of the receptor relative to the source and wind direction (m)

H = effective release height of emissions, in meters (m)

Q = mass flow of a given pollutant from a source located at the origin (mg/s)

μ = wind speed, in m/s

σ_y = standard deviation of plume concentration distribution in y plane (m)

σ_z = standard deviation of plume concentration distribution in z plane (m)

The results of the AERMOD dispersion calculations are the concentrations, given in micrograms per cubic meter (mg/m3), at receptors for each hour.

Dispersion Data Output: Modeling concentrations is a three-step process in EDMS. First, the user must select the meteorological data to be used via the AERMET Wizard, which is started from the Weather dialog, which is opened from the Airport menu heading. EDMS includes the optional use of AERMAP, which is the terrain preprocessor of AERMOD. AERMAP creates source (.SRC) and receptor (.REC) files for inclusion in AERMOD dispersion analyses. Next, the user must Generate AERMOD Input Files, under the Dispersion menu heading. This step pre-processes the emissions for every source for every hour in the weather data. The user also has the opportunity to select different averaging periods as well as the desired pollutant at this time. Finally, the dispersion calculations may be run by selecting Run AERMOD under the Dispersion menu heading.

As the dispersion algorithms execute, AERMOD displays its current status on the screen. Once AERMOD has finished, the AERMOD window will close and the user will be returned to EDMS. After AERMOD has run, EDMS will have created a directory for each scenario within the study directory. In each scenario directory there will be a directory for each airport. In each airport directory will be a file with the (.OUT) extension for each year and pollutant in the scenario-airport combination. These files contain both the list of inputs to AERMOD along with the concentrations for that scenario-airport-year-pollutant combination. These results can be viewed and printed in any text editor. Concentration (.CON) files can be viewed in the Concentrations View.

The example result of using EDMS tool: Results from the simulations of Zurich airport traffic in 2003 with the EDMS tool. The number of aircraft considered reached nearly 266000. Pollutants considered were CO, HC and NOx.

EDMS	CO (t/y)	HC (t/y)	NO_x (t/y)
Aircraft	1225.5	150.9	949.8
Stationary Sources	15.3	89.1	71.4
Roadways	1805.5	179.6	166.6
APU, GPU, Engine Start	65.7	5.4	46.3
GSE	54.2	8.1	77.3

3. Balanced Approach to Aircraft Noise Management

In 2001, the ICAO Assembly endorsed the concept of a "balanced approach" to aircraft noise management (Appendix C of Assembly Resolution A35-5 (ICAO, 2004). The Assembly in 2007, reaffirmed the "balanced approach" principle and called upon States to recognize ICAO's role in dealing with the problems of aircraft noise (Appendix C of Assembly Resolution A36-22 (ICAO, 2004). This consists of identifying the noise problem at an airport and then analyzing the various measures available to reduce noise through the exploration of four principal elements, namely reduction at source (quieter aircraft), land-use planning and management, noise abatement operational procedures and operating restrictions, with the goal of addressing the noise problem in the most cost-effective manner. ICAO has developed policies on each of these elements, as well as on noise charges. ICAO Balanced Approach to aircraft noise control consists of long-term and short-term measures. Long-term measures must force the solutions at regional level: reduction of noise at source and certification; phase-out of non-certificated airplanes; noise charges; and land-use planning and management. Short-term measure must facilitate the solutions at local level, like noise abatement procedures and mitigation of aircraft operation. In any case huge relationship between them must be provided, because of reaching more efficiency and advantages.

Reduction of Noise at Source: Much of ICAO's effort to address aircraft noise over the past 30 years has been aimed at reducing noise at source. Aircraft and helicopters built today are required to meet the noise certification standards adopted by the Council of ICAO. These are contained in Annex 16 (ICAO, 2007) - *Environmental Protection*, Volume I - Aircraft Noise to the Convention on International Civil Aviation, while practical guidance to certificating authorities on implementation of the technical procedures of Annex 16 is contained in the *Environmental Technical Manual on the use of Procedures in the Noise Certification of Aircraft* (ACAO, 1993). The first generation of jet-powered aircraft was not covered by Annex 16 and these are consequently referred to as non-noise certificated (NNC) aircraft (e.g. Boeing 707 and Douglas DC-8). The initial standards for jet-powered aircraft designed before 1977 were included in Chapter 2 of Annex 16. The Boeing 727 and the Douglas DC-9 are examples of aircraft covered by Chapter 2. Subsequently, newer aircraft were required to meet the stricter standards contained in Chapter 3 of the Annex. The

Boeing 737-300/400, Boeing 767 and Airbus A319 are examples of "Chapter 3" aircraft types. In June 2001, on the basis of recommendations made by the fifth meeting of the Committee on Aviation Environmental Protection (CAEP/5), the Council adopted a new Chapter 4 noise standard, more stringent than that contained in Chapter 3. Starting 1 January 2006, the new standard became applicable to newly certificated aircraft and to Chapter 3 aircraft for which re-certification to Chapter 4 is requested.

Technology solutions and the positive measures taken by airport authorities (restrictions on use of land, procedures for takeoff and landing, operating restrictions, compensating residents, ...), failed to reduce their impact near airports because of the growth in air traffic.

Operating Restrictions: Noise concerns have led some States, mostly developed countries, to consider banning the operation of certain noisy aircraft at noise-sensitive airports. In the 1980s, the focus was on NNC aircraft; in the 1990s, it moved to Chapter 2 aircraft; today, it has moved to the noisiest Chapter 3 aircraft. However, operating restrictions of this kind can have significant economic implications for the airlines concerned, both those based in the States taking action and those based in other States (particularly developing countries) that operate to and from the affected airports. On each occasion, the ICAO Assembly succeeded in reaching an agreement – contained in an Assembly resolution – that represented a careful balance between the interests of developing and developed States and took into account the concerns of the airline industry, airports and environmental interests.

In the case of Chapter 2 aircraft, the ICAO Assembly in 1990 urged States not to restrict aircraft operations without considering other possibilities first. It then provided a basis on which States wishing to restrict operations of Chapter 2 aircraft may do so. States could start phasing out operations of Chapter 2 aircraft from 1 April 1995 and have all of them withdrawn from service by 31 March 2002. However, prior to the latter date, Chapter 2 aircraft were guaranteed 25 years of service after the issue of their first certificate of airworthiness. Thus Chapter 2 aircraft which had completed less than 25 years of service on 1 April 1995 were not immediately affected by this requirement. Similarly, wide body Chapter 2 aircraft and those fitted with quieter (high by-pass ratio) engines were not immediately affected after 1 April 1995. Many developed countries including Australia, Canada, the United States and many in Europe, have since taken action on the withdrawal of operations of Chapter 2 aircraft at their airports, taking due account of the Assembly's resolution. This has had a substantial impact in reducing noise levels at many airports.

However, the benefits of removing Chapter 2 aircraft have now been largely achieved.

In the case of Chapter 3 aircraft, the ICAO Assembly in 2001 urged States not to introduce any operating restrictions at any airport on Chapter 3 aircraft before fully assessing available measures to address the noise problem at the airport concerned in accordance with the balanced approach. The Assembly also listed a number of safeguards that would need to be met if restrictions are imposed on Chapter 3 aircraft. For example, restrictions should be based on the noise performance of the aircraft and should be tailored to the noise problem of the airport concerned, and the special circumstances of operators from developing countries should be taken into account (Appendix E of Assembly Resolution A35-5).

In the frames of this work it is proposed to consider the following types of operating restrictions:

1) Global: to all traffic;
2) Aircraft specific: restriction on aircraft type or a group of aircraft based on individual noise performances.
3) Partial: restrictions applied for an individual time period or specific days.
4) Progressive: gradual decrease in the maximum level of traffic or noise energy:

   - 4.1 number of movements per period;
   - 4.2 quotas expressed as a combination of movements and aircraft noise characteristics.

Noise Abatement Operational Procedures: Noise abatement operational procedures are being employed today to provide noise relief to communities around airports from both arriving and departing aircraft. Noise abatement procedures are in-flight operations designed to minimize noise disturbances to residential areas along flight paths, as well as areas close to the airport. Noise abatement procedures enable reduction of noise during aircraft operations to be achieved at comparatively low cost. There are several methods, including preferential runways and routes, as well as noise abatement procedures for take-off, approach and landing. The appropriateness of any of these measures depends on the physical lay-out of the airport and its surroundings, but in all cases the procedure must give priority to safety considerations. ICAO's noise abatement procedures are contained in Annex 16, Volume I, Part V and Procedures for Air Navigation Services — Aircraft Operations (PANS-OPS, Doc 8168), Volume I — Flight Procedures, Part V. On the basis of recommendations made by CAEP/5, new noise abatement take-off procedures became applicable in November 2001.

Noise abatement operational procedures in use today can be broken down into three broad categories:

Noise abatement flight procedures

- Continuous Descent Arrival (CDA) – (Zaporozhest and Khardi, 2004 ; Eurocontrol, 2008 ; Khardi et al. 2010)
- Noise Abatement Departure Procedures (NADP)
- Modified approach angles, staggered, or displaced landing thresholds
- Low power/low drag approach profiles
- Minimum use of reverse thrust after landing

Spatial management

- Noise preferred arrival and departure routes
- Flight track dispersion or concentration
- Noise preferred runways

Ground management

- Hush houses and engine run up management (location/aircraft orientation, time of day, maximum thrust level)
- APU management

- Taxi and queue management
- Towing
- Taxi power control (Taxi with less than all engines operating)

Although noise abatement procedures may have quantifiable environmental benefits, effective implementation may be difficult: procedures must be developed, tested, and evaluated for benefits and ATC impacts; approved and accepted by the airport and the ANSP (Air Navigation Service Provider); and adopted by the airlines and other airport users.

Prior 2001 the PANS OPS guidance contained only two recommended procedures: ICAO A and ICAO B. In 2001 revision of PANS OPS development of noise abatement departure procedure (NADPs):

- engine thrust reductions cannot be made below 800' above the runway;
- the thrust reduction cannot be below the thrust level required by the certificated aircraft flight manual or approved manufacturers' operations manual;
- procedures for Air Navigation Services – Aircraft operations (PANS OPS) Volume 1 – Flight procedures (Doc 8168), Part 1, Section 7 (ICAO, 2007b):
    - NADP 1 - For noise areas sensitive near the airport;
    - NADP 2 - For more distant areas.

Noise Abatement Departure Procedure 1 (NADP 1): This procedure involves a power reduction at or above the prescribed minimum altitude and delaying flap/slat retraction until the prescribed maximum altitude is attained. At the prescribed maximum altitude, accelerate and retract flaps/slats on schedule while maintaining a positive rate of climb and complete the transition to normal en-route climb speed. The noise abatement procedure is not to be initiated at less than 800 feet AGL. The initial climbing speed to the noise abatement initiation point shall not be less than V_2 + *10 knots*. On reaching an altitude at or above *800 feet AGL*, adjust and maintain engine thrust in accordance with the noise abatement thrust schedule provided in the aircraft operating manual. Maintain a climb speed of V_2 + *10 to 20 knots* with flaps and slats in the take-off configuration. At no more than an altitude equivalent to *3000 feet AGL*, while maintaining a positive rate of climb, accelerate and retract flaps/slats on schedule.

At *3000 feet AGL*, accelerate to normal en-route climb speed.

Noise Abatement Departure Procedure 2 (NADP 2): This procedure involves initiation of flap/slat retraction on reaching the minimum prescribed altitude. The flaps/slats are to be retracted on schedule while maintaining a positive rate of climb. The thrust reduction is to be performed with the initiation of the first flap/slat retraction or when the zero flap/slat configuration is attained. At the prescribed altitude, complete the transition to normal en-route climb procedures. The noise abatement procedure is not to be initiated at less than 800 feet AGL. The initial climbing speed to the noise abatement initiation point *is* V_2 + *10 to 20 knots*.

On reaching an altitude equivalent to at least *800 feet AGL*, decrease aircraft body angle whilst maintaining a positive rate of climb, accelerate towards Flaps Up speed and reduce thrust with the initiation of the first flaps/slats retraction or reduce thrust after flaps/slats retraction. Maintain a positive rate of climb and accelerate to and maintain a climb speed equal to *Flaps Up speed + 10 to 20 knots till 3000 feet AGL*. At *3000 feet AGL*, accelerate to normal en-route climb speed.

SII-specific noise modeling system development: A specific noise modeling system has been developed, on the basis of the INM7.0 version. Its main characteristic is to better account for the airframe noise component (which varies with aircraft configuration - flaps/gear – and speed) during approach procedures, through the use of multi-configuration and multi-speed NPD data. The different elements of this noise modeling system are illustrated in Figure 8 below:

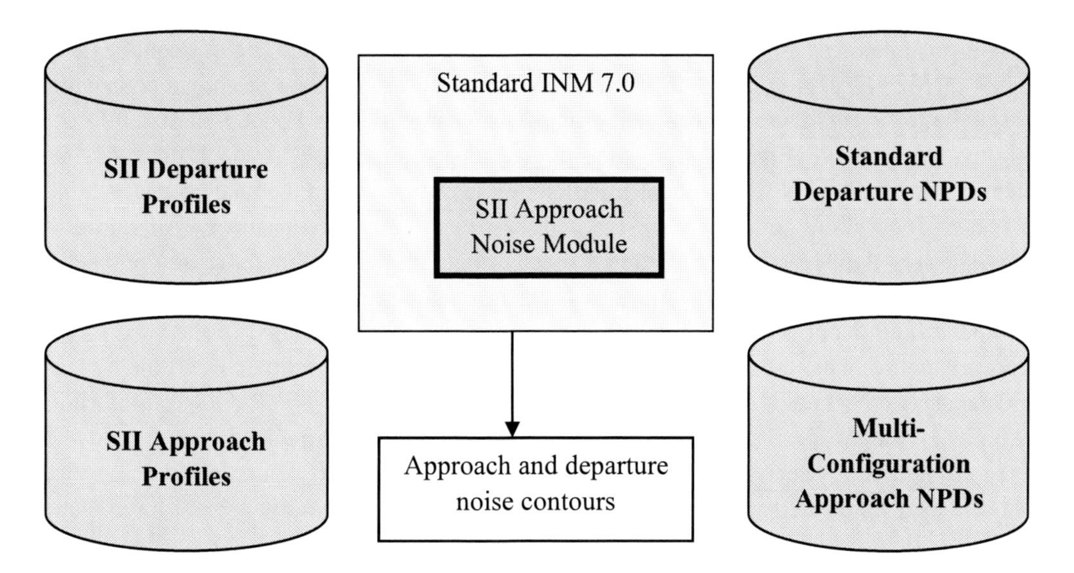

Figure 8. The SII noise modeling system.

Modifications of the INM single-event noise calculation method (INM segmentation process): To calculate the noise contribution of a given finite segment of the flight path, the segmentation process considers the aircraft configuration in which the segment is flown as an additional parameter, used to select the subset of NPD data associated to that configuration. It has to be noted that the aircraft configuration (flaps/gear state) is treated as a single, discrete state throughout the entire segment (Figure 9). In particular, the method does not model any transition between two configuration states (i.e. the fact that in reality, flaps deployment takes several seconds to be complete): configuration changes are here assumed to be instantaneous.

The noise contribution of a finite segment flown with configuration *Conf_N*, is determined using variants of equations below (respectively for maximum and exposure metrics), provided below:

The maximum noise level from a specific segment $L_{max,seg}$ is expressed as (CNELI, 2009):

$$L_{max,seg} = L_{max}(Conf\_N, P, V, d) + \Delta_I(\varphi) - \Lambda(\beta, l)$$

The contribution from one segment to the L_E is expressed as:

$$L_{E,seg} = L_{E\infty}(Conf\_N, P, V, d) + \Delta_V + \Delta_I(\varphi) - \Lambda(\beta, l) + \Delta_F$$

The only modified terms in the above equations (compared with the standard ones) are the 'baseline' noise levels L_{max} *(Conf_N, P, V, d)* and $L_{E\infty}$ *(Conf N, P, V, d)*, which are defined using *Conf_N* and speed *V* as additional input parameters. In particular, all the correction terms applied to the 'baseline' noise levels remain unchanged.

The above 'baseline' noise levels are interpolated for distance *d,* power settings *P* and speed *V* from the subset of NPD data associated to *Conf_N* (using respectively the LA_{max} and *SEL* NPDs). Values of *P*, *d* and *V* are computed exactly in the same way (for standard INM), as a function of the receiver-to-segment geometry (i.e. observer alongside or behind/ahead of the segment) and the noise metric (Figure 9).

Interpolations in configuration and speed-based NPD data: The 'baseline' levels L_{max}*(Conf_N, P,V,d)* and/or $L_{E\infty}$*(Conf_N, P,V,d)*, required to calculate the noise contribution of each finite segment, are interpolated from the configuration and speed-based NPD data, using the subset of NPD data associated to the configuration *Conf N* at which the segment is flown. As for standard NPDs, a linear interpolation is used between tabulated power-settings, whereas a logarithmic interpolation is used between tabulated distances. An additional interpolation is applied between tabulated speed values for which the configuration specific NPDs are provided (at least two values in order to enable the interpolations/extrapolations). This speed interpolation is linear, in a same way as for power settings. In the case of SEL (exposure metric) data, the noise calculation module preliminarily normalizes all the NPDs of the database to the same 160 knot reference speed (by adding *10log (V/160)* to the original noise levels), before performing the interpolations. The hence normalized NPDs include all the same duration effect (corresponding to the theoretical 160 knots reference speed), whatever their associated tabulated speed value is. Therefore, the above speed interpolation captures only the variations of the noise source state as a function of speed (through variations of the airframe noise component).

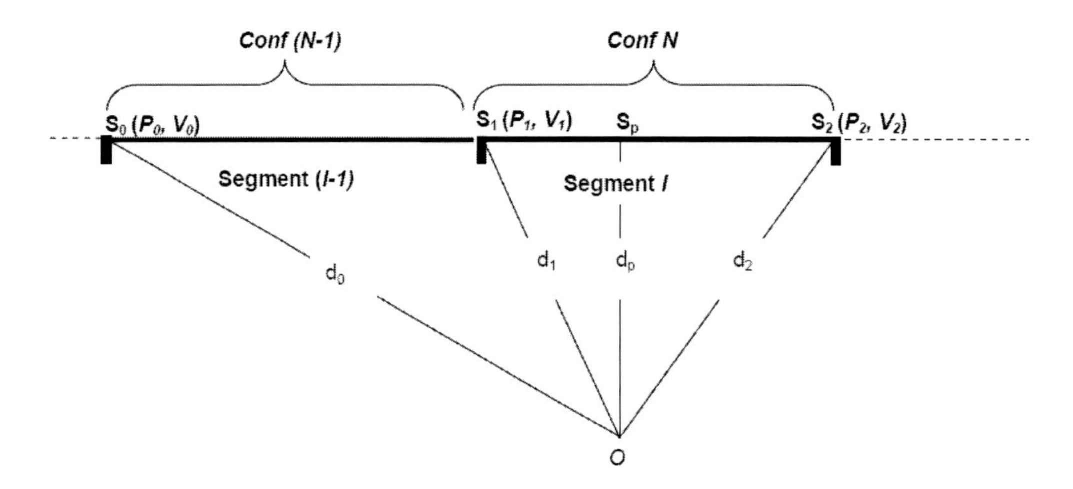

Figure 9. Multi-configuration NPD data segment geometry.

Additionally, whatever the speed effect on the overall noise (the decibel sum of the engine noise and airframe noise components), this linear interpolation may be considered as a reasonable assumption, given that the tabulated speed values for which the noise data are available are sufficiently close to each other. Moreover, large – uncontrolled – extrapolations for high speed values (associated to clean configuration) are limited, through the use of "duplicated" NPDs). How the noise level associated to a power *P*, speed *V* and distance *d* is interpolated between tabulated values of these parameters, is illustrated in Figure 10. The different steps are described below.

Interpolations at distance d: The noise level at engine power P_1, speed V_1 and distance d is given by:

$$L_{P1,V1,d} = L_{P1,V1,d1} + \frac{L_{P1,V1,d2} - L_{P1,V1,d1}}{\log d_2 - \log d_1}(\log d - \log d_1)$$

P_1 engine power value for which noise data are available in the database,
V_1 speed value for which noise data are available in the database,
d_1, d_2 distance values for which noise data are available in the database,
$L_{P1,V1,d}$ noise level at power P_1, speed V_1 and distance d,
$L_{P1,V1,d1}$ noise level at power P_1, speed V_1 and distance d_1,
$L_{P1,V1,d2}$ noise level at power P_1, speed V_1 and distance d_2.

The noise level at engine power P_2, speed V_1 and distance d is given by:

$$L_{P2,V1,d} = L_{P2,V1,d1} + \frac{L_{P2,V1,d2} - L_{P2,V1,d1}}{\log d_2 - \log d_1}(\log d - \log d_1)$$

P_2 engine power value for which noise data are available in the database,
V_1 speed value for which noise data are available in the database,
d_1, d_2 distance values for which noise data are available in the database,
$L_{P2,V1,d}$ noise level at power P_2, speed V_1 and distance d,
$L_{P2,V1,d1}$ noise level at power P_2, speed V_1 and distance d_1,
$L_{P2,V1,d2}$ noise level at power P_2, speed V_1 and distance d_2.

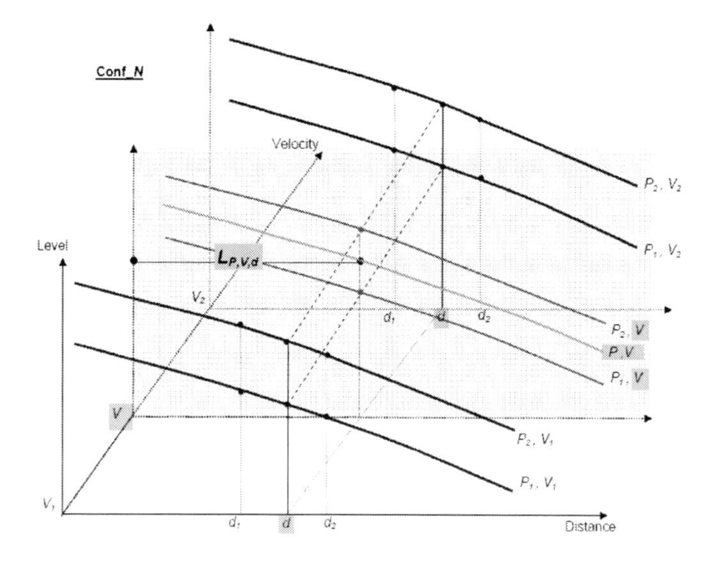

Figure 10. Interpolation in configuration/speed-based NPD curves.

The noise level at engine power P_1, speed V_2 and distance d is given by:

$$L_{P1,V2,d} = L_{P1,V2,d1} + \frac{L_{P1,V2,d2} - L_{P1,V2,d1}}{\log d_2 - \log d_1}(\log d - \log d_1)$$

P_1 engine power value for which noise data are available in the database,
V_2 speed value for which noise data are available in the database,
d_1, d_2 distance values for which noise data are available in the database,
$L_{P1,V2,d}$ noise level at power P_1, speed V_2 and distance d,
$L_{P1,V2,d1}$ noise level at power P_1, speed V_2 and distance d_1,
$L_{P1,V2,d2}$ noise level at power P_1, speed V_2 and distance d_2.
The noise level at engine power $P2$, speed $V2$ and distance d is given by:

$$L_{P2,V2,d} = L_{P2,V2,d1} + \frac{L_{P2,V2,d2} - L_{P2,V2,d1}}{\log d_2 - \log d_1}(\log d - \log d_1)$$

P_2 engine power value for which noise data are available in the database,
V_2 speed value for which noise data are available in the database,
d_1, d_2 distance values for which noise data are available in the database,
$L_{P2,V2,d}$ noise level at power P_2, speed V_2 and distance d,
$L_{P2,V2,d1}$ noise level at power P_2, speed V_2 and distance d_1,
$L_{P2,V2,d2}$ noise level at power P_2, speed V_2 and distance d_2.

Interpolations for speed V: The noise level at engine power P_1, speed V and distance d is given by:

$$L_{P1,V,d} = L_{P1,V1,d} + \frac{L_{P1,V2,d2} - L_{P1,V1,d1}}{V_2 - V_1}(V - V_1)$$

P_1 engine power value for which noise data are available in the database,
V_1, V_2 speed values for which noise data are available in the database
$L_{P1,V,d}$ noise level at power P_1, speed V and distance d,
$L_{P1,V1,d}$ noise level at power P_1, speed V_1 and distance d,
$L_{P1,V2,d}$ noise level at power P_1, speed V_2 and distance d.

In a similar way, the noise level at engine power P_2, speed V and distance d is given by:

$$L_{P2,V,d} = L_{P2,V1,d} + \frac{L_{P2,V2,d2} - L_{P2,V1,d1}}{V_2 - V_1}(V - V_1)$$

P_2 engine power value for which noise data are available in the database,
V_1, V_2 speed values for which noise data are available in the database
$L_{P2,V,d}$ noise level at power P_2, speed V and distance d,
$L_{P2,V1,d}$ noise level at power P_2, speed V_1 and distance d,
$L_{P2,V2,d}$ noise level at power P_2, speed V_2 and distance d.

Final interpolation at engine power P: The noise level at engine power *P*, speed *V* and distance *d* is finally obtained by:

$$L_{P,V,d} = L_{P1,V,d} + \frac{L_{P2,V,d} - L_{P1,V,d}}{P_2 - P_1}(P - P_1)$$

P_1 , P_2 engine power values for which noise data are available in the database,
$L_{P,V,d}$ noise level at power *P* , speed *V* and distance *d*,
$L_{P1,V,d}$ noise level at power P_1 , speed *V* and distance *d*,
$L_{P2,V,d}$ noise level at power P_2 , speed *V* and distance *d*.

Sourdine II noted that the methodology described above can be used for extrapolations as well (i.e. when *P, V* and/or *d* do not lie between tabulated values in the NPD data. However, extrapolations for high speed values (in clean configuration) are avoided, through the use of the "duplicated" NPDs.

Modification of the input flight path definition: The modified single-event noise calculation method uses the aircraft configuration in which each finite segment is flown as an additional parameter characterizing the aircraft noise state. As standard INM flight profiles do not contain flap/gear state explicitly (given that this type of information is not required in the standard INM noise calculation process), the format of *fixed-point* profile data has been modified to include flap/gear state as additional input parameter. With this modified format, the vertical profile database provides the configuration state of the aircraft at each point of the profile. All flap/gear labels used in the profile point's definition must match the configuration labels used in the multi-configuration NPD database. The 3-D flight path synthesis (where INM merges ground tracks and vertical profiles) has been adapted to account for this additional parameter. In particular, during the calculation of the contiguous straight flight path segments, each segment is assigned a single configuration (flaps/gear) label, treated as a discrete state throughout the entire segment. This section presents the results of series of tests done to obtain optimal flight procedure for take-off and climbing with the aim of minimizing noise and fuel consumption. Roll on the runway is excluded of the simulation; therefore the origin of local coordinates is located at the final point of roll on the runway. Initial conditions are presented in table 1. The analysis presented was done to compare standard take-off performed nowadays and optimal one (PANS OPS, 2000) for an Airbus-300 using Integrated Noise Model (INM) (Abdallah, 2007; Khardi et al. 2010). Altitude restrictions vary for different aircraft operations and are designed for both safety and noise abatement considerations. In general, departure procedures require jet aircraft to achieve an altitude above 3200 feet above ground level prior to initiating a turn.

For this comparison standard INM procedural profiles for take-off are used. Figures 11-13 present noise contours in terms of maximum noise levels L_{Amax} *(dBA)* for one take-off operation of A-300 performing standard INM take-offs procedures Standard 1 and Standard 5 and optimized procedure.

Table 1. Initial data for calculating take-off flight path.

Aircraft type	A-300
Maximum Gross weight, kg	165108
Maximum Gross Landing Weight. kg	133810
Maximum landing Distance, m	1636
Engine	jet
Number of engines	2
Static thrust, lb	52500
Initial altitude, m	0
Final altitude, m	3000
Initial speed at the final moment of roll, km/h	300
Final speed, km/h	500

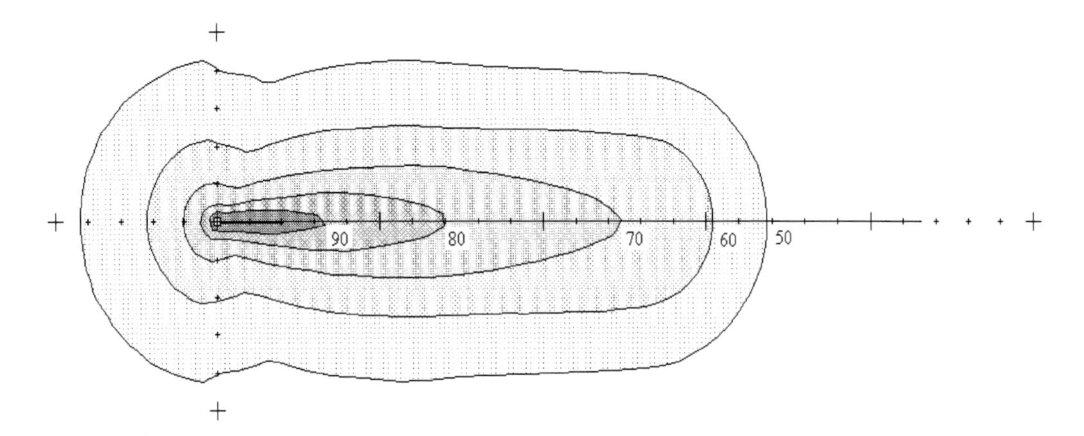

Figure 11. Noise optimized take-off procedure (case 5), maximum noise levels in dB(A).

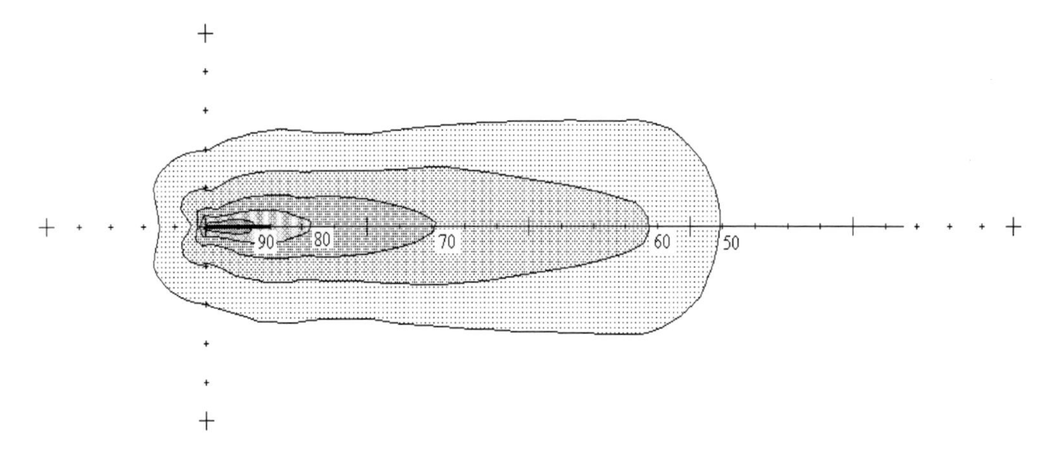

Figure 12. Standard take-off procedure 1, maximum noise levels in dB(A).

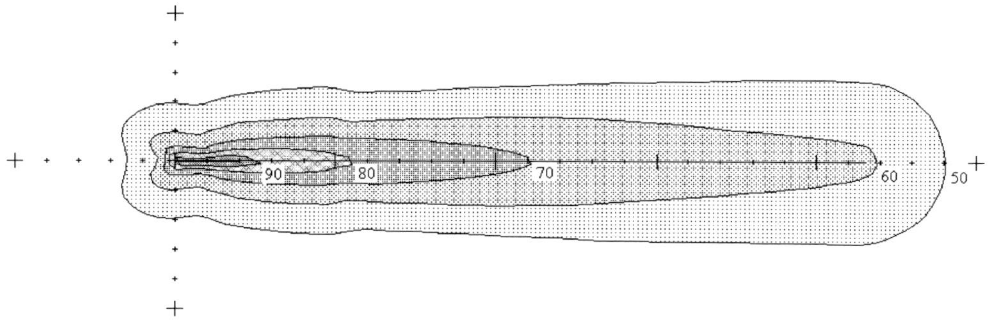

Figure 13. Standard take-off procedure 5, maximum noise levels in dB(A).

Approach procedures: Significant reduction in noise can be achieved through changes in aircraft operations that are enables by advanced approach flight path in addition to other measures of noise control. Operational changes include keeping arriving aircraft at their cruise altitude for longer than during conventional approaches. Aircraft is then performing a continuous descent to the runway at idle or near idle engine thrust with no level flight segments. Procedures with these approach flight path features are commonly referred to as continuous descent approach (CDA) procedures. In the absence of an internationally agreed definition of CDA in this report the definition proposed by Eurocontrol (INM) is used:

Table 2. Area of noise contours, SQ_KM.

L_{Amax}, dB(A)	Contour area, SQ_KM					
	stand 1	stand 2	stand 3	stand 4	stand 5	opt 5
50	289,29	312,12	331,97	374,08	427,74	536,90
55	197,51	214,46	229,51	260,62	300,06	375,33
60	122,68	134,06	143,43	163,99	190,06	262,87
65	62,94	68,34	73,48	83,45	96,21	175,05
70	32,82	35,60	38,07	43,12	49,44	104,95
75	16,67	18,21	19,50	22,26	25,71	57,34
80	7,98	8,84	9,49	11,06	12,98	30,09
85	4,06	4,43	4,71	5,40	6,28	14,38
90	1,71	1,86	1,92	2,24	2,63	6,18
95	0,19	0,14	0,47	0,55	0,65	2,19

"Continuous Descent Approach is an aircraft operating technique in which an arriving aircraft descends from an optimal position with minimum thrust and avoids level flight to the extent permitted by the safe operation of the aircraft and compliance with published procedures and Air Traffic Control (ATC) instructions." The term 'level flight' required to fulfill this definition should be locally determined for performance monitoring.

CDA starts ideally from the Top of Descent, i.e. at cruise altitude of about 10000 m, and allows the aircraft flying its individual optimal vertical profile down to runway threshold with engines at idle mode. CDA is based on the following operational concept:

- Replace the alternation descent / stable flyover / descent by a continue descent during the arrival phase in Visual Flight Rules (VFR).

- Avoid the sudden change of engine thrust variations, so decrease noise, fuel flow and emissions.
- Use modern avionics (Flight Management System) to manage the trajectory in vertical plan.

In operational terms, the CDA cannot be achieved from top of descent to touchdown point as to airspace or other operational constraints.

4. Aircraft Continuous Descent Approach Efficiency in Noise Control

The analysis presented was done to compare standard approach performed nowadays and optimal CDA (Khardi et al., 2010) for an Airbus-300 using Integrated Noise Model (INM) (Eurocontrol, 2008). For this comparison standard INM procedural profile for approach is used (table 3).

Table 3. Characteristics of standard approach procedural profile.

PT_NUM	Distance, m	Altitude, m	Speed, km/h	THR_SET, N
1	-34895,6	1828,8	511,0	2103,21
2	-17447,8	914,4	309,5	5845,46
3	-8723,9	457,2	283,5	7756,83
4	-5815,9	304,8	265,4	10304,32
5	0,0	0,0	261,5	9937,37
6	118,1	0,0	248,0	31500,00
7	1181,4	0,0	56,1	5250,00

Table 4. Characteristics of optimal CDA approach.

PT_NUM	DISTANCE, m	ALTITUDE, m	SPEED, km/h	THR_SET, N
1	-40000,0	2000,0	503,9	11474,00
2	-38317,0	2008,0	373,0	634,00
3	-25784,0	1351,0	322,1	707,00
4	-25382,0	1330,0	321,0	710,00
5	-23795,0	1247,0	313,0	1852,00
6	-21076,0	1105,0	310,0	3367,00
7	-17227,0	903,0	307,1	3378,00
8	-14935,0	783,0	305,0	3385,00
9	-11901,0	624,0	303,0	3393,00
10	-9265,0	486,0	301,0	3400,00
11	-7019,0	368,0	298,9	3407,00
12	-4046,0	212,0	297,1	3415,00
13	-362,0	19,0	290,0	830,00
14	0,0	0,0	288,0	830,00
15	954,0	0,0	150,0	830,00
16	1300,0	0,0	130,0	800,00

Features of the optimal CDA approach related to the altitude, speed and thrust versus distance from touchdown point, respecting flight safety rules and solving optimal control problem with the aim of minimizing noise on ground, were calculated using KNITRO (Khardi et al., 2010) solver and integrated in INM using fixed point profile procedure (table 4).

Figure 14 shows the altitude versus distance from touchdown point. The angle of descent is constant and equal to 3 degrees. It should be mentioned that this descent angle value is recommended by ICAO and accepted in the most civil aerodromes.

Figure 15 shows comparison of speed for both approaches including roll on the runway.

Figure 17 gives comparison of INM simulations of noise levels, in terms of maximum noise levels, at location points under flight path for an A-300 approach. Location points are situated up to the distance of 10 km from touchdown point along runway's longitudinal axe. For both approaches, at a relatively large ground distance from the touchdown point (about 10 km), noise levels have very close values which can be explained by large influence of noise propagation effects. In the middle of the graph and at close to runway noise levels differ, in mean, to more than 3 dB in favor of the optimal approach which reduces noise and therefore its impact.

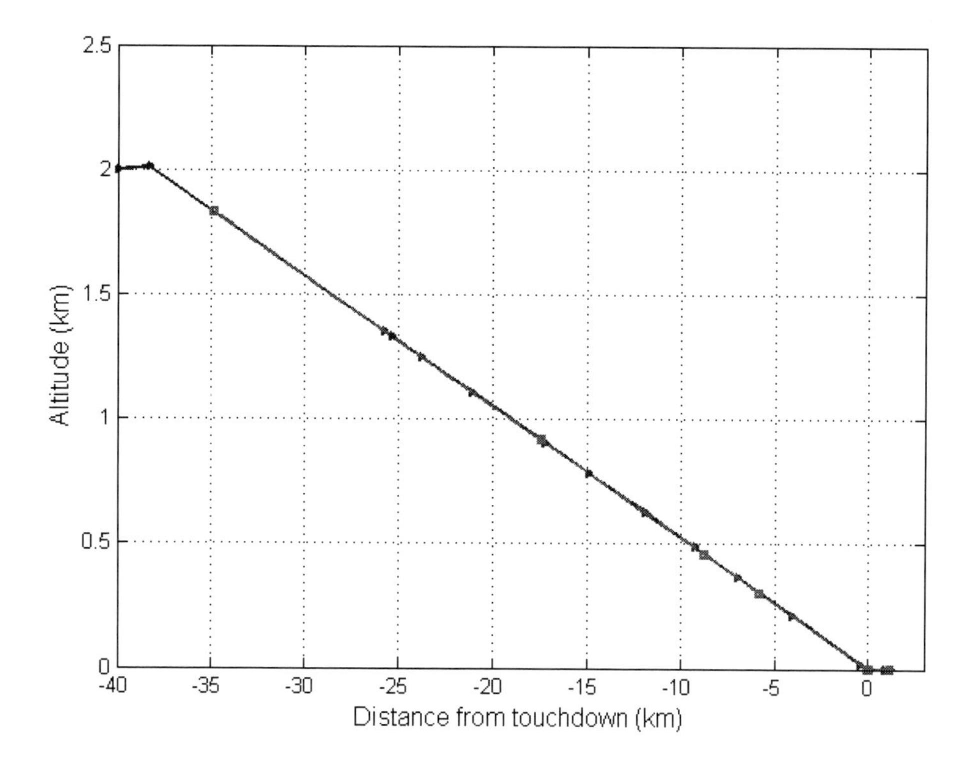

Figure 14. Characteristics of standard INM approach and optimal CDA approach. Altitude versus distance: —— optimal and – – – standard flight path.

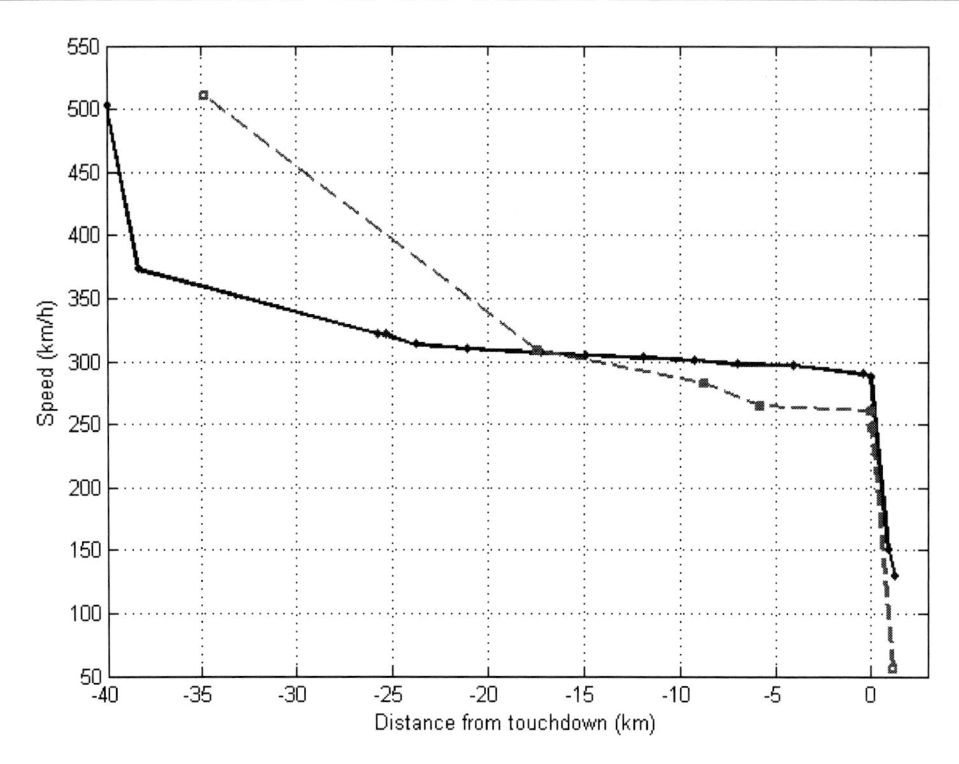

Figure 15. Characteristics of standard INM approach and optimal CDA approach. Speed of aircraft versus distance: —— optimal and – – – standard flight path.

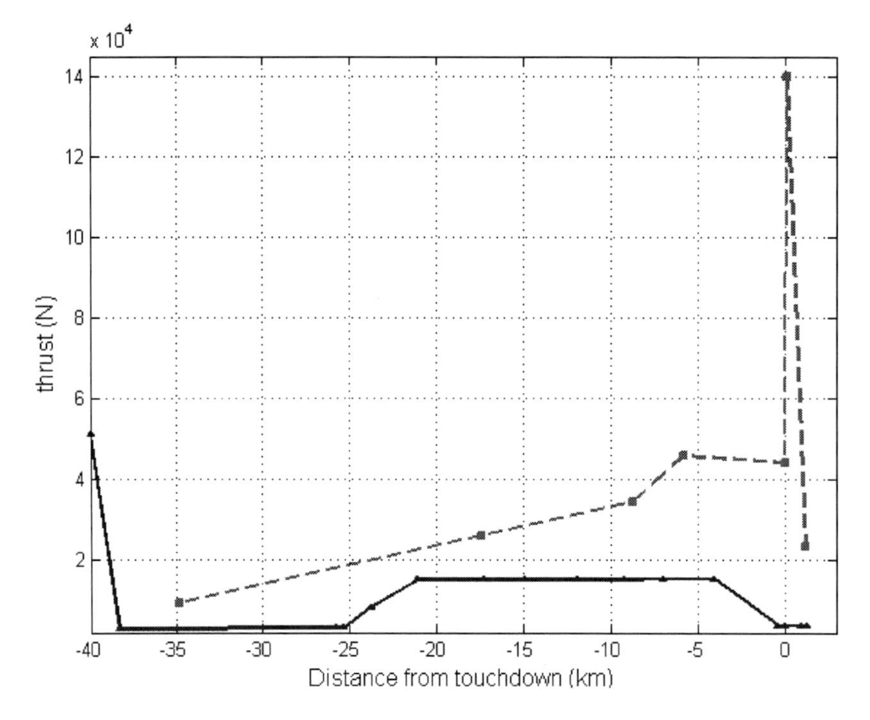

Figure 16. Characteristics of standard INM approach and optimal CDA approach. Thrust versus distance: —— optimal and – – – standard flight path.

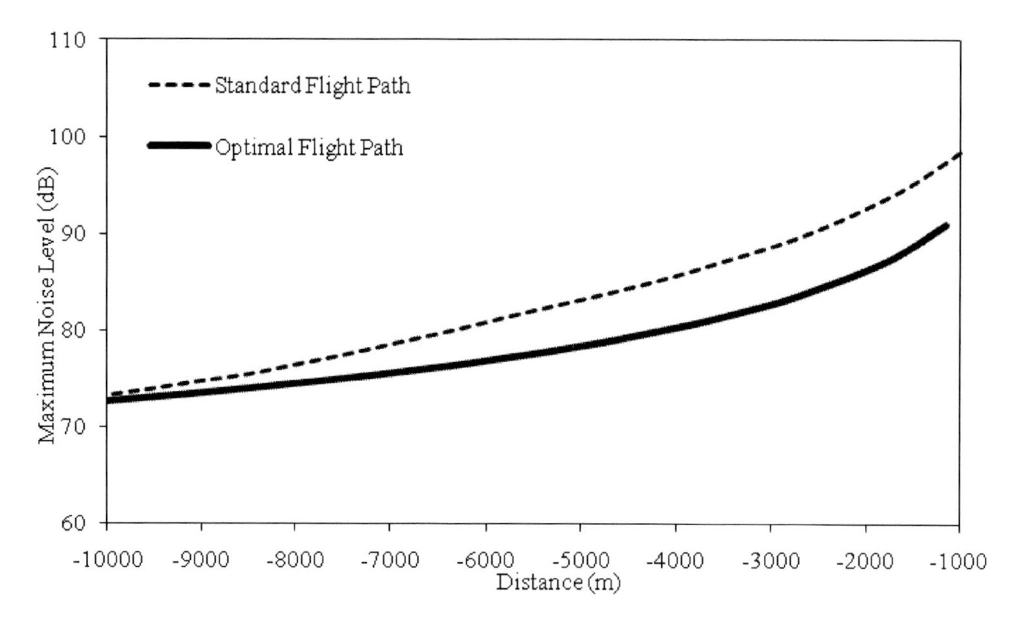

Figure 17. Comparison between noise levels at location points under the flight path.

Figure 18 presents noise contours in terms of maximum noise levels L_{Amax} *(dBA)* for one landing operation of A-300 performing standard INM approach procedure and optimal CDA procedure. This kind of shape of noise contours for standard INM approach procedure close to the runway can be explained because of using reverse of thrust at the phase of roll after touchdown. For optimal CDA procedure the reverse of thrust is not used. In the table 5 areas of noise levels presented in figure 18 is shown. Resuming the data it is possible to conclude that optimal CDA procedure generates smaller zones, from 15 to 30%, covered by noise levels.

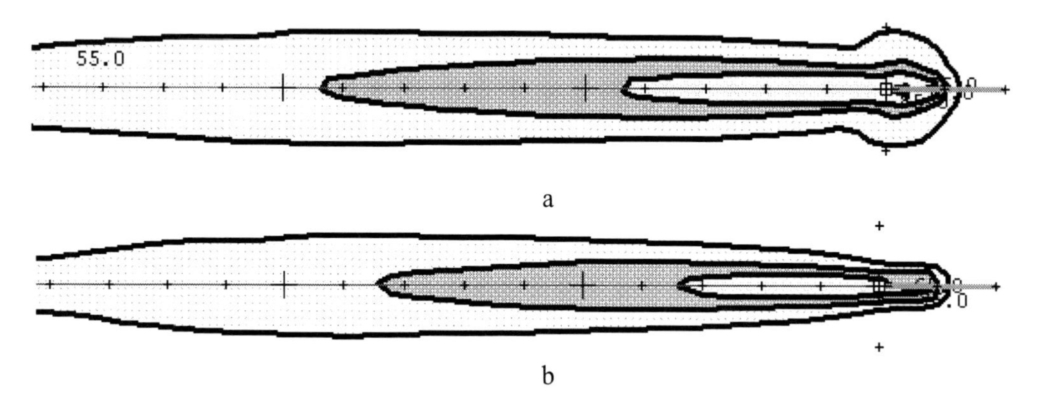

Figure 18. Contours of maximum noise levels L_{Amax} (dBA) for one landing operation of A-300: — - 55 dBA, — - 60 dBA, - 75 dBA, — - 85 dBA, a) standard INM approach procedure; b) optimal CDA procedure.

Maximum benefit from CDA is achieved when arriving aircraft closely follow an optimum continuous descent profile, whilst simultaneously minimizing thrust, avoiding sudden changes in thrust and reducing airframe noise by maintaining a clean aircraft configuration for as long as possible.

The nature and extent of the benefit from CDA will vary depending on the local situation but would typically include a significant noise, fuel and emissions reduction along the descent profile prior to the point at which the aircraft is established on the final approach path.

Table 5. Area of L_{Amax} *(dBA)* noise contours of one A-300landing operation.

Noise levels	Area, km^2	
	INM standard approach	Optimal CDA procedure
55	88,4	70,8
60	50,9	40,0
65	27,2	19,2
70	14,5	9,2
75	7,7	4,1
80	3,7	0,5
85	0,2	0,0

Experience with procedures that fit the definition of CDA indicate that the main noise benefit will be experienced on initial and intermediate approach segments (from approximately 8 to 25 nm from touchdown depending on local circumstances). In the open literature, analysis by Clarke et al. (2006), provided for some specific unrestricted CDA demonstrations, resulted in 5 to 6 db reduction in peak noise levels along some portions of the flight path (reduction of 30% in noise contours area). Also research by Zuijlen (2008) suggests a reduction of up to 50% of noise zones area over conventional approaches can be achieved. Therefore, the reliability of results presented fits with Clarke and Zuijlen, and Khardi et al. (2010) findings and confirm the significant advantage of CDA approach use in terms of noise and pollutant emissions, and fuel saving.

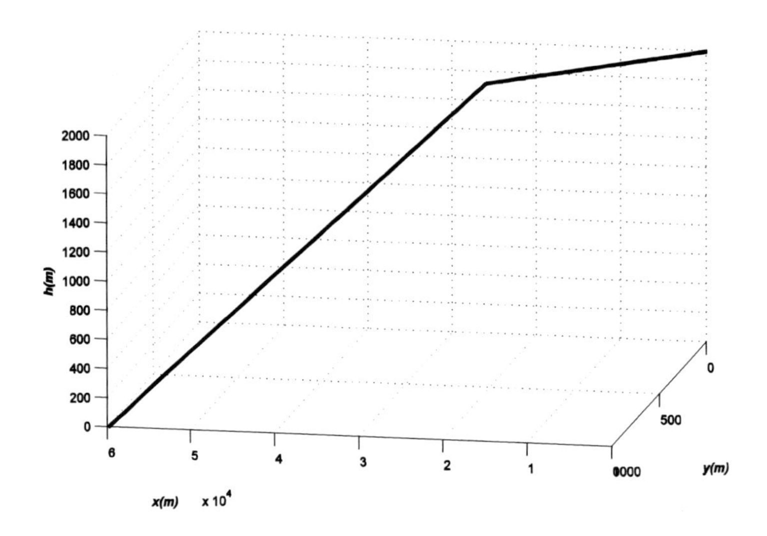

Figure 19. Optimized flight paths approach minimizing noise and fuel consumption in flight.

Aircraft Fuel Efficiency Improvements: In parallel with the aviation industry's natural vocation to develop high-performing products that respond to ever-increasing demands, market forces have always ensured that fuel burn and associated CO_2 emissions have been kept to a minimum for reasons of efficiency. Designing a product able to fulfill its mission safely with the lowest fuel consumption is a fundamental impetus behind reducing CO_2 (and other) emissions for each new aircraft type. This is the reason why aircraft engine and airframe manufacturers are always looking ahead at technological solutions that will enable significant environmental improvements. This is supported by extensive, continuous, and consistent research programs. In fact, generation after generation of aircraft has shown impressive weight reduction results due to improvements in: materials, manufacturing processes and systems, aerodynamics, engine performance, and advances in specific combustion and acoustic focused technologies. As a result, remarkable results have been achieved, not only without compromising the standards of safety and reliability, but by actually raising those to the highest levels. To give an order of magnitude of the reductions in CO2 emissions; a 70% reduction in fuel consumption was achieved between 1960 and 1990, thus more than tripling fuel efficiency during that period. From 1990 to 2004, that trend continued to the point where it is estimated that fuel efficiency in the commercial aviation fleet quadrupled from 1960 to 2004. Fuel efficiency gains result from a continuous flow of new and improved design features and technologies, which are hallmarks of the constantly evolving and innovative high-technology focused aviation industry. The mindset of continuous improvement through innovation is one that is deeply ingrained in the cultures of both manufacturers and operators in the aviation industry. Improved operational practices and enhanced air traffic management (ATM) systems and procedures also contribute to the overall efficiency improvements. Recent results by the International Air Transport Association (IATA) and IPCC (1999) show that the improvement trend continues. In addition to acquiring new modern high performance aircraft and engines, operators are investing in various fuel economy-related operational measures as another way to deal with challenging economic times in the industry. There is believed to be significant potential there, with manufacturers, ICAO and other stakeholders, striving to exploit all available means. An additional example of technological progress with respect to

Six Elements towards technology improvements: While the preceding text describes what technology has permitted industry to accomplish global globally to-date, this section focuses on six key ways (or strands of progress) in which technology plays a role in reducing aircraft emission impacts on global climate (AGAI, 2009). These key elements explain how past technological achievements were accomplished and also cover some of the potential areas for future improvements. The six key elements are:

1) Propulsion Systems
2) Materials
3) Structure, Aero & Systems Design & Methods
4) Manufacturing Processes
5) Aircraft Systems
6) Operational Procedures.

The subject of weight reduction is more complex than it first appears, for a number of reasons:

- Composites are not the only important factor involved in structural weight reduction. Experience shows that the optimum weight reduction depends on a combination of composites, advanced alloys, advanced processes, as well as improved overall structural architecture.
- Important contributors, other than materials and processes can also reduce the weight of an airplane. For example, load alleviation systems.
- Weight comparisons between different generations of airplanes are difficult to make and often not meaningful. This is because all technologies and design practices evolve over time, and the interpretation of gains estimated by projecting newer aircraft back to a past context is subject to interpretation. In addition, differences in design objectives at different points in time affect the characteristics and the optimization processes and trades. What really matters is the overall environmental performance resulting from the combination of aircraft concepts, integration, and optimization; in relation to the product requirements.

Active Controls: Active flight control can be used in many ways, ranging from the relatively simple angle of attack limiting found on airplanes such as the Boeing 727, to maneuver and gust load control investigated early with L-1011 aircraft, to more recent applications on the Airbus and 777 aircraft for stability augmentation. Reduced structural loads permit larger spans for a given structural weight and thus a lower induced drag. A 10% reduction in maneuver bending load can be translated into a 3% wing weight. This produces about a 6% reduction in induced drag. Reduced stability requirements permit smaller tail surfaces or reduced trim loads which often provide both drag and weight reductions. Such systems may also enable new configuration concepts, although even when applied to conventional designs, improvements in performance are achievable. In addition to performance advantages the use of these systems may be suggested for reasons of reliability, improved safety or ride quality, and reduced pilot workload, although some of the advantages are arguable.

New Airfoil Concepts: Airfoil design has improved dramatically in the past 40 years, from the transonic "peaky" sections used on aircraft in the 60's and 70's to the more aggressive supercritical sections used on today's aircraft. Continuing progress in airfoil design is likely in the next few years, due in part to advances in viscous computational capabilities. One example of an emerging area in airfoil design is the constructive use of separation. The examples below show the divergent trailing edge section developed for the MD-11 and a cross-section of the Aerobie, a flying ring toy that uses this unusual section to enhance the ring's stability.

Flow Control: Subtle manipulation of aircraft aerodynamics, principally the wing and fuselage boundary layers, can be used to increase performance and provide control. From laminar flow control, which seeks to reduce drag by maintaining extensive runs of laminar flow, to vortex flow control (through blowing or small vortex generators), and more recent concepts using MEMS devices or synthetic jets, the concept of controlling aerodynamic flows by making small changes in the right way is a major area of aerodynamic research. Although some of the more unusual concepts (including active control of turbulence) are far from practical realization, vortex control and hybrid laminar flow control are more likely possibilities.

Propulsion: Propulsion is the area in which most evolutionary progress has been made in the last few decades and which will continue to improve the economics of aircraft. Very high efficiency, unbelievably large turbines are continuing to evolve, while low cost small turbine engines may well revolutionize small aircraft design in the next 20 years. Interest in very clean, low noise engines is growing for aircraft ranging from commuters and regional jets to supersonic transports.

Advanced Airframe Concepts: Aerodynamic efficiency improvements such as higher lift/drag ratio (e.g., slotted cruise airfoil and natural laminar flow), new structural materials, and control system advances (such as fly-by-wire) could collectively improve fuel efficiency by about 10%, compared to current production aircraft. An aircraft representing some of these nearer term (2016) advanced airframe technologies (Condit, 1996). At the upper end of the airframe size scale (> 600 passengers), a more futuristic concept approach such as a Blended-Wing Bodied (BWB) could be developed. Studies have assessed the potential of the BWB design. The advantage of the BWB over conventional or evolutionary designs stem from extending the cabin span wise, thereby providing structural and aerodynamic overlap with the wing. This design reduces the total aerodynamic wetted area of the airplane and allows a higher span to be achieved because the deep and stiff center body provides "free" structural wingspan. Relaxed static stability allows optimum span loading. If engine and structural material technologies remain the same for the BWB, initial estimates show that fuel burn could be reduced significantly relative to that of conventionally designed large transports (Liebeck et al., 1998). In addition to the fuel burn and emissions reduction potential of this concept, the engine installation and airframe can help to minimize exterior noise: Inlets are placed above the wing so fan noise is shielded by the vast center body. An initial BWB concept could enter service after the year 2020. However, the passenger size and range of the initial design is not known at this time.

CONCLUSIONS

Noise abatement policies: they must be cooperatively developed and understood by aircraft and airport operators, engine and aircraft manufacturers and the local communities if such programs are to be effective.

Compared to the present-day approach procedures the environmental and economic benefits of the CDA procedure are:

- substantial reduction of community noise (smaller "noise footprint") as a result of
- higher altitude during larger part of the approach,
- lower power settings / clean aircraft configuration
- more flexibility in definition of (lateral) approach path geometry, due to curved approach application. This enables the procedure designer to design approach paths away from residential areas. Moreover, it has the additional advantage of reducing "the third party risk" (safety issue).
- less emission, due to application of "flight idle" thrust settings

- fuel conservation: less fuel is consumed during this procedure as compared to the conventional approach procedure,
- reduction of the overall approach time.

CDA will be most appropriately deployed where an aerodrome has existing or potential noise issues - typically where noise sensitive areas are being over-flown at medium altitude by arriving aircraft. However, the application of CDAs may be extended to all airports as a means of reducing the environmental impact of aviation. The local implementation of CDA shall always be subjected to a local safety assessment. As a result of noise levels comparison, it is obtained that the difference between is up to 4 dB confirming the new prospects of optimal CDA in aircraft operations. There can be significant fuel savings for the final arrival phase of flight with a optimal CDA, which also means that aircraft emissions will be reduced of about 30% to 35%. A system of flight procedures has to be only one part of a complete noise abatement program. Optimal utilization of the upcoming new systems for approach, navigation and flight management will lead to the introduction of more efficient flight procedures. These will contribute to:

- improvements in noise abatement,
- airport capacity enhancement.

Before these flight procedures become available for operational application, however, a lot of research work has to be carried out. This comprises not only flight- and ATC simulations studies but also in-flight demonstrations, to prove the operational feasibility of the concept of a particular procedure under real life conditions.

The Combined Effects of Aircraft Pollutant and Noise Emissions: Method for measurement, prediction and assessment of environmental problems such as Aircraft pollutant and noise emissions has been carried out. The use of certain methods will require justification and reliability that must be demonstrated and proven. Various methods have been adopted for the assessment of aircraft pollutant and noise emissions. The use of different and separate methodology causes a wide variation in results and there are some lacks of information. Because of these problem and there is no method of the combination for assessing between aircraft pollutant and noise emission, it is necessary to combine both of these method and that method should be integrated with the ICAO concept "Balanced Approach". But the questions are how to combine pollutant and noise emission, how about the balanced approach between operational, economic and environmental capacities, which method that would be reliable to use for quantifying the pollutant and noise emissions, how about the accuracy and the uncertainty of model that would be reliable and capable, how far the zoning area of the pollutant and noise emission that affect the people around airport, and why do we have to combine pollutant and noise emissions model.

In order to help Airport Authorities, Air Traffic Control, Political decision makers to decide, to manage and to get reliable information of the impacts of aircraft pollutant and noise emissions, it will be identify and comparison of various methods of pollutant and noise emissions assessment and evaluate the best methods to use in terms of accuracy, application, capability and problem of the uncertainty data and model which are an important part for improvement.

From a variety of assessment methods on both pollutant and noise emissions has shown the different assessment approach with different levels of uncertainty as well. This uncertainty factor has seen from the value of the index that is different from any method used for both assessment methods pollutant and noise emissions. Based on the problem of uncertainty factor of this index, it will be carried out the research on how to develop model of aircraft pollutant and noise emissions, which is the most suitable approach to be used, both for the assessment of pollutant and noise emissions and also the combination of those methods of assessment. Four objectives are necessary:

1) Identify and compare various methods of aircraft pollutant and noise emissions assessment.
2) Analyze and determine which method produces the best scenarios using data aircraft operations at the airport and evaluate the best methods to use in terms of accuracy, application, capability using Integrated Noise Model (INM) and the Emission Dispersion of Modeling System (EDMS) - Software Tools published by Federal Aviation Authority (FAA).
3) Develop and improve the combination method or model of aircraft pollutant and noise emissions and combine both emission zoning areas.
4) Develop model of the balanced approach between the economic, operational and environmental capacities to help Airport Authorities, Air Traffic Control, Political decision makers, to decide, to manage and to get reliable information of the impacts of aircraft operations

Currently it is proposed to make more deep analysis of operational procedures, grounding on solving of optimization tasks for flight paths at take-offs and approaches (Byrd, 2006; Khardi et al., 2010). For that an improved complex of models for particular dominant aircraft noise sources was developed. Among them the dominant sources of the engine – jet (including co-annular with by-pass ratio till 12-15), fans at inlet and outlet, combustion chamber, turbine, and the dominant sources of the airframe – wing with flaps and slats, landing gear. All improvements are reached on latest results of theoretical, numerical and experimental analysis of the sources from European and American aircraft noise programs. New models for the acoustic sources, their propagation and installation effects must be used for search of the new or for the improvement of currently used flight procedures. For that the optimization tasks must be formulated and solved. Criteria of optimization must include the flight event evaluation and overall scenario evaluation as well. Due to inclusion of propagation effects the routine meteorological and topographical conditions influence on results of optimization may be investigated. All the results for flight procedures must be examined on their relationship with long-term measures of the Balanced Approach to aircraft noise control, with zoning and land-use planning and management first of all. Resuming aforesaid the tasks have:

1) to make more deep analysis of operational procedures, grounding on solving of optimization tasks for flight paths at take-offs and approaches,
2) to search new or to improve currently used flight procedures,
3) to examine the results for flight procedures on their relationship with long-term measures of the Balanced Approach to aircraft noise control,

4) to analyze airport capacity in terms of its environmental characteristics,
5) to estimate airport environmental performances.

Land-use Planning and Management: Land-use planning and management is an effective means to ensure that the activities nearby airports are compatible with aviation. Its main goal is to minimize the population affected by aircraft noise by introducing land-use zoning around airports. Compatible land-use planning and management is also a vital instrument in ensuring that the gains achieved by the reduced noise of the latest generation of aircraft are not offset by further residential development around airports. ICAO guidance on this subject is contained in Annex 16, Volume I, Part IV and in the Airport Planning Manual, Part 2 - Land Use and Environmental Control (Doc 9184). A revised edition of this manual is being produced. The manual provides guidance on the use of various tools for the minimization, control or prevention of the impact of aircraft noise in the vicinity of airports and describes the practices adopted for land-use planning and management by some States. In addition, with a view to promoting a uniform method of assessing noise around airports, ICAO recommends the use of the methodology contained in Recommended Method for Computing Noise Contours around Airports (Circular 205).

Noise Charges: ICAO's policy with regard to noise charges was first developed in 1981 and is contained in ICAO's Policies on Charges for Airports and Air Navigation Services (ICAO, 2002). The Council recognizes that, although reductions are being achieved in aircraft noise at source, many airports need to apply noise alleviation or prevention measures. The Council considers that the costs incurred may, at the discretion of States, be attributed to airports and recovered from the users. In the event that noise-related charges are levied, the Council recommends that they should be levied only at airports experiencing noise problems and should be designed to recover no more than the costs applied to their alleviation or prevention; and that they should be non-discriminatory between users and not be established at such levels as to be prohibitively high for the operation of certain aircraft. Practical advice on determining the cost basis for noise-related charges and their collection is provided in the ICAO Airport Economics Manual (ICAO, 2008), and information on noise-related charges actually levied is provided in the ICAO Manual of Airport and Air Navigation Facility Tariffs (Callum, 2000).

Environmental Capacity of an Airport as an Element of Balanced Approach to Aircraft Noise Control: Aircraft noise disturbance is probably the most important factor (first of all, because it is the most geographically extensive form of impact) affecting the operation and development of airports around the world. Most of the world major airports have operational constraints or capacity limits based upon noise. But the future potential growths of air traffic imply that emission sources in the future will increase in importance.

The study of integrated airport impact shows that it is necessary to introduce the concept of airport traffic (operational) capacity according to environmental safety conditions. Evaluation of an airport impact on surrounding environment in terms of noise and air pollution could be realized by defining environmental capacity of an airport. Environmental capacity (Janic, 2001; Lieuwen, 2002; Zaporozhest et al., 2002) means such environmental performances of airport (both - operations and future development) that ensure fulfillments of normative conditions of noise at the control points around airport.

The main objective of this research is to use a concept of environmental capacity as it applies particularly to airports according to ICAO Balanced Approach to aircraft noise control and emissions control.

The Concept of Environmental Capacity: The concept of environmental capacity: as applied to an airport, the notion of "environmental capacity" C_{en} means reduction of an airport's (and by definition the air transport system's) capacity so as to ensure that airport environmental performances comply with the environmental rules Konovalova, 2004). According to this, airport's capacity C will be:

$$C = \min(C_{op} - C_{ec} - C_{en})$$

where, C_{op} - the operational capacity of an airport,

C_{ec} - the economic capacity of an airport,
C_{en} - environmental capacity of an airport.

The operational capacity C_{op} of an airport can be measured as the number of runway-taxiways slots, the terminal capacity or capacity of the apron areas. The operational capacity is limited only by means of flight safety. The economic capacity C_{ec} can be measured as the maximum number of passengers or aircraft, which can be accommodated on a particular day with a given amount of infrastructure under given economic conditions. In a short-term, the airport services load during peak and off-peak period determine these conditions. In the long-term, the availability of investments for airport expansion principally determines the economic conditions.

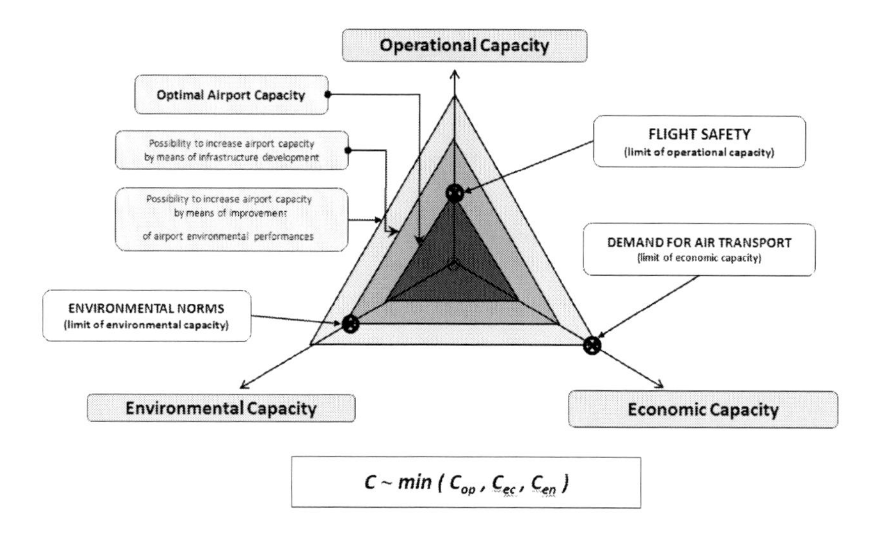

Interdependency of airport capacity types: The impact of the airports' operation upon the local environment is a major issue, which will affect both the capacity and the potential for future growth. This concept of "environmental capacity" as it applies to airports can be approached in at least two ways:

1) The first is that an airport's operational capacity is less the total sum of the individual environmental mitigation measures already in place at that airport.
2) The second is or could ever lead to an environmentally optimal solution:

   a) It is necessary to identify and separate short-run concerns which mainly affect quality of life (e.g., aircraft noise) from long-term issues which mainly affect the assimilative capacity of the environment to cope with what we are throwing at it (e.g., pollution and global warming);
   b) It is necessary to evaluate the viability of the environmental mitigation measures that are in the airport territory and in the vicinity. For example, many major airports have long-established night flight restrictions whose aim has been to protect local communities from excessive exposure from aircraft noise. From an environmental capacity perspective, such restrictions may be seen as a short-run, quality of life issue - and a successful mitigation measure –but with potentially more serious long-term environmental consequences.

The Methodology of defining airport environmental performances: A general definition is "Airport Environmental Performances are the individual for each airport characteristics which allow evaluating the interrelation of its operational, economic and environmental capacity". On the basis of this analysis it is possible to make the following conclusions:

1) Evaluation the environmental capacity starting from individual characteristics of the airport;
2) On the local level (environmental capacity defining from the environmental requirements on the airport territory and in the vicinity) the environment ability to assimilate this negative pressure varies widely and is dependent on different factors;
3) The airport is considered as the focus of environmental capacity definition; this can be explained so that there are many negative factors concentrated directly in the airport boundaries. Thereafter the levels of the influence of this factors and hence the number of people being subjected to it are individual for an airport;
4) Effective environmental management and long term planning are the key to ensuring that the environmental capacity and hence the operating capacity of an airport can be maximized.

Evaluation of an airport capacity according to noise and emissions is defined by three groups of conditions: technical, political and social.

Technical conditions:

- Monitoring and creation and survey noise and air pollution data bases.
  Integration these data bases with monitoring system for track keeping survey (for each landing-take-off cycle it is necessary to provide the following information - flight track used, time, aircraft type).

- Monitoring and survey for aircraft maintenance on ground (timetable of engine run-up operations for noise and emissions, aircraft anti-icing handling and painting for emissions).
- Modeling noise contours for various variants of operational conditions (intensity of aircraft operations). It is necessary to evaluate present day intensity of aircraft operations, to predict it for short-term and long-term period. But if the accuracy of such prediction is not sufficient, it is proposed to make noise contours assessment for intensity of aircraft operations equal to operational capacity of an airport.
- Defining sanitary-hygienic zone around an airport according to air pollutants emissions for all activities which generates emissions. Correlation noise contours and sanitary-hygienic zone to habitable areas around an airport.
- Analysis of probability that this airport has or can have in future constraints for its capacity or expansion according to its environmental performances.
- Analysis of probability that this airport can be influenced by any capacity constraints in associated airports.
- Land-Use Planning and Management: Planning (zoning, easement, etc.), mitigation (building codes, insulation, etc.) financial (tax incentives, charges, etc.); key to protecting noise reduction and abatement benefits; may involve "opportunity costs" for airports/local government.

Social conditions:

- Availability of information about ensuring noise and emissions standards for population living at the airport vicinity.
- Establishing special service for processing with complaints.

Political conditions:

- Establishing operational constraints according to noise and air pollutant emissions for short-term and long-term perspective.
- Ascertainment of conformity between, on the one hand, profits from airport operation and job placement for local community and, on the other hand, expenses for realization of environmental programs.
- Opportunity to make amendments to norms by means of implementation national, European or international recommendations.
- Limitations to noise during observation time T_{observ} are established in the fixed control points (zones) of noise control in the form of noise levels L_{AGoal} which, usually, serve as equivalent or day/night (which are also equivalent in fact, but include noise impact features during day and night time) levels. In Ukraine the equivalent noise level is established by national regulations as such criteria, for daytime $L_{AGoal}=65\ dBA$ and for nighttimes $L_{AGoal}=55\ dBA$. Limitations to air pollution are established in the fixed control points (zones) of air pollutants emissions control in the form of instantaneous and continuous (effective) concentrations.

Maximizing the environmental capacity of an airport: There are three variants of airport development according to the problem being discussed:

- Capacity changes at existing sites, without physical development.
- Physical development of airport including new sites or extensions to existing ones.
- Implication of redistributed air traffic to other airports.
- The key to maximizing the environmental capacity of an airport is the integration of environmental management into the corporate business planning process.
- It is a characteristic of environmental issues that they can require significant short-term investment in order to ensure long term return. Airport operators need to begin to plan now for the long term in order to meet anticipated infrastructure demand and environmental constraints. It is necessary to invest sufficient resources in environmental management and mitigation to ensure future capacity.
- And as applied to the air transport industry it means that not only airport operators are responsible for taking into account environmental factor, but also their service partners (airlines and other transport services). Of course, airport operator is the central figure who has to solve the problem of aircraft noise in the working area and in the vicinity (on the local level).

All organizations involved need to work jointly and where necessary airport operator has to enforce controls to ensure corporate environmental targets are met and that the environmental impact of the airports total operation is kept to a minimum. Airlines wishing to develop hub operations at a particular airport may actually take a proactive approach to encouraging the development of environmental best practice in order to secure their own future. In the long term it is to the advantage of all involved in the air transport industry that it continually strives to improve its environmental performance. Only through such action, coupled with planning for the long term that it will maximize its capacity within a climate of increasing environmental pressures. An important issue for airports is that the noise contamination can be minimized by:

- Minimizing noise disturbance using all technological, operational, and planning devices within its power;
- Considering mitigation and compensation for affected communities;
- Working within standards limits.

REFERENCES

Abdallah, L. *Minimization des bruits des avions commerciaux sous contraintes physiques et aérodynamiques. PhD.* thesis, Claude Bernard University, 2007

A.Bies, David and Colin H.Hansen, *Engineering Noise Control*, Theory and Practice, Spon Press, 2003

Airport Local Air Quality Studies, Information website: http://www.isa-oftware.com/ Alaqsstudy, [online] October, 2009

Antoine, N. E, *Aircraft Optimization For Minimal Environmental Impact*, Dissertation, 2004

AGAI - *Aviation and Global Atmosphere, Information Website* : http://www.grida.no/ publications/other/ipcc_sr/?src=/Climate/ipcc/aviation/094.htm, [online] October 2009

Burton, NJS, *Methods Of Assessment Of Aircraft Noise*, Master Thesis, 2004

Byrd, R.H., Nocedal, J. and Waltz R.A. *KNITRO: An Integrated Package for Nonlinear Optimization.* http://www-neos.mcs.anl.gov/neos/solvers. 2006

Clarke et al. *Development, design, and flight test evaluation of a continuous descent approach procedure for nighttime operation at Louisville International Airport. Final Report of the Partner CDA Development Team. Partnership for Air Transportation.* Noise and Emissions Reduction. An FAA / NASA / Transport Canada – Sponsored Center of Excellence. – January 2006

CNEL - *Community Noise Equivalent Level, Information website:* http://www.sfu.ca/ sonic-studio/handbook/Community_Noise_Equivalent.html, October, 2009

Callum, Th. *"Environmental capacity of airports – what does it mean?", in Wokshop Proceedings 2, Environmental Capacity.* The challenge for the aircraft industry, Heathrow airport, June, 2000

Directive 2002/30/EC, *On the Establishment of Rules and Procedures with Regard to the Introduction of Noise-Related Operating Restrictions at Community Airports*, 26 March 2002

Directive 2002/49/EC, *Relating to the Assessment and Management of Environmental Noise,* The European Parliament and of the Council, 25 June 2002

ECAC.CEAC Doc 29, *Report on Standard Method of Computing Noise Contours around Civil Airports,* 1997

e-CFR Data, *Information Website: http://ecfr.gpoaccess.gov [online]* November, 2009

Environmental Protection Agency (EPA), *Evaluation of Air Pollutant Emissions from Subsonic Commercial Jet Aircraft*, April 1999

Eurocontrol. *Continuous Descent Approach. Implementation Guidance Information.* European Organisation for the Safety of Air Navigation. May 2008

FAA, *Air Quality Procedures for Civilian Airports and Air Force Bases,* Appendix D: Aircraft Emission Methodology, April 1997

FAA, *Emission Dispersion Modeling System.* http://www.faa.gov/about/office_org/ headquarters_offices/aep/models, [Online], 15 Desember 2009

FAA, *Integrated Noise Model.* http://www.faa.gov/about/office_org/headquarters_ offices/aep/models, [Online], 15 Desember 2009

FAA, *Office Environment and Energy,* Aviation & Emissions A Primer, January 2005

FAA, *Noise Control and Compatibility Planning for Airports*, Advisory Circular 150/5020-1, 1983

Gardner, R.M et al., *The ANCAT/EC Global Inventory of NOx Emissions from Aircraft,* Paper, March 1997

ICAO. *Environmental protection. Annex 16 to the convention on international civil aviation. Aircraft noise*. Vol. 1, ICAO, 1993, 3rd Edition

ICAO. DOC 9082/6. *ICAO's Policies on Charges for Airports and Air Navigation Services.* ICAO Working paper. Montreal, 2002

ICAO. Doc. 9848. *Assembly Resolutions in Force*. Montreal. – 2004

ICAO. Doc 9501-AN/929. *Environmental Technical Manual on the use of Procedures in the Noise Certification of Aircraft*. Noise Certification Workshop. 3rd Edition. 2004

ICAO, *Doc 9562-Airport Economics Manual*, Second Edition, 2006

ICAO, *Doc-8168-Vol-1-5th-Edition-OPS-Aircraft-Operations-Flight-Procedures*, 5th Edition, 2006

ICAO. *Resolutions Adopted by the Assembly 36th*. Provisional Edition. 2007

ICAO. *Review of noise abatement procedure. Research and development and implementation results*. Discussion of survey results. Preliminary edition. 2007

ICAO, Environnemental Report, 2007

ICAO, *Airport Local Air Quality Guidance Manual*, 2007

ICAO. Doc. 9552. *ICAO Airport Economics Manual*. Conference on the Economics of Airports and Air navigation Services. Montréal, 15 to 20 September 2008

ICAO. Doc 7100. *Tariffs for Airports and Air Navigation Services*. Conference on the Economics of Airports and Air navigation Services. 2008

ICAO, *Annex-16-Vol-1-5th-Edition*, Aircraft Noise, July 2008

ICAO, *Annex-16-Vol-2-3rd-Edition,* Aircraft Engine Emissions, July 2008

ICAO, *Working Paper,* Conference On The Economics Of Airports And Air Navigation Services; September 2008

ICAO, *Doc 9082-Policies on Charges for Airports and Air Navigation Services*; Eight Edition 2009

INM. *Integrated Noise Model*. Office of Environment and Energy, Federal Aviation Administration (USA) http://www.faa.gov/about/office_org/headquarters_offices/aep/models/inm}

INM. *User's Guide and Technical Manual. Integrated Noise Model 7.0*. http://www.faa.gov/ .../inm/

IPCC Report, 1999

Janic M. *An Analysis of the Concepts of Airport Sustainable Capacity and Development // Ist Intern. Conf. "Environmental Capacity at Airports"*. – The Manchester Metropolitan University. – Manchester, Great Britain. 2nd and 3rd Apr. 2001

J. Crocker, Malcolm, *Handbook of Noise and Vibration Control,* Copyright © 2007 John Wiley & Sons, Inc, 2007

Kalivoda, Manfred T. and Kudrna, M. *Methodologies for Estimating Emissions from Air Traffic,* October 1997

Khardi S., Abdallah, L., Houacine, M. and O. Konovalova. *Optimal approach minimizing aircraft noise and fuel consumption*. Acta Acustica united with Acustica. Vol. 96, 68-75, 2010

Konovalova E.V. *Airport Acoustic Model // Science and Youth. Collected articles*. – NAU, Kiev, – 2004

Lieuwen, A. *The Concept Of Airport Environmental Capacity*. Manchester Metropolitan University, Department of Environmental & Geographical Sciences. October 2002

NASA, *Review of Integrated Noise Model (INM) Equations and Processes*, NASAICR-2003-2 124 14, May 2003

PANS OPS. *Procedures for Air Navigation Services – Aircraft operations* Volume 1 - Flight procedures (Doc 8168), Part 1, Section 7. 2000

Peeters P.M., Middel J., Hoolhorst A., *Fuel efficiency of commercial aircraft an overview of historical and future trends*, National Aerospace Laboratory, NLR,November 2005

Sourdine II WP5, Airport Noise and Emission Modelling Methodology, 2005

Zaporozhets A.I., Konovalova E.V., Kartishev O.A. *Environmental Capacity of an Airport According to Noise Conditions // Transport noise and vibration: Proceedings of the VIth Int.* Sumposium (4-6 June 2002), – St. Petersburg, Russia. – 2002

Zaporozhets O.I. and Tokarev V. *Predicted Flight Procedures for Minimum Noise Impact.* Applied Acoustics, Vol. 55, No. 2, pp. 129-143. 1998

Zaporozhets O.I. and Khardi S. *Optimisation of aircraft flight trajectories as a basis for noise abatement procedures around the airports.* Theoretical considerations and acoustical applications. INRETS Report. N° 257. 2004

Zuijlen, E. van. *Noise, Europe's airports' perspective.* Airport Environmental Colloquium. Airport Council International. – Cairo, 26-27 November 2008

ADDITIONAL REFERENCES

Adkins R. C., Elsaftawy A. S. - *A double Acting Variable Geometry Combustor.* ASME Gas Turbine Conference, San Diego, California, 12-15 March 1979

Airbus. (2005a). *"Aircraft families / A330 / A340 family / A330-300 The wealth generator."* Airbus S.A.S.

Airbus. (2005b). *"Aircraft families / A330 / A340 family / A340-300 Pathfinder to profitability."* Airbus S.A.S.

Albritton, D., Amanatidis, G., Angeletti, G., Crayston, J., Lister, D., McFarland, M., Miller, J., Ravishankara, A., Sabogal, N., Sundararaman, N., and Wesoky, H. (1997). *"Global atmospheric effects of aviation. Report of the Proceedings of the symposium held on 15-19 Aprille 1996 in Virginia beach Virginia, USA."* NASA CP-3351, NASA, Washington

Andrews G. E., Alkabie H. S., Abdul Hussain U.S., Abdul Aziz M. - *Ultra Low NO, Ultra Lean Gas Turbine Primary Zones with Liquid Fuels. Fuels and Combustion Technology for Advanced Aircraft Engines*, AGARD-CP-536, May 1993

ATA. (2005). *"2005 Economic Report. New thinking for a new centry."* Air Transport Association of America, Washington

ATAG. (2005). *"Aviation & environment summit discussion paper."* Air Transport Action Group, Geneva

Babikian, R., Lukachko, S. P., and Waitz, I. A. (2002). *"The historical fuel efficiency characteristics of regional aircraft from technological, operational, and cost perspectives."* Journal of Air Transport Management, 8(6), 389-400

Bahr D. W. *Turbine Engine Developers Explore Ways to Lower NO, Emission Levels.* ICAO Journal, August 1992, 14-17

Bahr D.W. - Aircraft Turbine NO, *Emissions Limits - Status and Trends. ASME International Gas Turbine and Aeronengine Congress,* Cologne, Germany, 1-4 June 1992, 87-GT-126

Bickerstaff, C. (2005). *"Aircraft Technological Developments."* AERONET III Workshop on Air Transportation Systems, Stockholm

Boeing. (2005). *"Airport Technology. Airplane Characteristics for Airport Planning."* Boeing Commercial Airplanes

Brunner D., Stahelin J. - *Planned Measurements of NO, NO2 and O3 from a SWISSAIR Airliner. International Scientific Colloquium on the Impact of Emissions from Aircraft and Spacecraft upon the Atmosphere*, Cologne, Germany, 18-20 April 1994

Bureau of Transport Statistics. (2005b). *"Historical Air Traffic statistics,* annual 1954-1980"

CE. (2000). *"ESCAPE: Economic screening of aircraft preventing emissions; main report."* Publ. code: 00.4404.16, CE/Peeters Advies/ADSE, Delft

Chevron Products Company. (2005). *"Aviation Fuels Technical Review (FTR-3)."* Chevron U.S.A. Inc., San Ramon, CA

Correa S. M. - *Carbon Monoxide Emissions in Lean Premixed Combustion*. Journal of Propulsion and Power, November 1992, 1144-1151

Goodger E. M. - *Jet Fuels Development and Alternatives*. Journal of Aerospace Engineering, 1995, 209, 147-156

Goyal A., Ekstedt E. E., Szaniszlo A. J. -*NASA Advanced Low Emissions Combustor Program*. ASME, 1983, 83.JPGC-10

Haberland Ch., Kokomiak M. - *A Strategy for Configurational Optimization of Aircraft with Respect to Pollutant Emissions and Operating Costs*. International Scientific Colloquium on the Impact of Emissions from Aircraft and Spacecraft upon the Atmosphere, Cologne; Germany; 18-20 April 1994

Haberland Ch., Kranz O., Stoer R. - *Impact of Operational and Environmental Aspects on Commercial Aircraft Design.* 19th Congress of the ICAS, Anaheim, California, 18-23 September 1994, ICAS-94- 1.3.1

Hiittig G. - *Estimation of Civil Air Traffic Exhaust Emissions Evaluation for the Terminal Area (ICAOLTO-Cycle) and for the Territory of the FRG*. International Scientific Colloquium on the Impact of Emissions from Aircraft and Spacecraft upon the Atmosphere, Cologne, Germany, 18-20 April 1994

IATA. (1957-2004). *World air transport statistics*, International Air transport Association, Montreal

IATA. (2000). *World air transport statistics*, International Air transport Association, Montreal-Geneva-London

IEA. (2004). *"2. Units and conversions."* Energy balances of OECD countries, 2002-2003 International Energy Agency, Paris

International Civil Aviation Organization. *ICAO Engine Exhaust Emissions Data Bank,* first edition, December 1993

Jamieson J. B. - *Twenty-First Century Aero-Engine Design: The Environmental Factor.* Journal of Aerospace Engineering, 1991, 204, 119-134

Kermode, A. C. (1972). *Mechanics of flight,* Pitman Publishing Limited, London.

Koff B. - *Aircraft Gas Turbine Emission Challenge Presentation*, Aero Engine and Environment Symposium, Berlin, Germany, 17 June 1992

Lee, J. J. (1998). *"Historical and future trends in aircraft performance, cost, and emissions,"* Massachusetts Institue of Technology, Cambridge

Lee, J. J., Lukachko, S. P., Waitz, I. A., and Schafer, A. (2001). *"Historical and future trends in aircraft performance, cost and emissions."* Annual Review Energy Environment, 26, 167-200

Lefebvre A. H. - *The Role of Fuel Preparation in Low Emissions Combustion*. ASME Gas Turbine and Aeroengine Congress, Houston, Texas, 5-8 June 1995, 95-GT-465

Lufthansa. (2005). *"Balance".* Key data on environmental care and sustainability at Lufthansa.

Marenco A., Nedelec P., Thouret V., Grouhel C. - *Measurement of Ozone and Water Vapor on Airbus In-Service Aircraft: The MOZAIC Program. International Scientific Colloquium on the Impact of Emissions from Aircraft and Spacecraft upon the Atmosphere*, Cologne, Germany, 18-20 April 1994

Miles D. E. - *The Asymptotic Impact of Aircraft Emissions. an Issue for Mutual EU Activities between Aeronautics and Environment. International Scientific Colloquium on the Impact of Emissions from Aircraft and Spacecraft upon the Atmosphere,* Cologne, Germany, 18-20 April 1994

Nicol D. G., Malte P. C.: Lai J., Marinov N. N., Pratt D. T., Corr R. A. - *NO, Sensitivities for Gas Turbine Engines Operated on Lean-Premixed Combustion and Conventional Diffusion Flames.* ASME Gas Turbine and Aeroengine Congress, Cologne, Germany, 1-4 June 1992, 92-GT-115

Nicol D. G., Steele R. C., Marinov N. M., Malte P. C. - *The Importance of the Nitrous Oxide Pathway to NO, in Lean-Premixed Combustion.* ASME Gas Turbine and Aeroengine Congress, Cincinnati, Ohio, 24-27 May 1993, 93.CT-342

Noda C. - *Development of a 2-D Atmospheric Model and Measurement of CO2 and CH, in Air sampled by Airline Flights.* International Scientific Colloquium on the Impact of Emissions from Aircraft and Spacecraft upon the Atmosphere, Cologne, Germany, 18-20 April 1994

OECD. (2003). *"Energy statistics odf OECD countries 2000-2001."* Organization for Economic Co-operation and Development, Paris

Penner, J. E., Lister, D. H., Griggs, D. J., Dokken, D. J., and McFarland, M. (1999). *"Aviation and the global atmosphere; a special report of IPCC working groups I and III."* Cambridge University Press, Cambridge

Prop-Liners of America Inc. (2005). *"Lockheed 1649A technical specifications."* Prop-Liners of America, Inc, Hartford, CT

Pulles, J. W., Baarse, G., Hancox, R., Middel, J., and van Velthoven, P. F. J. (2002). *"AERO main report. Aviation emissions and evaluation of reduction options."* Ministerie van V&W, Den Haag

Ralph M. O., Newton P. J. - *Experts consider operational measures as means to reduce emissions and their environmental impact.* ZCAO Journal, 1996, 51, no 2, Montreal

Reporting year 2004." Lufthansa, Cologne

Schumann U., ed. - AERONOX - *The Impact of NO, Emissions from Aircraft upon the Atmosphere at Flight Altitudes S-15 km, Final Report to the Commission of European Communities.* Publication EUR 16209 EN of the European Commission, Brussels, August 1995

Segalman I., McKinney R. G., Sturgess G. J., Huang L.-M - *Reduction of NO, by Fuel-Staging in Gas Turbine Engines - A Commitment to the Future. Fuels and Combustion Technology for Advanced Aircraft Engines*, AGARD-CP-536, May 1993

Sturgess G. J., McKinney R., Morford S. – *Modification of Combustor Stoichiometry Distribution for Reduced NO, Emissions from Aircraft Engines.* Journal of Engineering for Gas Turbines and Power, July 1993, 570-580

Tilston J. R., Wedlock M. I., Marchment A. D. - *The influence of Air Distribution on the Homogeneity and Pollutant Formation in the Primary Zone of a Tubular Combustor.*

Fuels and Combustion Technology for Advanced Aircraft Engines, AGARD-CP-536, May 1993

Wesoky H. L., Thompson A. M., Stolarski R. S. - *NASA Atmospheric Effects of Aviation Project: Status and Plans. International Scientific Colloquium on the Impact of Emissions from Aircraft and Spacecraft upon the Atmosphere,* Cologne, Germany, 18-20 April 1994

Wit, R. C. N., Dings, J., Mendes de Leon, P., Thwaites, L., Peeters, P. M., Greenwood, D., and Doganis, R. (2002). *"Economic incentives to mitigate greenhouse gas emissions from air transport in Europe."* 02.4733.10, CE, Delft

In: Advances in Mechanical Engineering Research, Volume 3 ISBN: 978-61209-243-0
Editor: David E. Malach ©2011 Nova Science Publishers, Inc.

Chapter 2

WELDING OF TITANIUM ALLOYS: FATIGUE AND RESIDUAL STRESSES

***Caterina Casavola*[*], *Luciano Lamberti and Carmine Pappalettere*[†]**
Dipartimento di Ingegneria Meccanica e Gestionale, Politecnico di Bari
Viale Japigia 182 - 71026 - Bari – Italy

ABSTRACT

Titanium alloys are very suited for highly specialized applications such as naval constructions where resistance to corrosion is the most important issue to be considered in materials selection, aerospace structures where the primary goal is to achieve high specific strength, and orthopedic prosthetic treatments where biocompatibility between implants and human tissues is mandatory.

Welding of titanium alloys is still considered a fairly uncommon process because industrial applications of these materials cover only the above mentioned niches where the unique properties of titanium and its alloys represent the best option in terms of minimizing the lifetime cost of the structure yet meeting stringent requirements on structural integrity and safety. However, whilst high quality welded components without inclusions and distortions and with a mild notch at the weld toe can be obtained through appropriate selection of welding process parameters, it should be considered that the weld cord is anyhow a geometric discontinuity in the structure, which modifies stress distribution. Furthermore, residual stresses are an unavoidable consequence of the thermal welding cycle.

Fatigue strength of welded joints, the most important information needed in aerospace applications of titanium alloys, is obviously lower than its counterpart for the base material. In spite of this, data available in literature are still not sufficient to make design procedures general and reliable. For this reason, in order to investigate the mechanical behavior of titanium welded joints and to build a technical database useful to designers, this chapter presents some fatigue strength curves recently obtained for butt joints made of Titanium grade 2 and grade 5. The joints tested in this research are welded

[*] Corresponding author. Tel.: +39 080 5962787; Fax: +39 080 5962777; E-mail address: casavola@poliba.it
[†] Member of the Permanent Committee of the European Association for Experimental Mechanics

by using different techniques. Experimental results of fatigue tests are discussed in terms of both nominal stress amplitude σ_a and local strain amplitude ε_a.

Fatigue curves determined experimentally are then compared with fatigue design curves reported in a very recent AWS standard that was specifically released for titanium welded joints. The role played by weld seam geometry is studied with respect to the fatigue strength reduction factor.

In order to complete the experimental analysis, residual stresses are measured by means of the hole drilling method and then correlated with mechanical response of welded components and microstructure modifications.

In addition to experimental tests, very detailed finite element analyses are carried out. Numerical results are in good agreement with the experimental evidence and confirm the importance of considering properly the interactions between weld seam geometry, defects eventually included in the weld and changes in microstructure induced by welding thermal cycles.

NOMENCLATURE

σ_a	nominal stress amplitude [MPa]	$\sigma_a = (\sigma_{max} - \sigma_{min})/2$
$\Delta\sigma$	stress range [MPa]	$\Delta\sigma = (\sigma_{max} - \sigma_{min})$
σ_{max}	maximum stress [MPa]	
σ_{min}	minimum stress [MPa]	
σ'_f	fatigue strength [MPa]	
b	Basquin's exponent	
σ_D	endurance limit [MPa]	
$f_{s,\Delta\sigma}$	safety factor on stress range	
$f_{s,N}$	safety factor on fatigue life	
ε_a	local strain amplitude [mm/m]	$\varepsilon_a = (\varepsilon_{max} - \varepsilon_{min})/2$
ε_{max}	maximum strain when σ_{max} is applied [μm/m]	
ε_{min}	minimum strain when σ_{min} is applied [μm/m]	
R	load ratio in fatigue test	$R = \sigma_{min}/\sigma_{max}$
N	number of fatigue cycles to failure	
K_t	stress concentration factor	
K_f	fatigue notch factor	
E	Young modulus [GPa]	
R_u	ultimate strength [MPa]	
R_{02}	yield strength [MPa]	

1. INTRODUCTION

The increasing use of titanium alloys in different industrial applications in chemical, aerospace, naval and biomedical engineering is motivated by the special properties of this material. In fact, titanium has excellent mechanical behavior and nearly perfect biocompatibility with human tissues. At room temperature, titanium grade 2 (known as "commercially pure titanium") has density of 4500 kg/m^3, while titanium grade 5 has density of 4400 kg/m^3 [1], about one half compared to steel. The tensile strength of titanium alloys can reach 1300 MPa depending on chemical composition and heat treatments. Furthermore, titanium shows excellent resistance to corrosion in aggressive environments because of the formation of a thin protective layer of TiO_2 oxide that can later be removed very easily.

In naval industry, the current trend is to explore the possibility of broadening the field of application of titanium alloys: high tensile strength and resistance to corrosion are required especially for hull ship structures that must operate at high-speed regimes. These working conditions induce large stresses that may facilitate corrosion caused by the marine environment. Protective titanium coatings of ship bottoms against corrosion are already realized, thus replacing the traditional protective paints utilized for steel hulls that can produce severe marine pollution if they are dissolved in the seawater.

Aeronautical industries are spending considerable effort in trying to improve the use of this homogeneous and isotropic metallic material in place of low-weight composite materials. This is motivated in most part by the high specific strength of titanium alloys, which is, for example, about 45% higher than that of aluminum alloys. Furthermore, composite structures are not very easy to be inspected and their mechanical behavior in the nonlinear regime (for example, the buckling strength of shells including geometric imperfections whose presence is almost unavoidable in real aerospace structures) may be extremely sensitive to the layup chosen in the design phase.

Welding represents a good alternative to mechanical joining of structures made of metallic alloys. However, welding of titanium is still considered a fairly uncommon process because titanium alloys are utilized only in highly specialized engineering applications. Considerations on the lifetime cost of a welded structure make indeed titanium the preferred option in view of its unique mechanical properties. However, special cares must be taken in carrying out the welding process because pure titanium and titanium alloys are highly susceptible to contamination from atmospheric gases. Shielding with inert gases may solve this problem but at the same time introduces complications in the technological process.

The following fusion-welding processes can be used for joining titanium and titanium alloys: Gas-Tungsten Arc Welding (GTAW); Gas-Metal Arc Welding (GMAW); Plasma Arc Welding (PAW); Electron-Beam Welding (EBW); Laser Beam Welding (LBW); Friction Welding (FRW); Resistance Welding (RW).

Gas-tungsten arc welding is massively utilized except for parts with thick sections. Square-groove butt joints can be welded without filler metal in base plates up to 2.5 mm thick. Gas-metal arc welding is used for joining parts more than 3 mm thick. The technological process, realized by means of pulsed current or in the spray mode, is less costly than gas-tungsten arc welding.

Laser beam welding is the most used technique for commercially pure titanium and titanium alloys. High-energy beams are focused onto a very narrow area. This allows the welded and heat affected zones to be much smaller in size than in other welding processes. Furthermore, a more uniform thermal distribution can be achieved. Consequently, high quality smooth welded joints without inclusions and distortions as well as lower residual stress can be realized. However, laser welding technology requires more expensive equipment. Technological parameters such as the laser beam power, welding speed and shielding gas purity govern the final geometry of the weld seam and the joint quality.

Weldability of titanium alloys depends on the welding process, preparation and cleanness of weld edges and welding pool shielding. Because of the high affinity between titanium and atmospheric gases as oxygen, nitrogen and hydrogen, special devices must be utilized in the welding operations in order to protect the melting pool with inert gases till temperature decreases below 700°C. In addition, surface dust and impurities can diffuse into the material thus generating porosity and causing embrittlement. In view of this, titanium alloy materials must be handled with the greatest care in order to avoid possible contaminations. For example, surface impurities can be removed by properly preparing the edges to be welded.

Titanium is a reactive metal which forms compounds that exhibit sub-optimal thermo-physical properties. Part surfaces heated in air contain brittle carbides, nitrides and oxides that can reduce fatigue strength and notch toughness of the welded joint and heat-affected zone (HAZ). The backside of the weld also must be carefully protected because it is as sensitive as the weld surface.

Design of welded joints is rather complicated because the weld toe can be considered as a geometrical discontinuity that affects stress distribution. Stresses near weld toe are higher than nominal stress and can even locally exceed the yield limit. Interactions between different factors (i.e. local and global geometry of the seam, micro-structural modifications induced by the welding thermal cycle, distortions and misalignment caused by material heating and cooling, residual stresses, etc.) affect the fatigue limit of welded structures by producing stress concentration near the weld toe. Therefore, to evaluate the fatigue life of welded joints is a very complicated task because of all these parameters involved.

In spite of the importance of titanium alloys in aerospace, naval and biomedical applications, the amount of experimental data on fatigue behavior available in literature is rather limited. Whilst there are several official standards that may provide designers with important indications on steel and aluminum structures, an AWS standard on titanium alloys (the first standard on titanium materials ever issued) was released just recently [2].

It should be pointed out, in fact, that official standards as Eurocode, ASTM or BS [3-9] refer generically to "metallic materials". Steel and aluminum alloys can be well covered by the above mentioned classification. Conversely, titanium alloys have a hexagonal close-packed (hcp) crystal structure, which is different from the body-centered cubic (bcc) and face-centered cubic (fcc) crystal structures typical of "traditional" metals like steel and aluminum. This fact results in a significantly different behavior in terms of mechanical response as the hcp structure has fewer slip systems and remarkable constraint against intergranular plastic deformation than bcc and fcc structures [10].

Mechanical behavior of titanium welded joints under fatigue was investigated just recently [11-14] and therefore there is the strong need to include in the official standards some sections, always supported by a clear and extensively proven experimental evidence, specifically devoted to titanium alloys (i.e. taking into account, should the later mentioned factors affect fatigue strength, the influence of different welding techniques, the effect of welding defects, the effect of main plate thickness, etc.).

As a very important consequence of the "history" of the manufacturing process undergone by the welded components, residual stresses also should be considered. The heat supplied by a welding cycle, besides the fact that modifies material microstructure, causes distortions and hence generates residual stresses in welded parts [15-16]. Although residual stresses are self-equilibrated systems, their effect on mechanical and fatigue strength [17] of materials is often relevant and must be contemplated in the design process of the joint. The residual stress field, superposed to a field of in-service stress generated by the applied loads, may produce local overloading that can even raise the resulting stress level above the yield limit of the material.

Distribution, magnitude and direction of residual stresses must be known in order to properly take into account their effect on structural efficiency. In the design of a welded joint, there is nothing more questionable than to idealize material conditions: it is well known, in fact, that compressive residual stresses reduce the buckling strength and that tensile residual stresses alter the load ratio during fatigue cycles and consequently affect component life. However, residual stress is still an open question with many debated issues.

Titanium and its alloys, because of their low thermal conductivity, usually present a heat affected zone (HAZ) limited in size. However, this fact does not prevent by itself the development of residual stresses. The role of residual stresses on titanium welded components still remains not clearly explained. For this reason, in any novel industrial application where it may be very convenient to use titanium, the problem of residual stresses, which seems to become more critical as the material has very good mechanical properties like titanium alloys do have actually [18], discourages its use. In order to solve this inherent contradiction and spread out the use of titanium alloys, comprehensive experimental data and/or reliable prediction models are needed [19].

This chapter aims at resolving as most as it is feasible the above mentioned lack of experimental information on welded components made of titanium alloys. For that purpose, two different titanium alloys (Titanium grade 2 and Titanium grade 5) are studied. In particular, butt joints welded by means of laser, hybrid (laser and MIG) or electron beam techniques are investigated. Results of static and fatigue tests carried out on titanium welded joints are presented and discussed in the paper. Fatigue strength curves are reported in terms of both local strain amplitude, according to the WEL.FA.RE. method [20-25], and nominal stress amplitude. Macrographs and shape contouring of butt joints are carried out in order to study the microstructure and the local geometry of the joint.

Fatigue resistance is discussed also in terms of safety factors with respect to fatigue design curves of welded joints provided by official standards.

Finite element analyses including information obtained from experimental tests are carried out in order to investigate the stress/strain field in the heat affected zone with a great deal of detail. Nine different finite element models are developed in this research in order to analyze interactions between weld geometry, material properties and level of porosity.

Finally, residual stresses values measured by the incremental hole drilling method (HDM) are presented both for titanium grade 2 and grade 5, with respect to different welding techniques [26].

The chapter is structured as follows. Following this Introduction section, details on the welding processes used for realizing the joints and their influence on the resulting mechanical properties are given in Section 2. Results of the experimental static and fatigue tests are presented in Section 3. The trade study conducted on the nine different FE models and the corresponding numerical results are described in Section 4. In Section 5, fatigue data are critically compared with official standards. Section 6 discusses experimental data on residual stresses measured on welded plates by means of the hole drilling method. The last Section briefly summarizes the main findings of this research.

2. Properties of Titanium Alloys, Welding Techniques and Characteristics of Welded Joints

2.1. Chemical and Physical Properties

Two different types of titanium alloy were considered in is research: titanium grade 2 and titanium grade 5. Titanium grade 2 is the commercially pure titanium (CP). At room temperature, it has a hexagonal close-packed (hcp) structure called α phase. At about 882°C, an allotropic transformation takes place and the crystalline structure changes into the β phase (up to 1670°C) which is characterized by a body centered cubic lattice (bcc). The addition of alloying elements may modify the transition temperatures and determines which microstructure (α, α–β, β) is predominant at the working temperature. Titanium grade 5 microstructure has the α–β phase, where aluminum is a α–stabilizing element while vanadium is a β–stabilizing element.

Material properties are directly related to microstructure: α-phase titanium has very high corrosion strength and excellent weldability with good ductility, but has a relatively low strength; α–β titanium alloys possess better mechanical properties because the bcc structure is stronger than hcp. Both of these alloys can be used for structural purposes.

Table 1 reports the chemical composition of the above mentioned materials. Table 2 compares chemical, physical and mechanical properties of titanium with those possessed by steels used in structural applications. All properties listed in the table are taken from data sheet provided by manufacturers. The Young modulus and mass density are about one half of the corresponding values typical of steel. The coefficient of thermal expansion (CTE) and thermal conductivity of titanium are considerably smaller compared with steel.

Table 1. Chemical composition of commercially pure titanium and titanium grade 5 (Ti-6Al-4V).

	Ti	N	C	Fe	O	Al	Sn	Zr	Mo	V
Ti grade 2	99.2	0.03	0.1	0.3	0.25	-	-	-	-	-
Ti grade 5	90	0.05	0.1	0.3	0.2	6	-	-	-	4

Table 2. Mechanical, chemical and physical properties of titanium alloys compared to structural steels (as they are reported in the data-sheets provided by manufacturers).

Material	R_{02} [MPa]	R_u [MPa]	E [GPa]	Elongation at break %	Hardness Vickers	Density [kg/m^3]	α (CTE) at 20°C [1/°C]	Thermal conductivity [W/m·K]
Ti grade 2	275	345	105	20	145	4510	$8.6 \cdot 10^{-6}$	16.4
AISI 1010	305	365	205	20	108	7870	$12 \cdot 10^{-6}$	50
Ti grade 5	1100	1200	114	10	396	4430	$8.6 \cdot 10^{-6}$	6.7
AISI 4340	862	1282	205	12	384	7850	$12 \cdot 10^{-6}$	44.5

2.2. Determination of Mechanical Properties: Tensile Tests on the Base Metal

Mechanical characterization under static loading was carried out on titanium grade 2 specimens of thickness 1.5 mm and on titanium grade 5 specimens of thickness 3 mm. Specimen dimensions and testing procedures were selected according to RINA indications [9] for naval applications since specific guidelines on the execution of tensile tests on titanium alloys are not available.

The static tensile tests were performed on a Instron 4467 testing machine equipped with a load cell capacity of 100 kN. The testing speed was set as 6 mm/min.

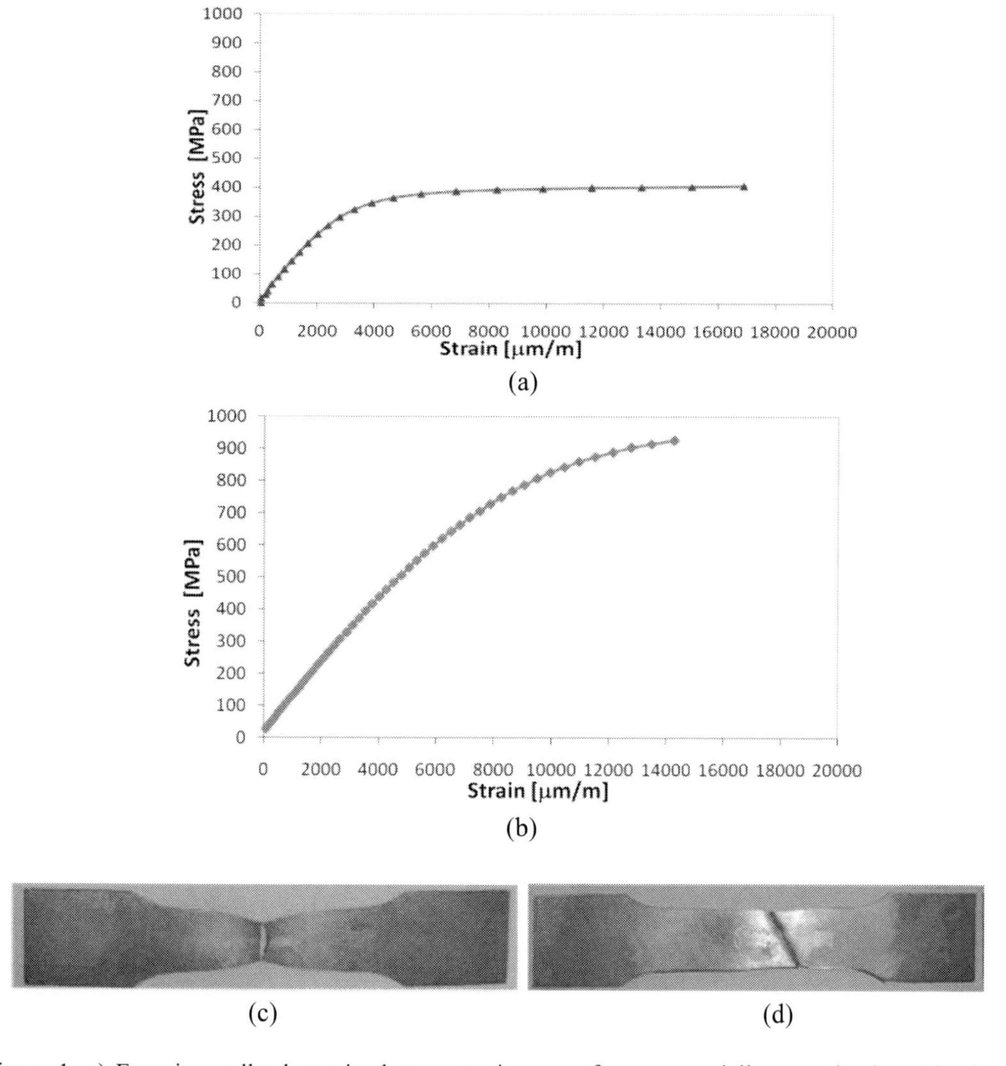

Figure 1. a) Experimentally determined stress-strain curve for commercially pure titanium (titanium grade 2); b) Experimentally determined stress-strain curve for titanium grade 5; c) Localized necking at failure in titanium grade 2 specimens; d) Slant fracture in titanium grade 5 specimens.

Stress-strain curves were obtained by recording strain values developed during the tensile tests by means of electrical strain gauges bonded on both sides of each specimen. Typical constitutive curves obtained from experiments are shown in Figure 1 while mechanical properties (Young modulus, yield stress, ultimate strength and elongation at fracture) are summarized in Table 3. It can be seen that yield and fracture strength of titanium alloy grade 5 are two times as high as their counterpart in the case of titanium grade 2. The latter material undergoes ductile fracture with localized necking (Figure 1c) and its constitutive behavior is nearly perfectly elastic-plastic (Figure 1a). Elongation at break of titanium grade 5 is 20% smaller than for titanium grade, but still considerably large, 14%. Fracture paths developing before instable crack propagation and failure finally occur are inclined by about 45° with respect to the direction of the applied load (Figure 1d). Nonlinear elastic deformation before yielding was observed for both materials (Figures 1a and 1b).

Table 3. Mechanical properties of grade 2 and grade 5 titanium alloys determined with tensile tests.

Material	E [MPa]	R_{02} [MPa]	R_u [MPa]	Elongation at break
Ti grade 2	105.3	358.3	422.8	18%
Ti grade 5	110.1	765.0	965.1	14%

2.3. Description of the Welding Process

Three joining processes were considered in this study: laser beam welding (LBW), hybrid laser-MIG welding and electron beam welding (EBW). The most important parameters characterizing each welding process are summarized in Tables 4 through 6.

An important aspect in laser welding is represented by the final configuration taken by the weld that appears narrow in size and goes deeply through the specimen thickness. This behavior is due to the high-energy concentration entailed by the joining process and the huge welding speed that concur to reduce the heating in the specimen. The heat affected zone hence becomes narrow in size thus producing smaller modifications in microstructure and generating less residual stress. The thermal energy supplied to the specimen is very concentrated, actually at one of the largest extents among the different welding processes currently available in industry. Such a large density of thermal energy allows welds with a high depth to width ratio and minimal thermal distortions to be obtained [27]. The welding process is also pretty fast with a clear benefit in terms of productivity. However, besides the many advantages deriving from the deep and narrow shape of the weld, accurate machining and careful positioning of the workpieces are required. In summary, the strength points of laser welding are: narrow and deep weld seam of good quality, high completion rates, no requirement of additional material (filler), limited amount of heating in the workpiece that hence results in small distortions easily predictable, reduction of post-weld processing, and feasibility of joining highly dissimilar materials.

However, LBW presents also some drawbacks: high costs in terms of equipment and maintenance, poor gap bridging ability which leads to severe requirements on joint preparation, poor electrical efficiency intended as the ratio between the thermal power actually supplied to the material and the electrical power given in input to the welding head (for example, 2% for CO_2 lasers and 10% for Nd–YAG lasers), and metallurgical problems caused by high cooling rates.

Table 4. Characteristics of the laser welding process.

LASER WELDING	Ti grade 5 - THICKNESS 3 mm
Equipment	Laser Nd–YAG - HAAS HL 2006D
Focal length of the beam focusing apparatus [mm]	f=100
Device for gas plasma suppression	standard nozzle f=6mm; special nozzle: Scarpeta TINEA
Gas for plasma suppression	He 4.8, delivery capacity 25 Nl/min
Shielding gas	Ar 5.0
Edge preparation	right (gap 0)
Power of laser beam [kW]	1.6
Feed velocity [m/min]	3

Table 5. Characteristics of the hybrid (laser – MIG) welding process.

HYBRID WELDING	Ti grade 5 - THICKNESS 3 mm
Laser CO_2 power [kW]	3
MIG power [kW]	7.2
Distance laser-MIG [mm]	0
Welding speed [m/min]	2
Wire velocity [m/min]	4
Gas for plasma suppression	He
Shielding gas	Ar

Table 6. Characteristics of the electron beam welding process.

ELECTRON BEAM WELDING	Ti grade 5 - THICKNESS 3 mm
Equipment	Techmeta CT4
Electron beam gun power [kW]	50
Voltage [kV]	50
Welding current [mA]	73
Welding speed [m/min]	2.7

Electron Beam Welding [28] is a fusion joining process where a beam of electrons traveling at very high velocity hit the materials to be welded. The basic principle of this technique is that the workpieces melt as the kinetic energy of the electrons is transformed into heat upon impact against the specimen surface. The electron beam is always generated in a high vacuum environment to achieve the best level of purity as it is feasible and to realize welds with high depth to width ratio.

The main problem with EBW is the vacuum chamber required in the welding process: this equipment is costly and the working dimensions of the chamber limit the size of the parts that can be welded. Moreover, X-rays are generated during welding operations. On the other hand, both LBW and EBW create a narrow heat affected zone and are hence characterized by high density power (with no need of filling metal), thus achieving a very deep penetration. For example, EBW allows welding very tick joints in a single pass.

Finally, because of the high risk of contamination of the melted metal, specific devices must be designed in order to protect the weld until its temperature drops below 700°C.

In the hybrid laser-arc welding technique, a laser power source (CO_2 or Nd–YAG) is used in combination with an arc process (TIG, MIG, MAG or plasma). This makes it possible to benefit from the advantages of both processes. The laser beam allows deeper welds to be produced in just one pass while the arc energy serves to increase welding speed and to fill the fit-up defects between the pieces to be joined. Therefore, when the laser and the arc operate in the same melting pool, higher welding speeds can be obtained with even deeper penetration and greater tolerance to fit-up compared with the standalone laser process. Hybrid laser-MIG welding allows higher completion rates to be obtained with respect to standard laser welding processes: the laser power required in the joining process drops down and welding reliability is improved. Hybrid welding is less costly than laser beam welding and maintains, or even increases, technical benefits peculiar to laser welding. Furthermore, the hybrid welding technique minimizes the weak points of both processes and hence allows the joining process to be optimized.

2.4. Distortion and Misalignment of Welded Plates

In order to evaluate the effect of secondary bending in fatigue tests, the angle α formed between welded plates was measured for all specimens; a centesimal comparator was used for that purpose. Furthermore, the seam profiles of the butt specimens realized via laser and hybrid welding techniques were contoured with a high precision Mahr® tactile device (±1 μm accuracy) in order to determine precisely the local geometry of the weld cord (height, opening and fillet radius). Such information may turn very useful in the estimation of stress concentrations caused by the geometric discontinuity introduced by the weld cord. The seam profile was measured in the direction orthogonal to weld cord.

The average misalignment angles measured are 0.03° for LBW plates and 0.33° for hybrid plates.

Therefore, LBW plates are practically planar. The laser process utilized and/or the thermo-physical properties of the titanium allowed in all likelihood misalignment values to be strongly reduced with respect to welded joints obtained with laser and MIG process combined together. Steel joints built with electric-arc welding usually show much more misalignment [22].

The stress concentration factor due to misalignment was calculated using the Maddox's formula [8]: a mean value of Kt=1.35 was found for hybrid welded plates.

Figures 2 and 3 show the profiles measured respectively for the LBW and hybrid welds by the tactile device, a high precision mechanical coordinate measurement machine (CMM), in the direction orthogonal to the weld seam for both right and reverse sides of the specimen.

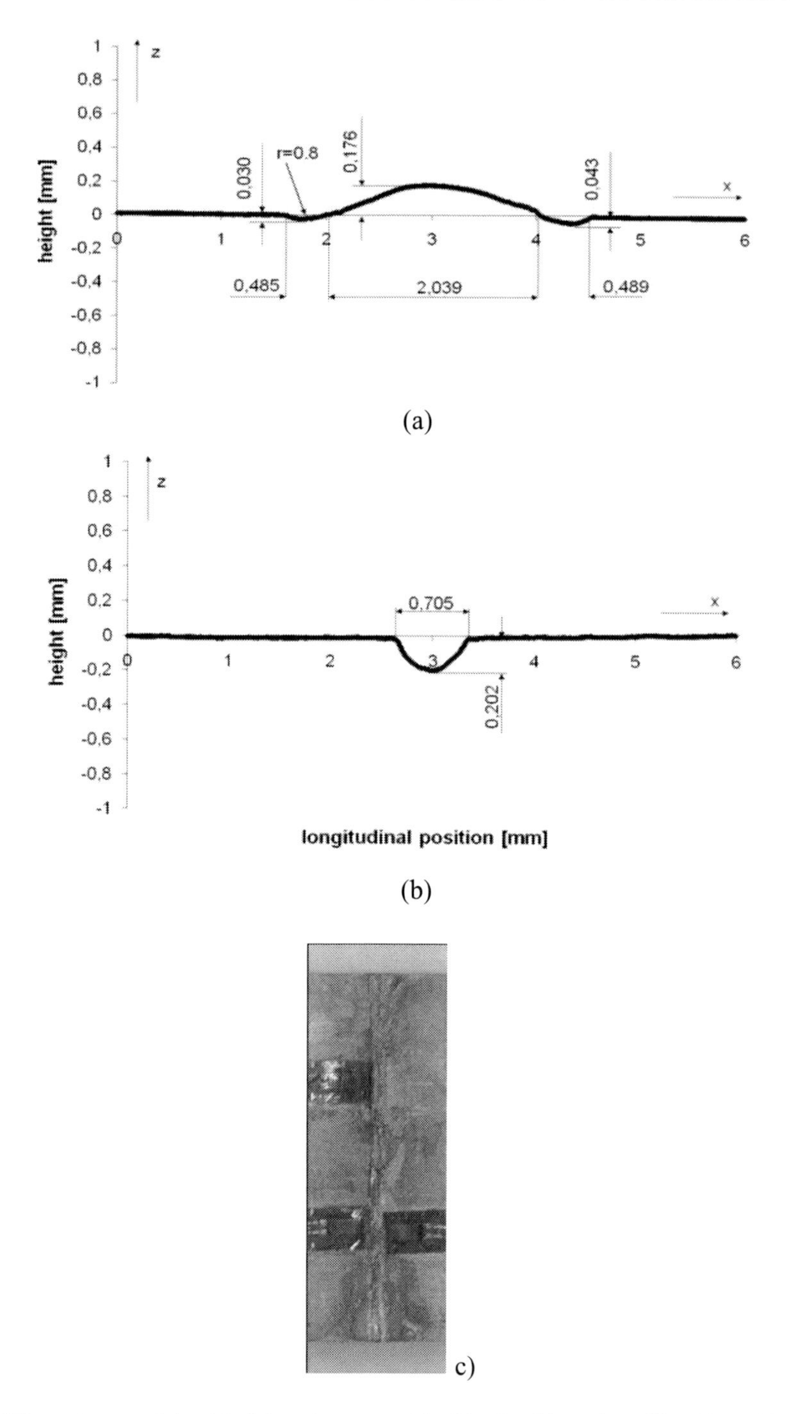

Figure 2. LBW specimens: right side (a) and reverse side (b) of weld cord profile; (c) Weld cord.

The seam profile appears regular. The connections between the weld cord and the base material also are rather smooth on both right and reverse sides of the joint. The small undercut observed on the right side of laser beam welded specimens is a direct consequence of the high density of thermal energy involved by the LBW technique. The width of the LBW weld seam is about 2 mm on the right side and 0.7 mm on the reverse side. The corresponding width increased to more than twice in the case of hybrid welding.

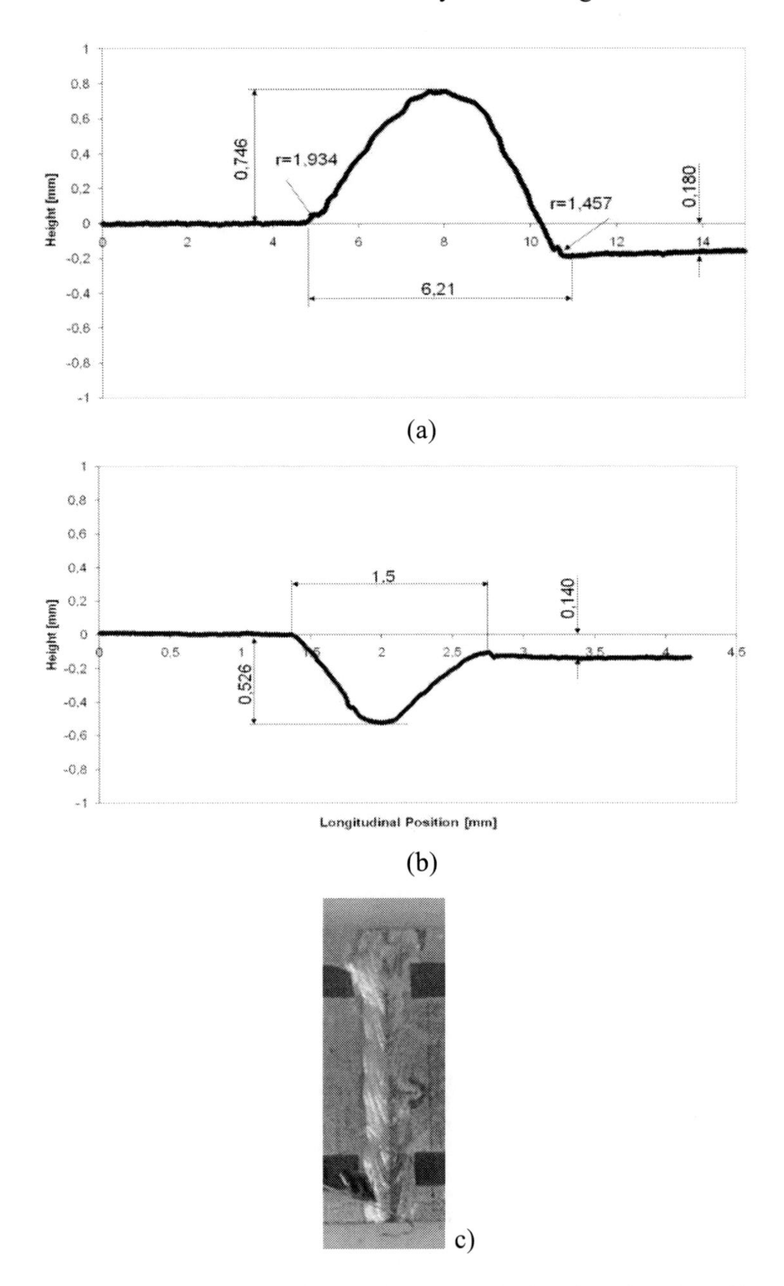

Figure 3. Hybrid weld specimens: right side (a) and reverse side (b) of weld cord profile; (c) weld cord.

2.5. Metallurgical Properties

Mechanical strength of welded joints depends on the quality of the welding process and on the final characteristics possessed by the component. Information on hardness of the fused zone, heat affected zone and base material can clarify the effects produced by the thermal welding cycle. Investigations on welded plate misalignment and residual stress field developed in the weld seam may also help us to fully understand the mechanical behavior of welded joints.

In order to study the profile taken by the weld seam and to assess microstructure modifications caused by the welding cycle, photo-macrographs of the specimens extracted from the welded plates were taken.

Specimen preparation for macrography involves the following steps. First, the welding joint is cut by a cropper machine and the cut piece is embedded into black epoxy resin. Then, the specimen is lapped and polished by means of nylon cloth and diamond paste. Finally, the specimen surface is chemically attacked by a HNO_3 27% / HF 3% / H_2O solution.

Figure 4 shows some macrographs of the welded plates analyzed in this research. Images were recorded by a Leica DMRME microscope with the magnification set as 25X. The analysis of the micrographs revealed that the weld seam is well connected to plates on both right and reverse sides. However, a small amount of porosity was observed in the case of LBW process (Figure 4c). The fused zone (where size of crystalline grains is larger) and the heat affected zone can be clearly distinguished (see, for example, the detail view of Figure 4a). A little and smooth undercut is present in both LBW and EBW specimens but is more pronounced in the latter case.

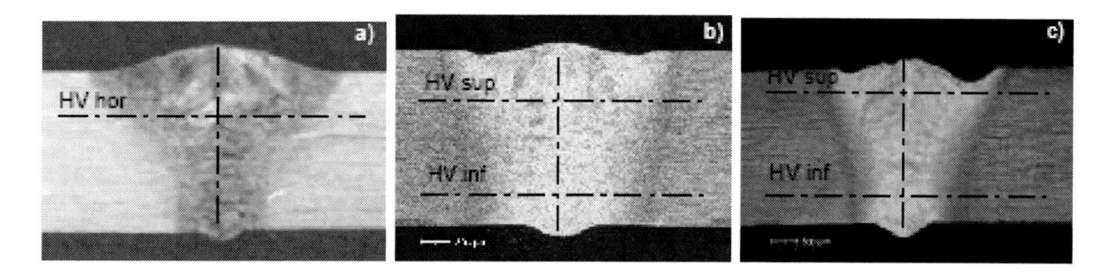

Figure 4. Photo-macrographs of joints built with different welding techniques: a) HYBRID Ti grade 5 of 3 mm thickness; b) EBW Ti grade 2 of 3 mm thickness; c) LBW Ti grade 5 of 3 mm thickness.

Mechanical properties of titanium alloys depend strongly on the temperature to which the material is exposed as well as on the duration of the exposure. In the regions where temperature exceeds the crystal phase transition threshold value, microstructure of $\alpha-\beta$ alloys like titanium grade 5 is entirely comprised of β grains. The laser welding technique allows high quality titanium welded joints to be realized: in fact, unlike it happens for example in the case of stainless steel joints, the size of grains in the fused and heat affected zones of titanium joints grows by only a small amount with respect to the initial grain size of the base material (Figures 5a-c) [13,19,29]. This is because the time of exposure at high temperature is short and consequently the thermal gradient developed between the heated/fused zones and the base material is rather small. Mechanical properties throughout the welded joint hence are fairly uniform compared to the case of traditional materials such as, for example, stainless steel.

Vickers micro-hardness was measured along control paths respectively normal and parallel to the weld seam (see control paths sketched in Figure 4).

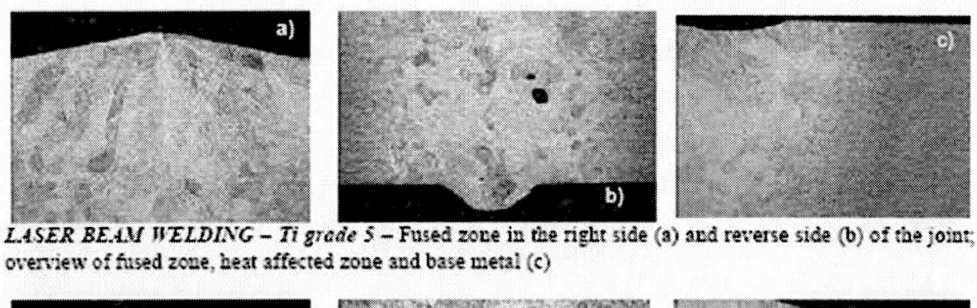

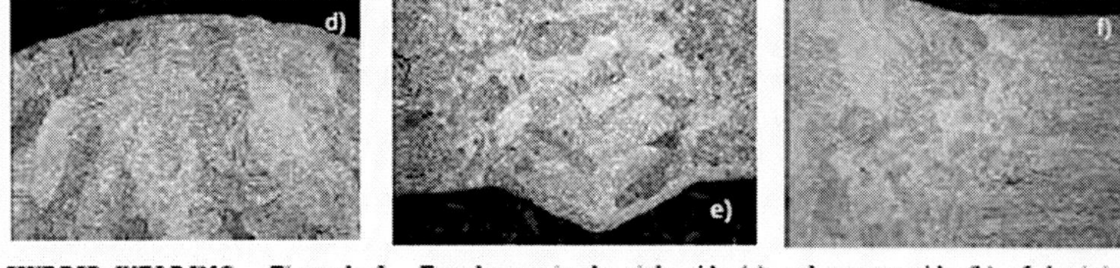

Figure 5. Photo-macrographs of LBW and HYBRID welded joints.

It can be seen from Figures 6-8 that the hardness values measured in the direction of the weld cord are higher for LBW and hybrid joints than for EBW joints. Moreover, in the direction orthogonal to the weld cord, hybrid welded joints are harder than LBW and EBW: the difference in micro-hardness becomes more pronounced as the distance from the weld cord midline increases. This was expected because the weld cord formed in the hybrid welding process is twice larger in width than in the case of specimens joined via laser beam welding (see Section 2.4)

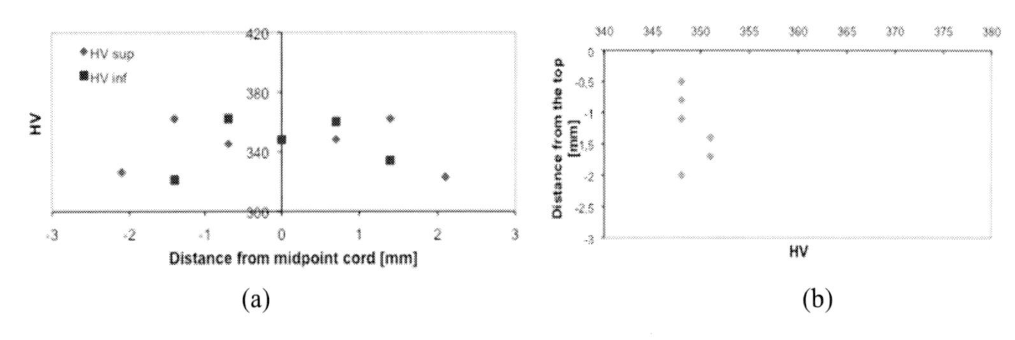

Figure 6. Vickers micro-hardness of EBW joints: a) Control paths orthogonal to weld cord; b) Control path in the direction of the weld cord.

From micro-hardness distribution, it is possible to locate precisely the bounds of the different regions included in the weld. Figures 6 through 8 show that micro-hardness in the vicinity of the fused zone is higher than its counterpart in the base material. However, micro-hardness changed on average only by about 15%: from 320 HV in the base material to 368 HV in the fused zone. Micro-hardness of the heat affected zone was instead about 350 HV for all welding techniques considered in this study: this value was hence practically the same as the corresponding micro-hardness of the fused zone.

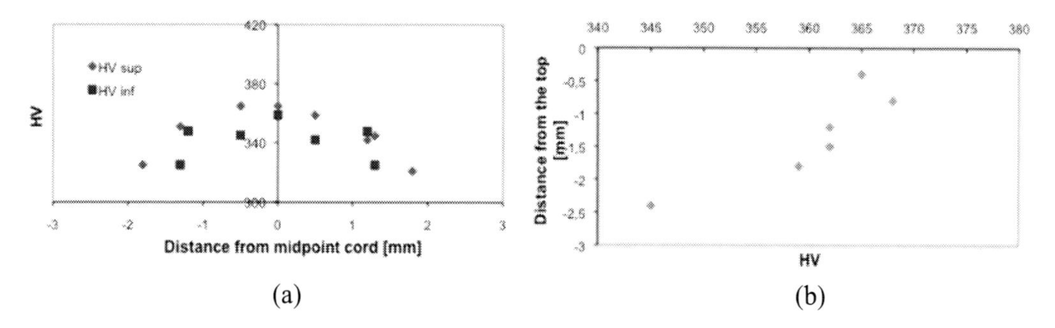

(a) (b)

Figure 7. Vickers micro-hardness of LBW joints: a) Control paths orthogonal to weld cord; b) Control path in the direction of the weld cord.

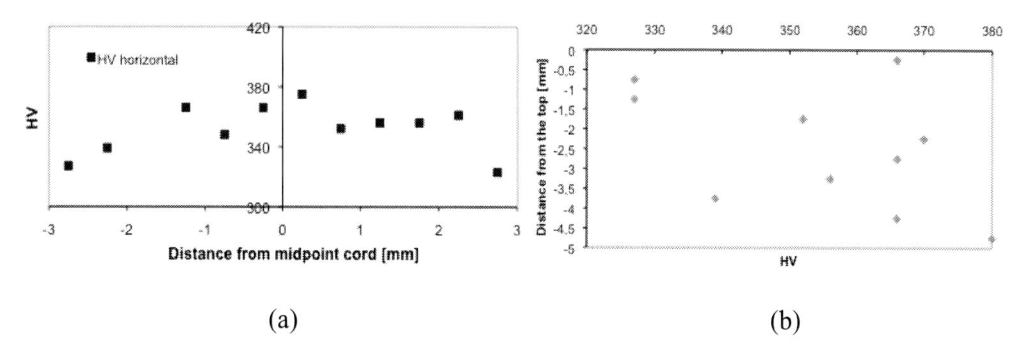

(a) (b)

Figure 8. Vickers micro-hardness of HYBRID welded joints: a) Control path orthogonal to weld cord; b) Control path in the direction of the weld cord.

3. Fatigue Tests

Assessment of fatigue life of welded joints is very difficult because the endurance limit depends on many factors such as, for example, the irregularity of weld seam geometry and material inhomogeneity induced by welding operations. In the case of structural steels, the type of material and the corresponding static strength as well as the type of welding process marginally affect the resistance to fatigue expected for a specific structural detail. This fact has simplified much the process of developing design rules for components subject to fatigue loads since it is not necessary to dispose of data for all of the different types of steel.

The most common approach to fatigue design and assessment of metal structures is very well established and relies on the traditional S–N (stress vs. cycles) curves reported in Official Standards [3-9]. However, for all details not covered by standards or cases like those studied in [30], new methodologies for designing welded structures were developed and extensively discussed in [31]. Guidelines on design of titanium welded joints subject to fatigue are provided by a very recent AWS standard [2]. This section presents some experimental data on titanium joints welded by means of different techniques. Two titanium alloys, grade 2 and grade 5, are considered. Experimental fatigue results are presented in terms of both nominal stress and local strain amplitude. This was done in order to overcome the limitations of traditional S–N based approaches that may result inappropriate for designing welded structures subject to fatigue as they do not account properly for all factors concurring in the fatigue strength of material (for example, global geometry of the joint, local geometry of the cord, distortions and misalignment, type of load, defects, plasticization, residual stresses, etc.) [30,31].

The WEL.FA.RE. method, developed by Pappalettere *et al.* [20-25], can be used to predict fatigue life of several welded joints of different shapes (cruciform, angular, butt, overlap and T with full penetration). WEL.FA.RE. is based on the local strain amplitude ε_a measured at the weld toe. This is the strain value averaged over the area of the specimen surface where electrical strain gauges are bonded: that is, where the fatigue failure finally occurs [24,32]. It is assumed that the ε_a parameter can account for the different factors affecting fatigue life of welded joints. The local strain amplitude is measured by strain gauges before the fatigue test at nominal load amplitude is executed. Consequently, fatigue life curves at the level of safety of 50% are expressed, at fixed load ratio R, in terms of local strain amplitude $\varepsilon_a=(\varepsilon_{max}-\varepsilon_{min})/2$ instead of the nominal stress amplitude $\sigma_a=(\sigma_{max}-\sigma_{min})/2$.

Strain gauges with grid length of 3 mm were bonded to the specimen surface with their transverse axis located at the distance of 2.5 mm from the real weld toe. Selection of the strain gauge position and grid length mediated between two conflicting aspects: (i) the need to correctly evaluate the local strain field in the critical zone that implies measurements be carried out in the region very close to the weld toe; (ii) the use of a reliable and user-friendly experimental technique which may be directly transferred to the industrial environment and utilized without difficulty [23,33]. The particular location of the strain gauge grid adopted in the WEL.FA.RE. method allows us to capture local effects such as joint global geometry, misalignments, welding local geometry and plasticity that are relevant for fatigue phenomena and can affect the strain field at the weld toe.

Specimens submitted to fatigue tests were obtained from the welded plates by removing the outer sides that included defects and other welding irregularities (the seam was gold in color because of partial oxidation caused by insufficient protection of plate edges). Specimen geometry and loading modality are shown in Figure 9. All specimens have the same dimensions: 310 mm length and 40 mm width.

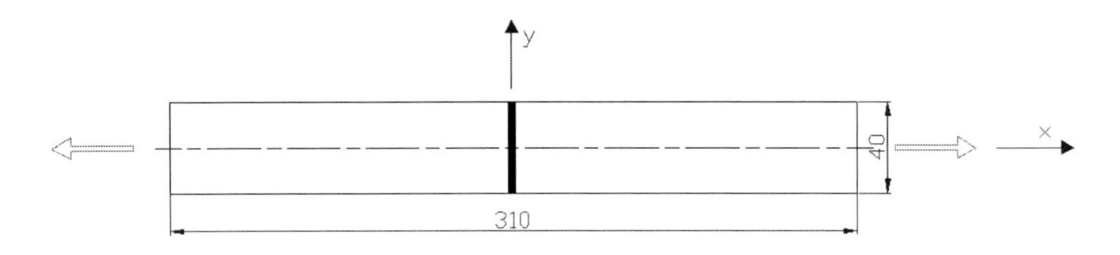

Figure 9. Geometry and type of load for fatigue test specimens.

Fatigue tests were executed on butt welded joints. The first set of experiments was carried out to compare fatigue strength of titanium grade 2 and 5 alloys; rectangular plates of thickness 1.5 mm for titanium grade 2 and 3 mm for titanium grade 5 were considered. The second set of experimental tests investigated the effect of different welding processes on the fatigue behavior of titanium grade 5. Finally, the third set of fatigue tests was conducted on erased hybrid welded joints.

Before executing fatigue tests, all specimens were visually inspected in order to evaluate the quality of the weld seam. No macroscopic defects were observed. The seam was silver in color (this indicates a good gas protection without contaminations); surface is smooth and does not present spray.

Eight electrical strain gauges were bonded onto each specimen to be tested following the WEL.FA.RE. guidelines: 4 on the weld right side and 4 on the reverse side. The amplitude of local strain was measured at the weld toe. Strain values measured by gauges were recorded by the System 5000 (Micro Measurements Inc., USA) equipment. All fatigue tests were performed under load control on a 250 kN Schenck servo-hydraulic testing machine (at about 25 Hz frequency). Load ratio was always set as R=0.1. The experimental plan is summarized in Table 7.

Table 7. Experimental plan of fatigue tests carried out on butt joints for different materials and welding techniques.

Welded plates	Thickness [mm]	Material	Welding technique	Load ratio R
Butt A	1.5	Ti grade2	LBW	0.1
Butt B	1.5	Ti grade2	LBW	0.1
Butt C	3	Ti grade5	LBW	0.1
Butt D	3	Ti grade5	HYBRID	0.1
Butt E	3	Ti grade5	HYBRID	0.1
Butt F	3	Ti grade5	HYBRID	0.1
Butt G	3	Ti grade5	HYBRID	0.1
Butt H	5	Ti grade5	HYBRID	0.1
Butt I	5	Ti grade5	HYBRID	0.1
Butt L	5	Ti grade5	HYBRID	0.1

Table 8. Results of fatigue test on titanium butt joints.

Material	Fatigue strength at $N=2 \cdot 10^6$ cycles from σ_a–N curve [MPa]	Fatigue strength at $N=5 \cdot 10^6$ cycles from σ_a–N curve [MPa]	Fatigue strength at $N=10^7$ cycles from σ_a–N curve [MPa]
Ti grade2 - LBW	85	77	71
Ti grade5 - LBW	127	116	108
Ti grade5 - HYBRID	87	80	75
Material	Fatigue limit at $N=2 \cdot 10^6$ cycles from ε_a–N curve [mm/m]	Fatigue limit at $N=5 \cdot 10^6$ cycles from ε_a–N curve [mm/m]	Fatigue limit at $N=10^7$ cycles from ε_a–N curve [mm/m]
Ti grade2 - LBW	861	762	694
Ti grade5 - LBW	1217	1143	1089
Ti grade5 - HYBRID	943	873	823

Experimental results are reported in Table 8. Fatigue curves ε_a–N expressed in terms of local strain amplitude and fatigue curves σ_a–N expressed in terms of nominal stress amplitude are shown in Figures 10 and 11, respectively for the laser beam welding and the hybrid welding techniques.

Figure 12 shows that the fatigue fracture paths finally observed at failure are always localized in the base metal zone, away from the weld toe for all LBW specimens (except butt C2), and at the weld toe for hybrid welded specimens.

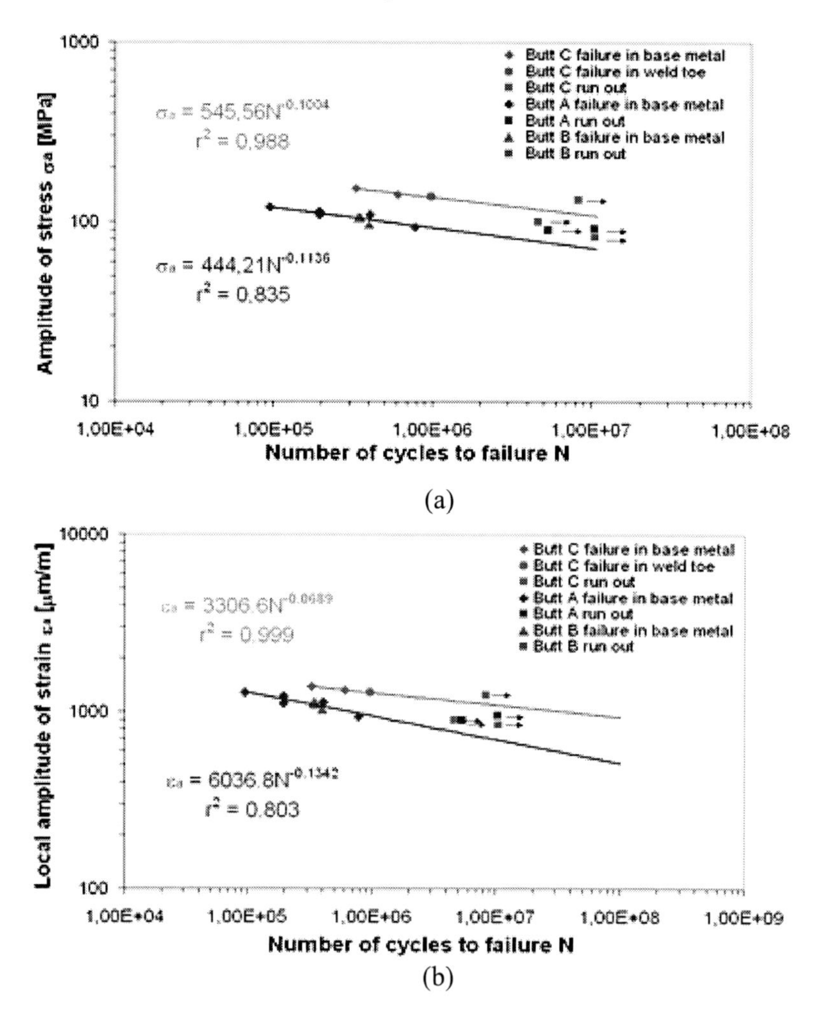

Figure 10. Fatigue curves for LBW joints: a) ε_a–N; b) σ_a–N.

Results of experimental tests carried out on titanium welded joints suggest some considerations. First, using LBW it is possible to limit the extension of the heat affected zone as well as the dimension of the grains interested by the welding thermal cycle [19]. This allows mechanical behavior and fatigue strength to be improved. Second, since both LBW and HYBRID joints were not misaligned, secondary bending effects, that can be captured by strain gauges and are actually included in the value of ε_a but are missed by the nominal stress amplitude σ_a approach, should not be considered in such cases a discriminating element in the assessment of fatigue behavior as it is instead done for steel joints.

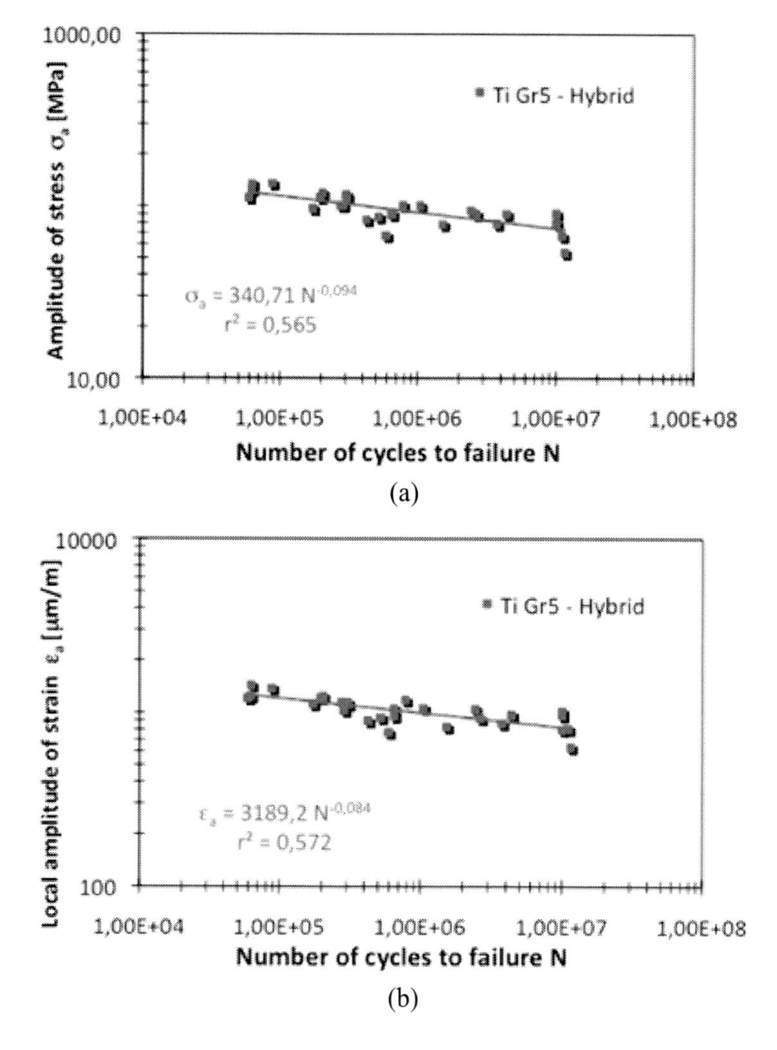

Figure 11. Fatigue curves for HYBRID joints: a) ε_a–N; b) σ_a–N.

Third, the welding seam profile (Figures 2 and 3) is fairly regular and the stress concentration factor is quite small. Consequently, the local stress field is practically the same as the nominal one. Although there may be some local plasticization caused by stress concentration at notch, it is well known that strain gauges can capture only a portion of this effect which is integrated over the grid length. The effect of plasticization may hence not be so important in governing the final rupture of the joint.

Remarkably, fatigue behavior of titanium welded joints seems to be well described both by "nominal" (i.e. σ–N) and "local" (i.e. ε–N) fatigue curves, where the latter plot was built from strain gauge measurements that are just marginally affected by notch effects.

The last aspect to be considered is the location of fatigue crack. Fatigue strength depends on many parameters, in particular on the magnitude of local peaks of stress or strain. The experimental tests carried out in this paper showed that fatigue cracks of hybrid welded joints are localized at the weld toe, that is, in the heat affected zone. This behavior is similar to that usually observed for welded parts made of steel. In the case of titanium joints built via laser beam welding, fatigue cracks were always localized far away the seam except in one case, the butt C2 specimen (see Figure 12).

This result seems to indicate that microstructural transformations experienced by titanium (manufactured with different welding techniques) are different from those of steel welded joints. In particular, while the transition from $\alpha-\beta$ to β phases during welding operation (i.e. at increasing temperature) is rather well known, the reverse transformation which should occur in the cooling phase still is not very clear and it is not necessarily similar to what is expected in the case of traditional materials. Indeed, the heat affected zone could be a region highly resistant to fatigue loads [11].

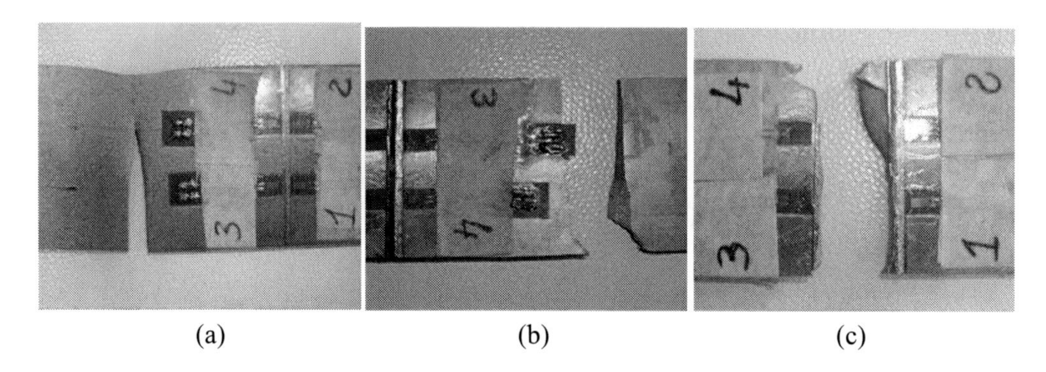

(a) (b) (c)

Figure 12. Fatigue fracture patterns observed for LBW joints at failure: a) Titanium grade 2 (butt B4); b) Titanium grade 5 (butt C4); c) Titanium grade 5 (butt C2).

4. Finite Element Analysis of Titanium Welded Joints

The results of experimental tests conducted on titanium welded joints were somehow surprising as fatigue failures occurred in the base metal instead of the notch region as it usually happens in "traditional" materials like stainless steel. For this reason, detailed finite element analyses of the welded joint were carried out in order to better understand failure mechanisms. The general purpose finite element software ANSYS® Version 11.0 [34] was utilized.

4.1. The Effect of Weld Cord Geometry on LBW Joint

The FE model of the joint was built on the basis of topographic data gathered on the real specimens by means of a profilometer. This allowed the real behavior of the welded joints under investigation to be reproduced as more accurately as it is feasible. Special care was taken in modeling geometric discontinuities produced by the presence of the weld seam and radius. The numerical simulations carried out in this study are relative to the titanium grade 5 butt welded joints - series C. The only specimen that experienced fatigue failure at the weld toe belongs to butt C welded plates.

Since fatigue strength of welded joints depends significantly on seam geometry, the seam profile modeled in the finite element analysis must be very similar to the real weld profile in order to correctly evaluate stress concentration effects at the weld toe. For this reason, the weld seam profile of titanium grade 5 butt C welded joints was contoured with a high precision tactile device (±1 μm accuracy). Different profiles were considered at some positions taken along the weld direction. The cross-section analyzed in FE analysis is located in correspondence of the deepest undercut (Figure 2), which hosts the very critical stress concentrations. The general 3D problem can be transformed in a much simpler 2D problem because of the small thickness of the joint: the plane strain was hence assumed (see [31,35-37] and the references cited therein). Since experimental measurements on weld shape indicated that the mean value of the misalignment angle is less than 0.3°, the FE model did not account for this kind of geometric imperfection.

Figure 13a shows the mesh of the welded joint: details of top and bottom surfaces of the weld seam are shown in Figures 13b and Figure 13c, respectively. The FE model is comprised of 127644 PLANE82 elements (these are quadratic elements including 8 nodes and 2 degrees of freedom for each node, the translations in X and Y-directions of the global reference system) and has 386933 nodes. The size of the smallest element in the mesh is 13.3 μm. Mesh was progressively refined through convergence analysis. This process ended when displacement values became mesh independent. The left edge of the model is clamped in order to reproduce the experimental conditions. The tensile load acting in the experimental tests was simulated by applying a uniform pressure to the "free" edge opposite to the constrained side of the model.

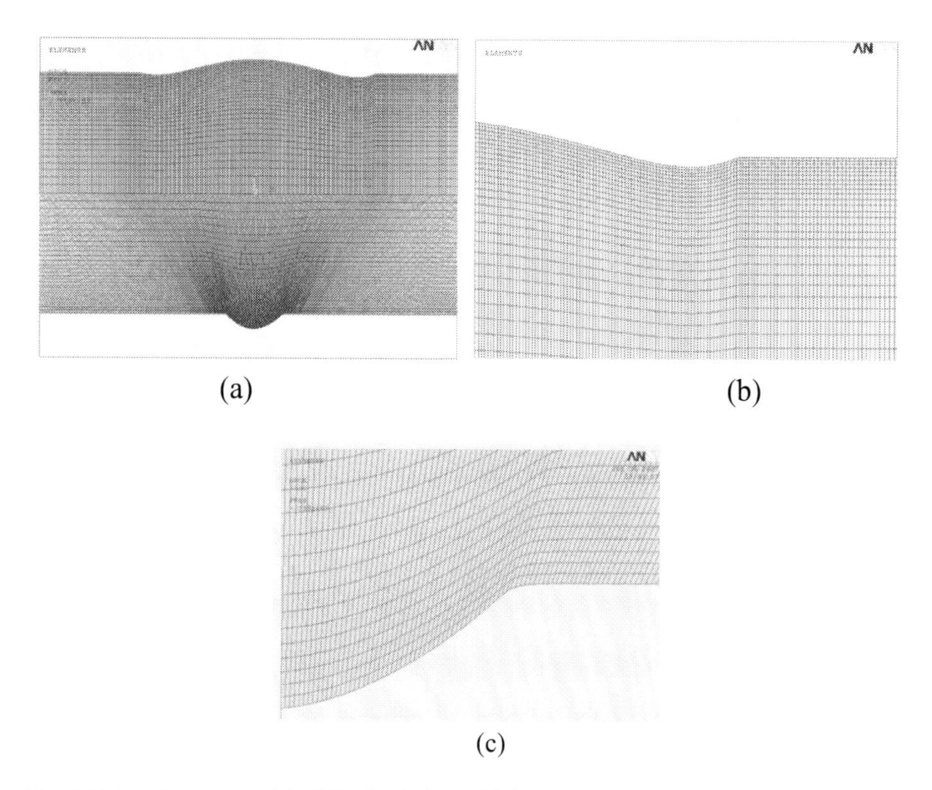

Figure 13. a) Finite element model of the LBW butt C joint; b) Undercut on the right side of the joint; c) Reverse side of the joint.

Structural analysis was carried out in the elastic-plastic range. The corresponding material properties defining the stress-strain curve are derived from experimental results of static tests on titanium grade 5 specimens (Section 2.2). In order to ensure smooth convergence behavior, the load was progressively increased through 20 substeps. Since nonlinear elastic behavior was observed in the tensile tests carried out on titanium grade 5 specimens, this feature was included in the FE model. In fact, the load applied to the model was gradually increased (20 steps) in order to capture the transition from nonlinear elasticity to the plastic field. The "MELAS" option (where the nonlinear elastic stress-strain curve is assigned as a data table) was initially assumed as the material constitutive behavior of all elements included in the model. At the end of each load step, nodal stresses in each element were checked in order to verify that none of them went beyond the yield limit. Each time this happened, the constitutive model assumed for the material was switched from the "MELAS" option to "MISO" option. The latter corresponds to the classical multi-linear isotropic hardening stress-strain curve available from experimental results.

Table 9. Comparison between finite element results and experimental data.

FEM right side			EXPERIMENTAL right side			Error FE vs. Exp.		
ε_{max} [μm/m]	ε_{min} [μm/m]	ε_a [μm/m]	ε_{max} [μm/m]	ε_{min} [μm/m]	ε_a [μm/m]	Err.% ε_{max}	Err.% ε_{min}	Err.% ε_a
1932	179	877	2011	213	899	4.1	16.0	2.4
2979	273	1353	3021	286	1367	1.4	4.5	1.0
2771	255	1258	2964	321	1321	6.5	20.6	4.8
2583	238	1173	2793	304	1244	7.5	21.7	5.7
2666	245	1210	2869	317	1276	7.1	22.7	5.2
FEM reverse side			EXPERIMENTAL reverse side			Error FE vs. Exp.		
ε_{max} [μm/m]	ε_{min} [μm/m]	ε_a [μm/m]	ε_{max} [μm/m]	ε_{min} [μm/m]	ε_a [μm/m]	Err.% ε_{max}	Err.% ε_{min}	Err.% ε_a
1928	179	875	1967	199	884	2.0	10.0	1.0
2972	273	1349	3066	312	1377	3.1	12.5	2.0
2764	255	1255	2710	255	1228	-2.0	0	-2.2
2577	238	1170	2527	226	1151	-2.0	-5.3	-1.6
2659	246	1207	2638	241	1199	-0.8	-2.1	-0.7

Strain and stress values at the weld toe calculated by the ANSYS program were compared with experimental results. In order to have a homogeneous basis of comparison, strain values at weld toe computed by the FE model were integrated along the length of the strain gauge grid utilized in the experimental tests and the local strain amplitude was hence evaluated.

Table 9 reports experimental data and finite element computations for both right and reverse sides of the joint. The percent difference on maximum strain never exceeded 7.5% while raised to 22.7% in the case of minimum strain which obviously is more sensitive to uncertainties on experimental data and modeling assumptions. However, the error on the local strain amplitude ε_a resulted always smaller than 6% as strain values were integrated over the grid length. Remarkably, all σ_a–ε_a curves present the same slope regardless if they were determined experimentally or computed by ANSYS (Figure 14). Therefore, the FE model was able to accurately reproduce the local effects experimentally observed for the real joint.

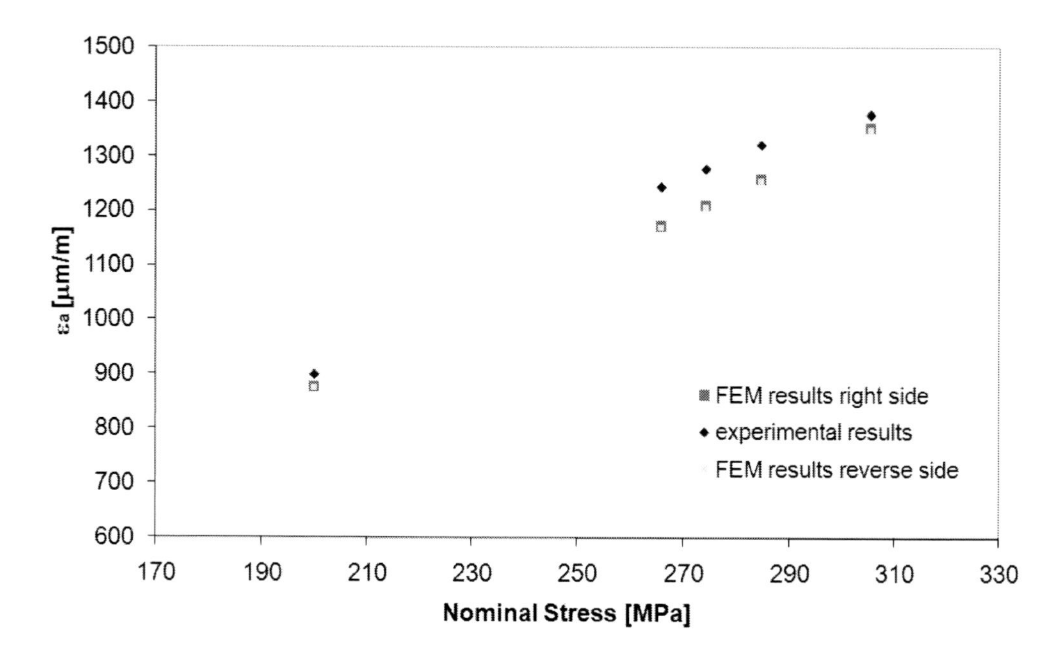

Figure 14. Comparison between experimental results and finite element computations for the titanium grade 5 butt joint.

The largest values of strain and stress on the reverse side of the weld seam profile are localized at the weld toe, in the region of transition between the weld metal and the base metal. This fact can be explained in view of the presence of the notch effect. Maximum stress/strain values depend on the notch effect resulting from the weld seam elevation with respect to the base metal and notch radius. Since these characteristic parameters of the joint were measured experimentally by the profilometer, FE computations depend directly on seam geometry.

Both macrographic inspection and mechanical contouring with the ± 1 μm precision CMM machine revealed the presence of a deep undercut on the right side of the titanium grade 5 butt joint. This feature is very common when laser beam welding or any other joining process involving high thermal flux is utilized. Strain and stress peak values were observed in correspondence of the deepest undercut (Figure 15). Peak values dropped down as the control location moves away from the undercut and became even smaller than their counterpart in the nominal zone. This local effect is due to the particular geometry of the weld seam and is related also with stress distribution. In fact, isostatic lines are highly concentrated near the undercut but become coarser at increasing distance from the undercut region (Figure 16).

Figure 17 shows that strain gauges could never capture strain peak values in the same position as that computed by the numerical model. However, FE predictions and experimental local strain amplitude measurements integrated over the 3 mm grid length practically coincided and were very close to the nominal strain value. The average level of stress near the weld seam could be estimated at a distance of 2 mm from the weld toe on both right and reverse sides of the joint: the average stress value was 301 and 244 MPa, respectively.

Conversely, in the nominal zone where the development of the final fatigue crack is observed, the maximum stress reached 340 MPa. The same behavior was observed regardless of the level of nominal load considered in the FE analysis.

The maximum load applied to the FE model did not produce any plasticization as the peak stress never went beyond the yield limit. In order to gather further evidence, another FE analysis was carried out considering a load for which the material can surely enter the plastic regime. However, the trend of total strain resulted again very similar to those shown in Figure 17 for the 40 kN load.

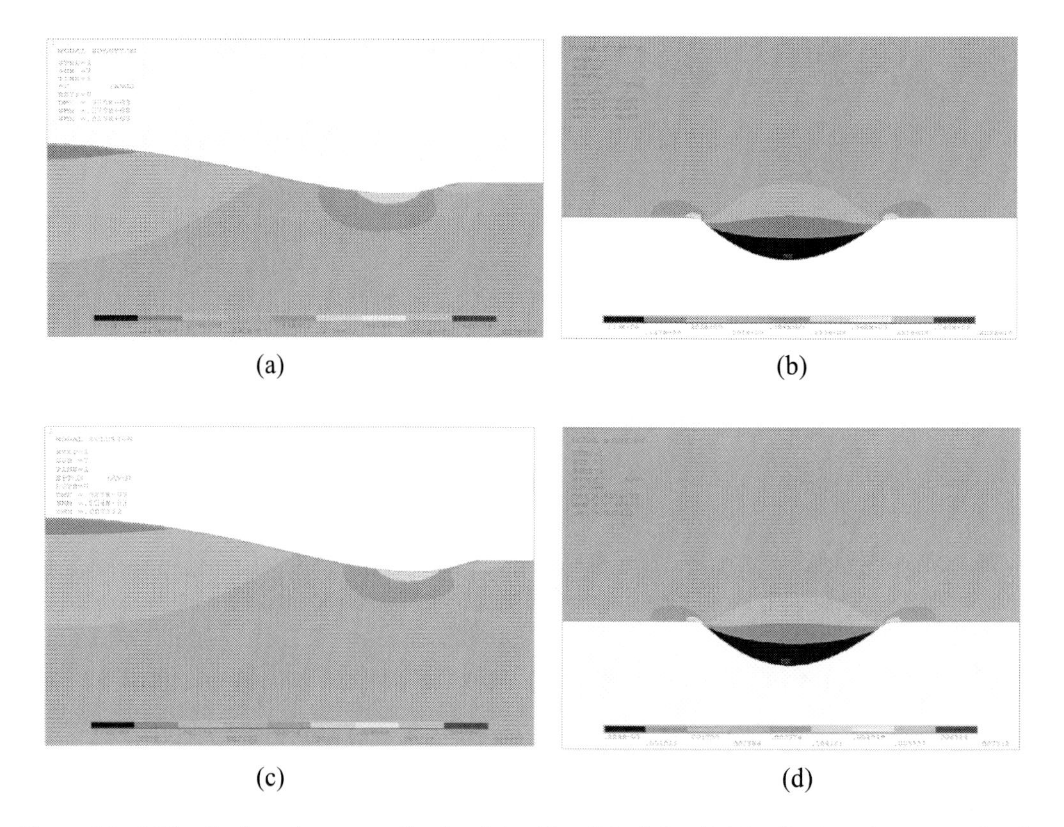

Figure 15. a) Distribution of stress σ_x for the right side of the joint; b) Distribution of stress σ_x for the reverse side of the joint; c) Distribution of total strain ε_x for the right side of the joint; d) Distribution of total strain ε_x for the reverse side of the joint.

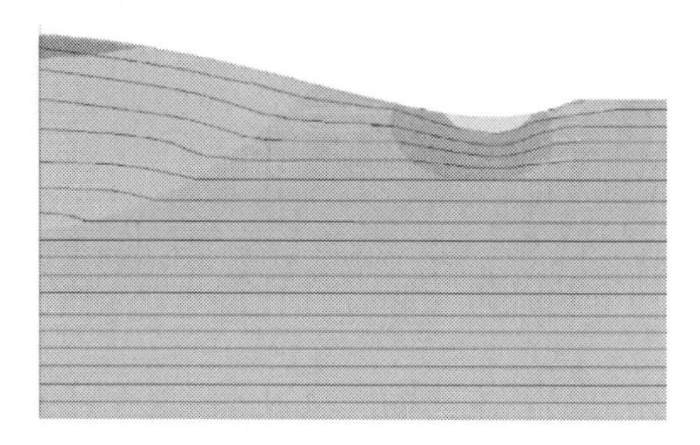

Figure 16. Distribution of isostatic lines computed by ANSYS for the Titanium grade5 butt joint.

Since the material properties given in input to ANSYS coincided with those of the base material, variations in microstructure and mechanical properties of the welded zone and heat affected zone of the joint were not considered in the FE simulation. This simplification is justified by the fact that LBW joints always failed in the nominal zone while the maximum stress was localized at the weld toe. Therefore, for the titanium alloy considered in this study, the welding process and the corresponding high thermal flux supplied to the metal seem not to affect mechanical properties in terms of microstructure modifications.

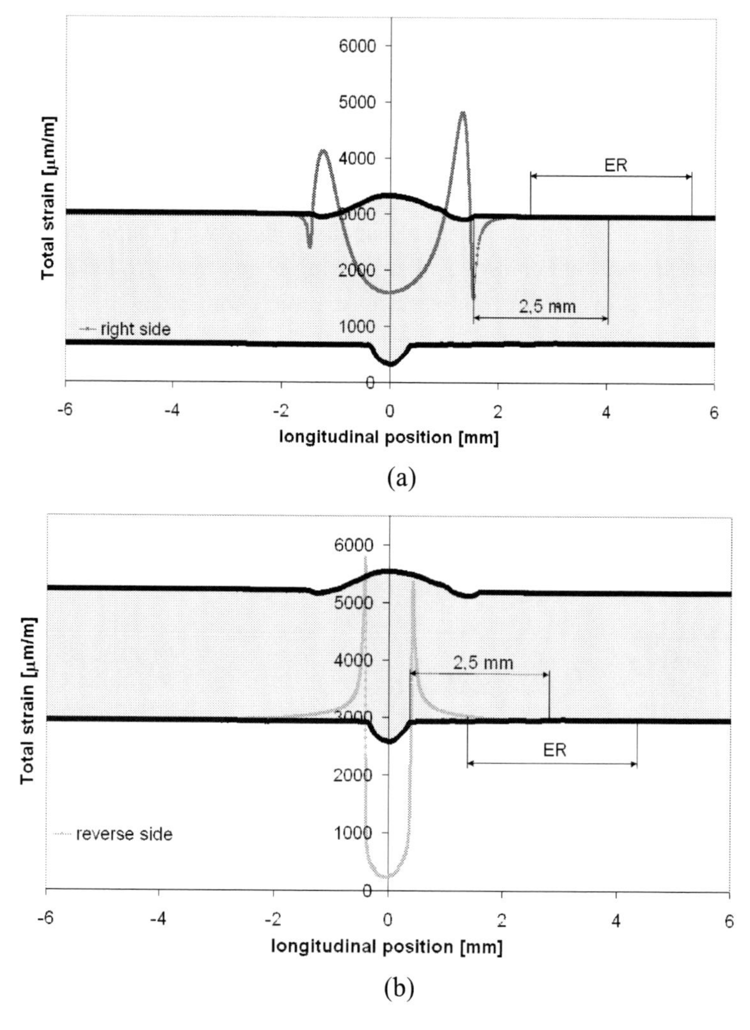

Figure 17. Total strain on the right side (a) and reverse side (b) of the butt C joint and corresponding strain gauge position for the nominal applied load of 40 kN.

4.2. The Effect of Porosity on LBW Joints

Most of the FE studies documented in literature include simplifications in order not to complicate the computational phase. For example, hypotheses on perfectly regular geometry (no imperfections or defects, circular weld radius, etc.) and linear elasticity are very common modeling assumptions used as "black boxes" even in presence of fully 3D finite element models. However, an important aspect to be properly considered in numerical analysis of titanium welded joints is the presence of defects in the weld seam. These defects, such as porosity, are caused by the welding process and may affect the fatigue resistance of the joint. The presence of defects often is not modeled in FE analysis of welded joints made of traditional metals because welding processes utilized in those cases are highly standardized and very well controllable. This may not be completely true for innovative materials like titanium alloys that are highly reactive with atmospheric gases at temperatures well below the melting temperature. Therefore, shielding systems must be designed to protect the welded component also in the cooling phase in order to avoid the formation of defects.

Porosity is caused by the high viscosity of welding pool that entraps gas particles and impurities inside the weld seam. Since aluminium and vanadium alloy elements included in titanium grade 5 (Ti6Al4V) vaporize above the vaporization temperature of pure titanium, the presence of the observed pores may be explained with the fact that shielding gas particles were entrapped in the fused zone. Pores are usually spherical in shape and very variable in size. Radiographic inspection could even miss single cavities. However, the weld seam usually includes clusters of cavities possessing the same geometrical characteristics: the size of these clusters may range from 60 to 160 μm.

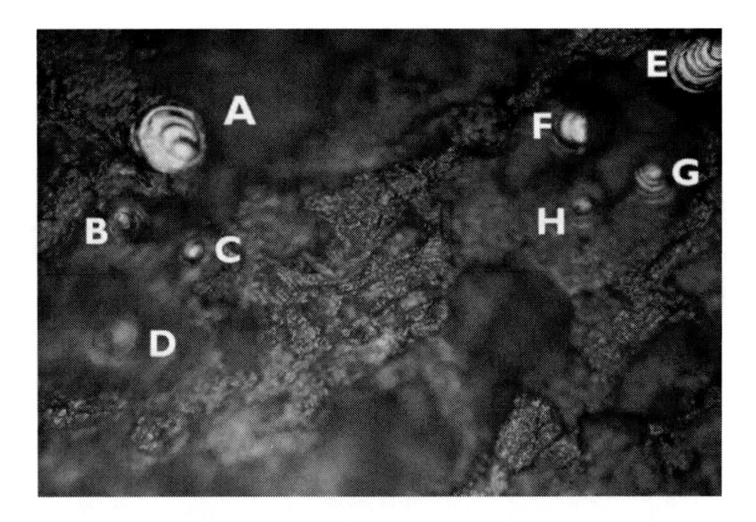

Figure 18. Fracture surface of the LBW titanium specimen broken at the weld toe (butt C2 specimen).

The fracture surface of the only LBW specimen, the butt C2 joint, broken at the weld toe (Figure 18) was inspected by means of a Nikon Eclipse TE200U optical microscope with 100X magnification. Pores were found to be present inside the weld seam. This surface map was utilized to determine precisely the dimensions and the position of the cavities. The corresponding elements of the FE mesh localized inside the region limited by each pore were killed in order to simulate the presence of material discontinuity: the stiffness assigned to each killed element was automatically scaled down by a factor 10^{-6} by the ANSYS program.

The butt C2 welded joint was again modeled with 8-nodes PLANE82 elements. The new FE model now included 263100 elements and 793109 nodes. Meshes were progressively refined via convergence analysis taking care to have enough elements in the region corresponding to each pore: hence, each pore "covered" about 10 elements in both horizontal and vertical directions. Finite element discretization was obviously finer near the weld seam in order to capture local stress gradients.

Figure 19 shows the FE models of the joint without porosity (Figure 19a) or including pores (Figure 19b). Figures 19b shows also the two control paths Γ_1 drawn along the weld profile (the dark curved line) and Γ_2 crossing the pore locations (the light straight line). Distortions due to misalignment were again not modeled as they are not significant (about 0.3°).

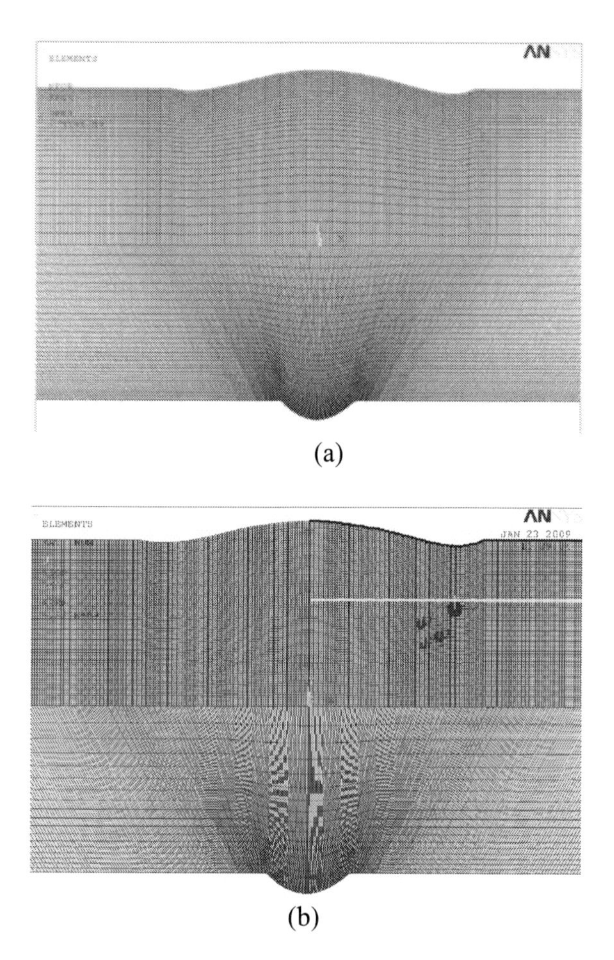

Figure 19. Finite element models of butt C2 titanium welded joint: a) without porosity; b) with porosity in the fused zone.

Table 10 summarizes the most important features of the FE models developed in this study. The level of stress concentration developed in the perfect joint with no pores because of the presence of the weld seam was analyzed by considering four different values of the ratio between the nominal load and the yield limit ranging from 0.85 to 1.15. This was done in purpose to simulate the two possible conditions of plasticization localized near the weld toe or diffused over the nominal zone (base material).

Table 10. Details of the different FE models developed to study the effect of porosity.

Weld seam profile	Thickness [mm]	Nominal stress [% yield limit]
Real	3	85 – 95 – 105 – 115
Porosity in weld seam	3	43.8
Porosity in nominal zone	3	43.8
Porosity in weld seam and nominal zone	3	43.8

Including pores in the FE model allowed to analyze stress concentration caused by the presence of those cavities. Porosity was then included also in the nominal zone where titanium welded joints usually exhibited fatigue failures. It should be noted that, since cavities are generated by the intrusion of shielding gases while defects in the nominal zone derive only from fabrication processes, porosity cannot be present in the nominal zone. However, including porosity also in the nominal zone may help to separate the amount of stress concentration caused by pores from the stress concentration caused by the notch effect near the weld seam. FE models with porosity were loaded at the nominal stress of 330 MPa corresponding to the failure load measured experimentally for the only titanium joint broken at the weld toe.

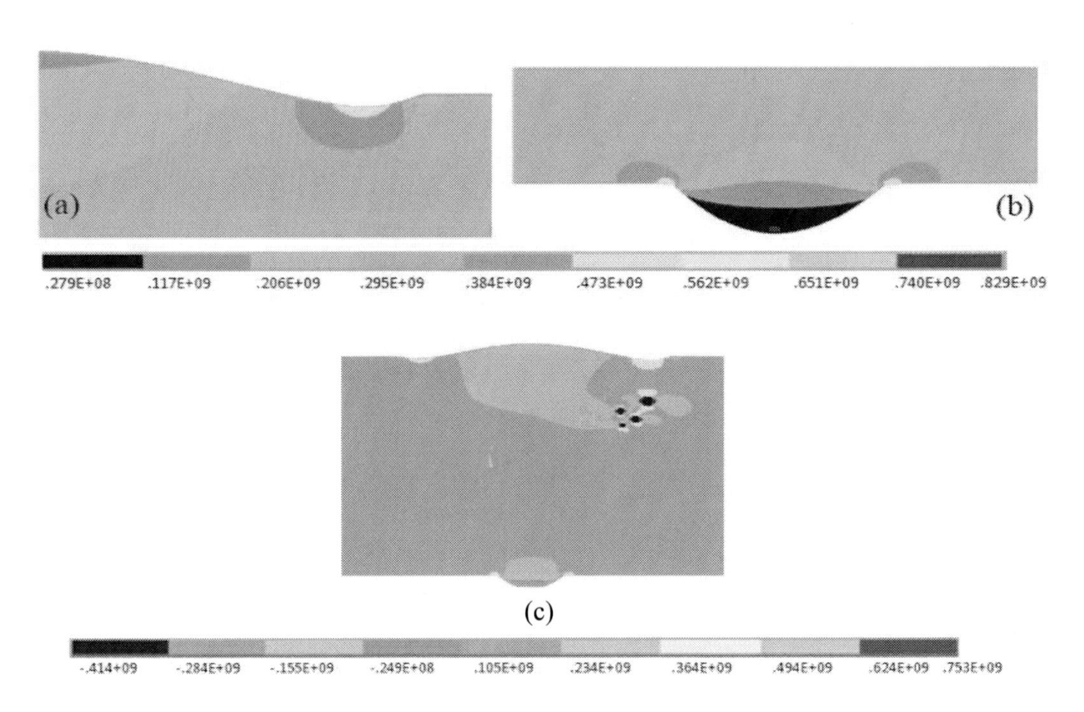

Figure 20. Stress field near the weld toe of titanium grade 5 butt joint: a) Right side of the perfect specimen; b) Reverse side of the perfect specimen; c) Specimen including porosity.

Figure 20 shows the distribution of stress σ_x in the different models with or without pores. This stress component is very significant in terms of the risk of crack initiation and weld joint failure. As expected, the highest value of stress computed by ANSYS for the joint without porosity is localized in correspondence of the deepest undercut (Figure 20a). By moving along the Γ_1 control path defined on the weld seam profile, one can observe that stress drops down to a value even lower than its counterpart in the nominal zone (see Figure 20a and Figure 21 where the latter illustration presents the results obtained for the perfect joint at different load levels). Remarkably, the same behavior was observed for both cases of concentrated and diffused plasticization (Figure 21). This local effect can be explained in view of the particular geometry of the weld seam and contributes to reduce the average level of stress and hence to prevent the risk of failure in the weld toe region.

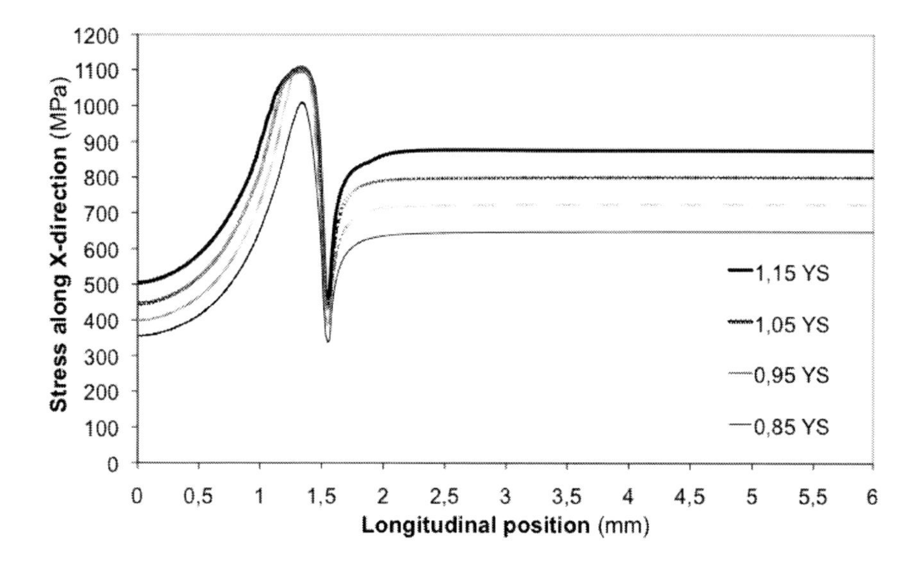

Figure 21. Perfect joint: stress distribution along path Γ_1 computed by ANSYS for different load levels.

Unlike in the case of perfect joint, the highest stress values were localized near pore cavities. Stress concentration caused by porosity is more important than that caused by the presence of the weld seam: respectively, 645 MPa vs. 530 MPa (Figure 22a). These results indicate that the combination of geometric effect (i.e. the presence of the weld cord) and porosity may increase the risk of failure.

Figure 22b shows that the 645 MPa stress peak exhibited by the titanium joint near cavities is always the same regardless of having modeled pores only in the fused zone or also in the nominal zone. No stress concentration occurred in the fused zone in correspondence of the control path Γ_2 when pores were included only in the nominal zone (i.e. where the computed stress is just equal to the applied nominal stress). Finally, Figure 22c shows the stress distributions computed in the nominal zone as one moves along Γ_2. It can be seen that the presence of defects in the fused zone does not affect at all the stress field in the nominal zone. Remarkably, the same value of maximum stress of 645 MPa was observed near pores for both fused and nominal zones. This indicates that there is the same risk of failure near pores regardless of where the pores are located. Therefore, failures observed in the nominal zone can be explained as long as there are enough pores that may interact thus forming a large region of high stress. Should this condition not occur, stress concentration near weld toe returns to be the primary cause of failure.

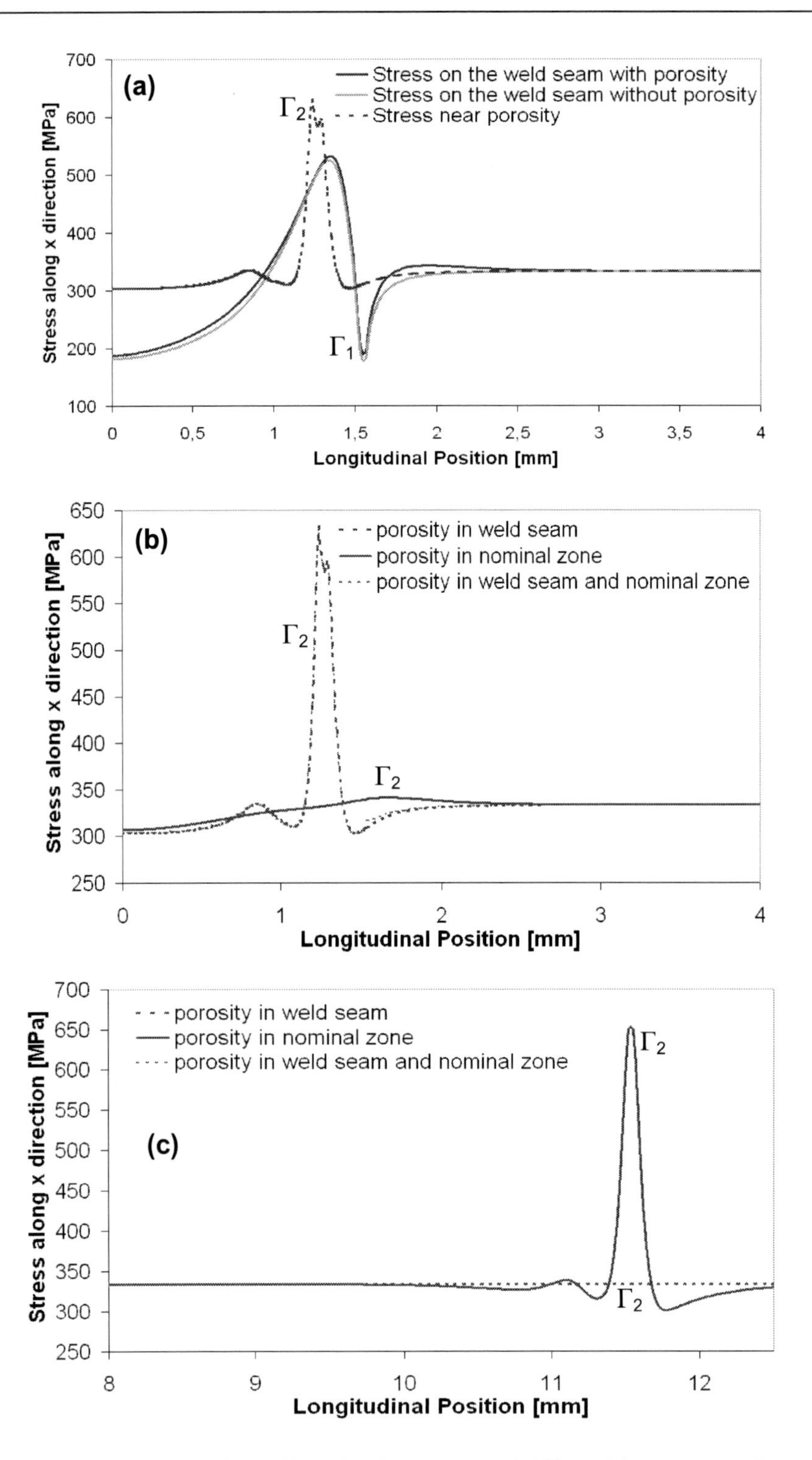

Figure 22. Stress distribution in the weld cord and near pores: a) Effect of the presence of pores in the fused zone; b) Effect of the presence of pores in the nominal zone on the local stress field in the fused zone; c) Effect of the presence of pores in the fused zone on the local stress field in the nominal zone.

Fatigue behavior of titanium joints is hence very sensitive to the presence of pores. In order to assess the relative influence of stress concentration produced by pores and/or local geometry of the weld cord, Table 11 summarizes the computed data. Stress concentration with respect to nominal load was 1.543: although this may indicate higher risks of failure near the weld toe, the experimental evidence is that fatigue fractures started in the nominal zone. The presence of pores increases considerably stress concentration: in fact, the stress concentration factor raises to about 2 either in the weld seam region and in the nominal zone. This result agrees with experimental evidence discussed in Section 3 of this chapter.

Table 11. Stress concentration factors for welded joints with or without defects.

Defect	Weld cord height [μm]	$\sigma_{max}/\sigma_{nom}$
no pores	176	1.543
pores	176	1.890 weld toe 1.960 nominal zone

A parameter that may clarify the interpretation of these results is the microstructural support length a: this quantity identifies a small material volume at the point of maximum stress that controls crack initiation [31]. The ratio between the microstructural support length and the notch radius ρ can be evaluated for titanium joints using the classical relationship between K_f and K_t given by Peterson [38]. Stress concentration factors were evaluated by carrying out new finite element analyses in the elastic regime. It was found that at the notch the a/ρ ratio is equal to 0.154; in the vicinity of pores included in the welded joint, the a/ρ ratio tends to 0. These results confirm that stress concentrations at pores observed in titanium joints may really be the cause of crack initiations.

The El Haddad-Smith-Topper parameter a_o supports the hypothesis that the mechanical strength of the titanium joint depends on the presence of defects rather than on weld cord local geometry. Under the assumption of crack radiating from the pore, which represents a circular hole in the plate, the Topper parameter [39] can be evaluated from the equation $\Delta K_{th}=1.12\cdot(\Delta\sigma_e)\cdot\sqrt{\pi a_o}$. In presence of pores, a stress concentration factor of 3 should be included in the previous expression (the factor 3 can be derived from the classical Inglis' equation on the stress value at the crack tip when the ratio between the crack length and the local curvature radius at the crack tip approaches 1 as it happens in the case of a circular hole [40]). The ΔK_{th} threshold value was estimated as 4.5 MPa $m^{1/2}$ [41-42] while the fatigue strength at $N=2\cdot10^6$ cycles was determined experimentally. The following values of a_o were found: 19.8 μm for the titanium joint with pores and 180 μm for the titanium joint without pores. With respect to the average measured diameter of defects 2a=160 μm, the El Haddad-Smith-Topper parameter is critical only in the case of joints with defects.

Although the FE models developed in this work were very detailed and refined on the basis of experimental evidence, the present trade study had some limitations. In the first place, no information on crystalline structures were included in the finite element models. Secondly, the titanium grade 5 alloy was considered isotropic and homogenous: however, mechanical behavior of titanium alloys actually depends on the direction of lamination as well as on the presence of interstitial elements within the crystalline structure. These factors could concur in causing fatigue failures in the nominal zone although it may not be easy to properly assess them in numerical analyses because of at least two practical reasons: (i) it is usually difficult to measure precisely microstructure transformations; (ii) should even detailed information on microstructure be available, the finite element model would include too many elements and hence FE analysis would be computationally unaffordable.

Finite element analyses did not account for the presence of residual stresses. Detailed information on residual stresses in terms of their distribution, magnitude and sign allow the real loading state of welded components to be correctly understood. However, experimental tests on stainless steel joints [26] showed that relaxation of residual stress already occur at about 10% of the joint fatigue life. Residual stresses were measured also for titanium grade 5 joints [43]. Since the ratio between residual stress magnitude and material yield limit (about 0.3) was the same both for stainless steel and titanium joints, the same mechanism of stress relaxation induced by fatigue was hypothesized to occur for both materials. For this reason, the effect of residual stress was not included in the finite element analyses carried out in this study.

5. Safety Factors Associated with Fatigue Resistance of Titanium Welded Joints

This section presents a critical analysis on fatigue strength of welded joints made of titanium grade 2 and grade 5 and welded by laser or hybrid process. Fatigue strength curves obtained for each alloy and each welding technique are compared in terms of safety factors with fatigue design curves of welded joints provided by a recent standard [2].

5.1. Influence of Material on Fatigue Resistance of Titanium Alloys

Fatigue test results can be plotted in terms of the stress range $\Delta\sigma$ versus the number of cycles to failure (Figures 23 and 24). Experimental data are then fitted by a power law of the Basquin's type where the number of cycles to failure increases as a negative power function of the stress range:

$$\Delta\sigma = \sigma'_f \cdot (N)^b \quad (1)$$

Titanium grade 5 has higher fatigue resistance than titanium grade 2. In the $N=10^3$–10^7 range and for the same life duration, the ratio between applied stress ranges is in fact 1.40–1.20. However, the ratio of yield stress is 2.14. It should be considered that fatigue resistance of titanium alloys depends on the chemical composition of the material. This aspect will be properly accounted for when experimental results are interpreted with respect to the fatigue design curves provided by the standard that do not include any dependence on material. Figure 23 reports the fatigue curve according to the AWS code [2] and other two curves obtained for titanium grade 2 and 5 from experimental data.

Table 12 reports values of fatigue strength and Basquin's exponent together with the endurance limit conventionally defined at $5 \cdot 10^6$ cycles. It can be noted that the empirical law used for steels, fatigue resistance equal to ultimate strength, is not satisfied for titanium joints. Values of Basquin's exponent are close to b=0.1 as common value.

Table 12. Fatigue resistance parameters of welded joints with respect to welding process and material.

		σ'_f [MPa]	-b	σ_D [MPa] ($N=5 \cdot 10^6$)
Welding (titanium grade 5)	LBW	1068	0.0991	232
	Hybrid	681.4	0.0940	160
Material (LBW)	Grade 2	609.5	0.0840	167
	Grade 5	1068	0.0991	232

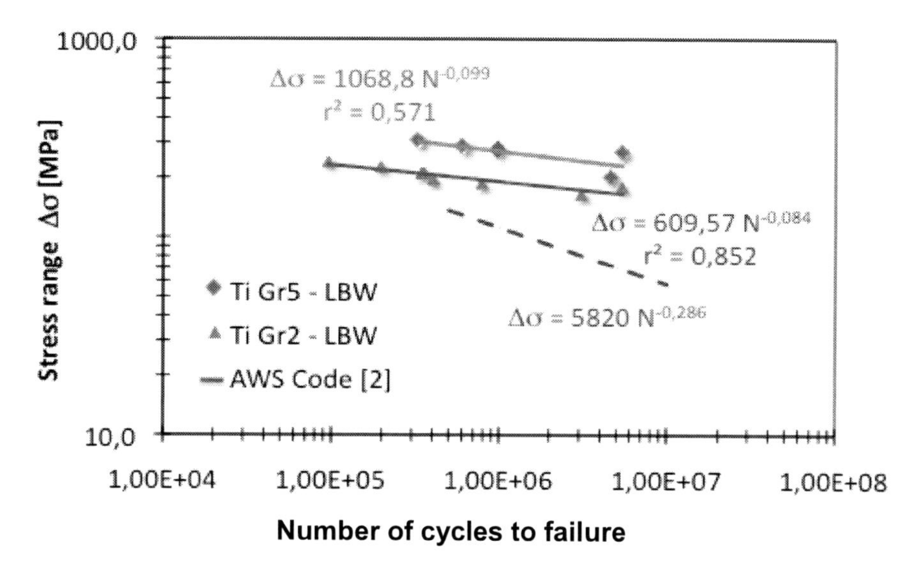

Figure 23. Fatigue resistance curves obtained for laser beam welded joints made of Ti grade 2 and Ti grade 5 alloys.

5.2. Effect of Welding Process

The effect of welding process on fatigue strength was analyzed only for the titanium grade 5 alloy by comparing results obtained for laser and hybrid laser-MIG specimens. Experimental results are presented in Figure 24. A power law of Basquin's type was again utilized to fit data. Fatigue resistance, Basquin's exponent and the endurance limit also are reported in Table 12.

Fatigue resistance is higher in the case of laser welding. In fact, in $N_r = 10^3 - 10^7$ range, for the same life duration, the ratio of applied stress range ranges between 1.43 and 1.49. It should be considered that fatigue resistance of welded joint made in titanium alloys is also sensitive to welding process.

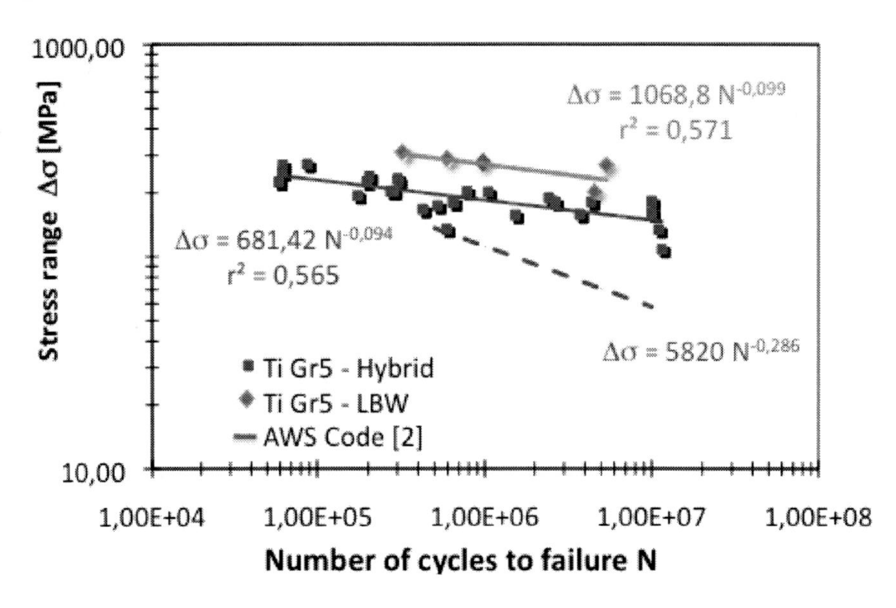

Figure 24. Fatigue resistance curves Δσ–N for laser and hybrid welded joints made of Ti grade 5.

5.3. Design of Welded Joints against Fatigue

Welded joints (steel, aluminum and titanium alloys) are designed against fatigue by using basic fatigue design curves that statistically incorporate the influence of residual stresses, stress ratio and safety factors. All design specifications for fatigue present a series of Δσ–N curves for particular weld details. Each curve is represented by a detail category corresponding to the limit value of stress range at $2 \cdot 10^6$ cycles. All fatigue curves drawn for the different stress range values are hence parallel. The basic fatigue design curves Δσ–N are expressed in the form:

$$\log_{10}N = \log_{10}K - m \cdot \log_{10}\Delta\sigma \qquad (2)$$

where K is the fatigue resistance limit of the welded joint and m the slope coefficient of the fatigue curve. In the case of steel, basic fatigue design curves include two segments with different slope. For short fatigue life, less than 10^7 cycles, a slope coefficient m=3 is used. For longer fatigue life, a slope coefficient m=5 is used until 10^8 cycles. For very long fatigue life, an endurance limit is expected. However, recent research work [44] found that fatigue life continuously decreases in the so-called “endurance limit” regime or giga-cycles regime and proposed to replace the endurance limit by a line with a slope coefficient m=45 for steel and m=22 for aluminum alloys.

Basic fatigue design curves of titanium alloys [2] are similar to the corresponding curves of steel but the slope coefficient is different: in particular, it holds m=3.5. Furthermore, endurance limit exists for long fatigue life, which corresponds to more than $5 \cdot 10^6$ cycles.

The slightly higher slope observed for fatigue crack growth curve was confirmed by fatigue tests on welded joints [45] and used to justify the change in slope with respect to steel alloys. The modified slope implies that many weld configurations will have very similar design allowable stress ranges for both titanium and steel alloys if the number of loading cycles is large enough: for example, more than 1 million of cycles.

The above mentioned AWS standard [2] includes a set of 14 original fatigue strength curves for different weld details. In [45], the fatigue strength curve obtained for a butt joint made of titanium alloy (FAT3.5 class 80) was compared with experimental results provided by technical literature. Derivation of fatigue curves is illustrated in Figure 25. All data found in literature are reported in one single graph. A safety factor of 10 is then applied to life duration: this corresponds to a safety factor on stress range close to 2. These rounded values indicate that deterministic safety factors were used for that purpose.

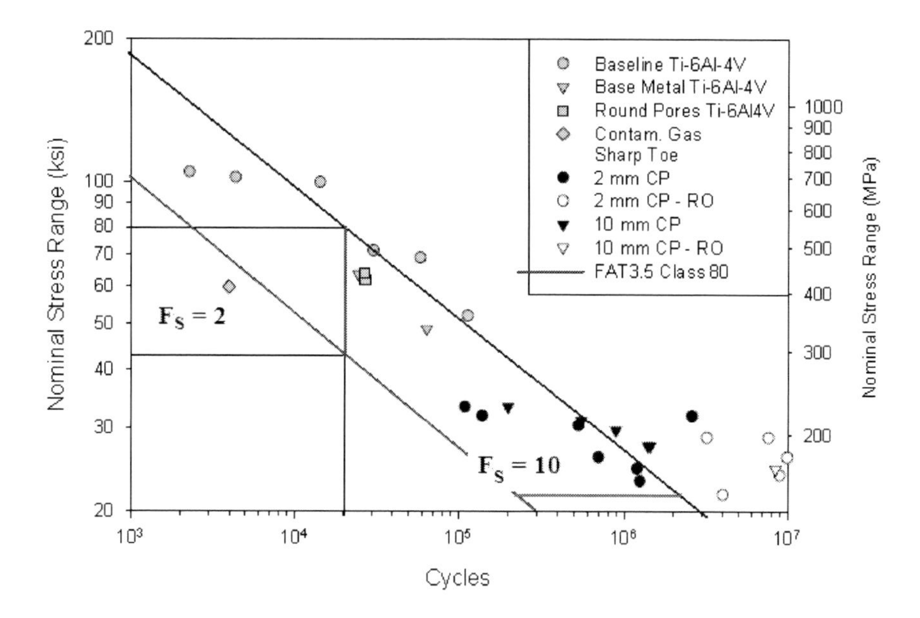

Figure 25. Derivation of fatigue design curves for welded titanium joints [45].

However, the current trend in design codes is to use probabilistic safety factors associated with a given level of probability P_r that fatigue failure may occur. The safety factor is thus defined as the ratio between the mean stress value and the corresponding limit value for the given probability P_r^*:

$$f_S = \frac{\sigma_a(P_r = 0.5)}{\sigma_a(P_r = P_r^*)} \tag{3}$$

The above mentioned approach requires the distribution of fatigue life duration to be known for a given stress range or the stress range distribution to be known for a given life duration. It is well known that for life duration shorter than 10^7 cycles and for a given stress range, the distribution of number of cycles to failure is either Lognormal or Weibull's type. For a given life duration, the distribution of stress range is instead Normal.

In many standard codes, the design curve is defined as that corresponding to 2.3% probability of failure. This is equivalent to 2 standard deviations of $\log_{10}N$ below the mean

$\Delta\sigma$–N curve if fatigue endurance is log-normally distributed. This corresponds to an approximate safety factor on fatigue life of about 3.

The probability density function for the Weibull distribution can be expressed as:

$$p(x) = c \cdot m \cdot x^{m-1} \cdot e^{-c \cdot x^m} \tag{4}$$

where m is the Weibull modulus and c is the normalization factor. The coefficient of variation is:

$$c_{V,x} = \sqrt{\frac{\Gamma(1+2/m)}{\Gamma^2(1+1/m)-1}} \tag{5}$$

where Γ is the Gamma-Euler function. The Weibull modulus can be estimated using the empirical relationship $m = c_{V,x}^{-1,09}$.

There are several values indicated in literature for the coefficient of variation. For example, British Defence standards suggest a value of 0.11 for aircraft grade aluminium alloys [46]. This value is intended to incorporate data scattering caused by material and manufacturing variations. Forgues [47] proposed a value of 0.0992 based on the analysis of 2451 fatigue tests on three aircraft grade alloys (Al7075, Al2024 and Al7475). Data for titanium alloys presently are not available but a good estimate can be made by assuming $C_{V,x}=0.1$.

For a given probability of failure, the safety factor is given by the following expression:

$$f_S = \frac{\Gamma(1+1/m)}{\left[Ln(1/P_S^*)\right]^{1/m}} \tag{6}$$

For the titanium butt joint class 71, the fatigue strength curve is described by the equation:

$$\Delta\sigma = \sigma'_f \cdot (N)^{1/3.5} \tag{7}$$

where the design fatigue strength σ'_f is 5820 MPa [2]. At $5 \cdot 10^6$ cycles, the corresponding stress range computed with Eq. (7) is equal to 71 MPa [2].

By combining the fatigue curve that contains experimental data and the fatigue curve reported in the code [2], two safety factors can be obtained. First, the safety factor on stress range $f_{s,\Delta\sigma}$ is defined as:

$$f_{s,\Delta\sigma} = \Delta\sigma_{exp}(N) / \Delta\sigma_{code}(N) \tag{8}$$

where $\Delta\sigma_{exp}$ and $\Delta\sigma_{code}$ are respectively the stress range relative to experimental results and the stress range indicated in the design fatigue curve for the same number of cycles to failure.

Second, the safety factor on fatigue life $f_{s,N}$ is defined instead as:

$$f_{s,N} = N_{exp}(\Delta\sigma) / N_{code}(\Delta\sigma) \quad (9)$$

where N_{exp} and N_{code} are respectively the limit number of cycles for the stress range relative to experimental tests and the limit number of cycles indicated in the design fatigue curve for the same stress range.

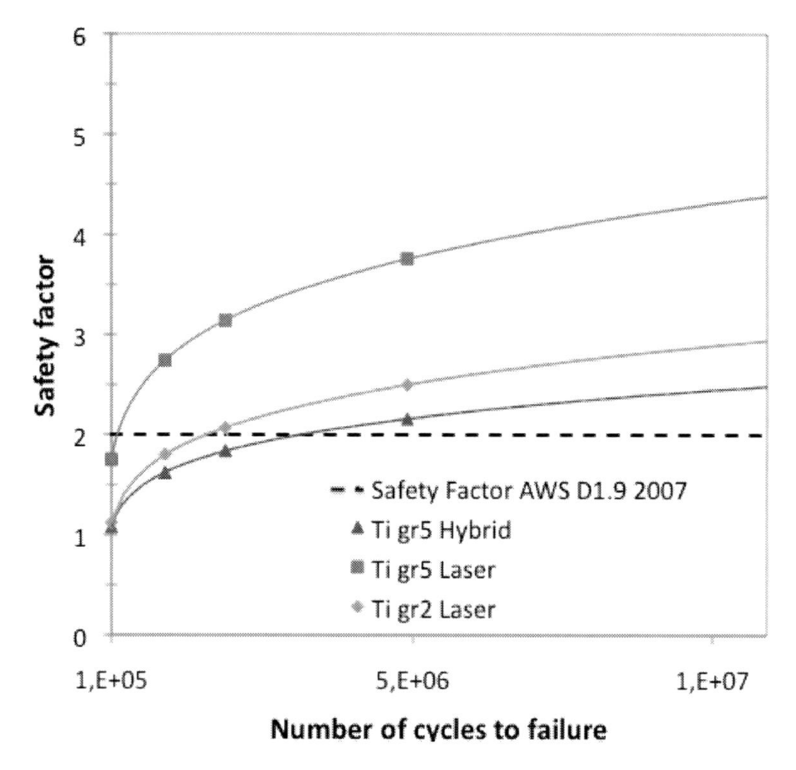

Figure 26. Safety factors for laser beam welded titanium grade 2 and 5 and for hybrid welded titanium grade 5 joints.

In Figure 26, safety factors on stress range are plotted vs. fatigue life duration for laser welded titanium grade 2 and 5 joints and hybrid welded titanium grade 5 joints, respectively. The following aspects should be underlined:

- Safety factors are never constant and become larger as life duration increases. The fact that for the Fat 3.5 Class 71 (71MPa) an endurance limit was expected for life duration longer than $5 \cdot 10^6$ cycles has been taken into account.
- Safety factors depend significantly on the joint material and are largest for laser beam welded joints made of titanium grade 5.
- For short life durations, safety factor is below the value of 2 indicated by the AWS D 1.9 2007 code [2].

Probability of failure associated with the safety factor values computed at different cycles by means of Eq. (6) is reported in Table 13. It can be seen that for long life durations the probability of failure is extremely low and stays below the conventional value of 10^{-6} used for assessing risks of human life. This level of probability corresponds to the safety factor of 2.95. For a safety factor of 2, the probability of failure increased to $1.2 \cdot 10^{-4}$.

Table 13. Probability of failure for laser welded Ti grade 2 and Ti grade 5 and for hybrid welded Ti grade 5.

Nr	P_f Ti grade 2 - Laser	P_f Ti grade 5 - Laser	P_f Ti grade 5 - Hybrid
10^5	$1.19 \cdot 10^{-2}$	$1.01 \cdot 10^{-4}$	$1.09 \cdot 10^{-2}$
10^6	$4.03 \cdot 10^{-5}$	$5.23 \cdot 10^{-7}$	$7.11 \cdot 10^{-5}$
$2 \cdot 10^6$	$7.26 \cdot 10^{-6}$	$1.07 \cdot 10^{-7}$	$1.56 \cdot 10^{-5}$
$5 \cdot 10^6$	$2.93 \cdot 10^{-7}$	$1.99 \cdot 10^{-8}$	$6.35 \cdot 10^{-7}$

The safety factor on life duration is more difficult to evaluate because the Basquin's exponent determined for the present material is much different from the corresponding values indicated in the Fat 3.5 Class 71 design fatigue curve. Furthermore, fatigue curve was limited to $5 \cdot 10^6$ cycles in view of the assumption that an endurance limit exists. However, comparisons can be made for the stress range of 167 MPa corresponding to 10^5 cycles on the Fat 3.5 Class 71 fatigue design curve: these data are listed in Table 14. Finally, it appears that the safety factor provided by the design fatigue curve Fat 3.5 Class 71 is close to 2 for short life duration (10^5 cycles) but highly conservative for long fatigue life.

Table 14. Comparison of fatigue safety factor on life duration ($\Delta\sigma$ =167 MPa) for laser welded joints.

	Fat 3.5 Class 71	Ti grade 2	Ti grade 5
N_r	10^5	$15 \cdot 10^6$	$40 \cdot 10^7$
$f_{s,\Delta\sigma}$	2	150	4000

6. Residual Stress

Residual stresses can be measured with different techniques. In the Hole Drilling Method (HDM) a small hole is drilled in the specimen at sequential steps. The resulting changes in strain distribution are measured by means of 3-grid electrical strain gauge rosettes. As the drilling operation proceeds, the residual stresses are gradually relaxed and the strain gauges allow the magnitude and orientation of in-plane principal stresses to be evaluated.

The experimental setup utilized in this research is basically an automatic system (RESTAN, REsidual STresses Analyzer, Sint Technology s.r.l, Italy) that utilizes the hole drilling method as it is indicated in the ASTM E837 standard [48]. The equipment includes a high speed turbine (300,000 rpm) powered by compressed air that moves the tungsten carbide mill, an electric step by step motor to control the vertical displacement of the mill so to perform measurements at sequential steps, an optical microscope to precisely line up the end mill with the hole axis, two centesimal comparators to measure the hole diameter and the amount of eccentricity, and three grid strain gauges to measure strains.

The RESTAN device can precisely execute the drilling process [49-50]. This limits the generation of additional stresses that would derive from some unavoidable plasticization around the hole. Furthermore, the initial point of drilling can be univocally defined. The eccentricity and effective diameter of the hole can be measured with optical techniques.

By measuring eccentricity it is possible to properly correct the computed values of residual stresses [51]. In order to account for variations in residual stress measured through thickness, the power series method was used in this study [52-53].

Strain gauge rosettes were bonded to the weld toe and positioned in such a way that the RESTAN system could drill holes centered at the maximum distance of 1.5 mm from the weld toe.

Residual stresses were measured on titanium grade 5 butt welded plates built with both LBW and hybrid techniques. Figures 27 and 28 show the distributions of residual stress calculated from experimental data for LBW and HYBRID, respectively. Stress components acting in the direction normal (i.e. normal direction) to the weld cord or in the direction parallel (i.e. longitudinal direction) to the weld cord are presented. Residual stresses are plotted with respect to hole depth. Each curve corresponds to a different location on the welded plates.

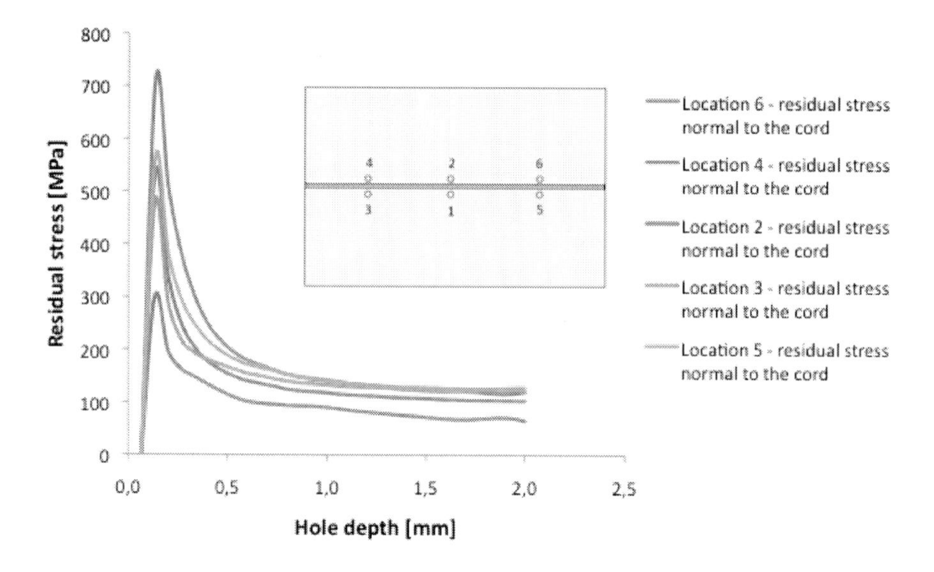

Figure 27. Titanium grade 5 laser butt welded plate: residual stresses normal to the weld cord.

Some out-of-scale values were measured in correspondence of the early drilling steps. This could be due to the presence of stresses generated during the surface preparation for strain gauge bonding. At higher depth (about 1 mm), residual stresses followed a more regular trend.

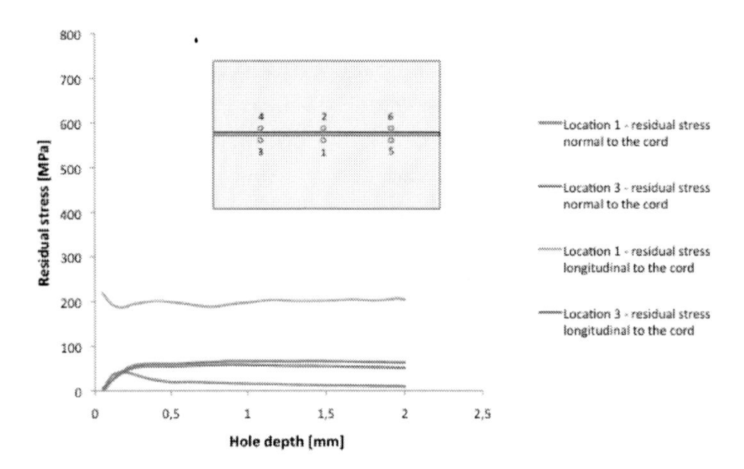

Figure 28. Titanium grade 5 hybrid butt welded plate: residual stresses normal to the weld cord.

The analysis of the distribution of residual stresses suggests the following considerations. Magnitude and distribution of residual stresses found in titanium welded plates depend on the material properties that govern temperature changes, thermal expansions and consequent shrinkages. In general, residual stresses are caused by thermal stresses and plastic strains generated during welding. The thermal stress $\sigma=\alpha ET$ for titanium plates is smaller in magnitude than for steel plates (both the Young's modulus E and thermal expansion coefficient α of titanium are lower than for steel while only the melting temperature T_f is higher for titanium). Furthermore, since thermal conductivity of titanium is low, heat propagates slowly through the plates: this probably explains why HAZ is very narrow. Since cooling also is slow, distribution of residual stresses along the cord for titanium welded plates is more regular than in the case of steel plates [54-57].

Residual stresses in the longitudinal direction were found to be higher than their counterpart in the transverse direction (Figure 28). Such a behavior is similar to that usually seen in the case of steel joints.

Residual stresses of Ti grade 5 are well below 50% of the material yield point: the average values of residual stress normal to the cord are 120 MPa for LBW joints and 50 MPa for hybrid weld joints, respectively.

Recent progress in the field of welding techniques and applications of titanium alloys in naval constructions suggest to deeply understand not only the process of generation of residual stress caused by welding and the corresponding stress peak values, but also to analyze the possible interactions with other phenomena such as fatigue. Effective prediction models should be developed in order to properly account for residual stresses already in the design stage.

CONCLUSIONS

This chapter presented a comprehensive experimental-numerical study on Titanium grade 2 and Titanium grade 5 butt joints welded by means of the laser beam and laser-MIG hybrid welding techniques. Preliminary investigations were conducted on the titanium specimens. Micro-hardness measurements and macrographs served to identify the different regions (i.e.

fused zone, heat affected zone, base metal) of the joint and to detect the eventual presence of defects. The geometry of the joint was accurately determined by measuring the weld cord profile, misalignments and distortions included in the weld: for that purpose, a CMM tactile device with ±1 μm accuracy was utilized. Static and fatigue tests were then carried out on the different specimens.

From the static tests, it was found that titanium grade 5 (also denoted as Ti6Al4V) has very high strength and is highly suited for applications where components may undergo high stresses. However, this titanium alloy is less machinable with tools than titanium grade 2, the commercially pure titanium.

Fatigue tests allowed the σ_a–N and the ε_a–N curves to be obtained. Surprisingly enough, those curves presented a very similar coefficient of correlation. Although this fact is rather unusual, it can be explained with the limited presence of factors that modify the local stress/strain field with respect to the nominal field.

Fracture patterns observed in the LBW broken specimens were localized in the base metal region, that is far away from the weld toe where failures of welded steel joints usually occur.

In order to analyze experimental results more deeply, a finite element model of the welded joint geometry was developed. To make FE analysis as most reliable and accurate as possible, the real geometry of the weld seam and the material properties of the different regions were determined experimentally and then given in input to the numerical model. Strain values at the weld toe predicted by FE analysis (numerical results were integrated over a 3 mm path corresponding to the length of the strain gauge grid) were in good agreement with experimental measurements. Furthermore, local stress concentration was correctly detected by the FE model. However, both experimental data and numerical results obtained for LBW joints indicated that the average level of stress near the weld seam is lower than in the nominal region of the joint where fatigue fracture finally develops. This suggests that the local stress/strain peaks, found by finite element analysis but not detected by strain gauges, not necessarily were the primary cause of failure of laser welded titanium alloy plates.

Although the FE model implemented in this research was very detailed and refined on the basis of experimental evidence, the model itself could not cover completely all of the complex mechanisms behind initiation, development and conclusion of the fatigue fracture process. In spite of the fact that there is an isolated local peak of stress correctly captured by the FE model, the experimental evidence revealed that fractures occur in regions hosting a lower level of stress. Therefore, one should not conclude "blindly" that failure is always localized where the largest stress peak is predicted by the FE model, but must carefully consider the possible relationships between stress values and the attitude and capacity of material to withstand such a stress level. As is clear, this would imply the exact knowledge of other important material properties deeply connected with microstructure (very difficult to measure experimentally) and rarely accounted for in numerical analysis because of the serious hardware constraints (CPU time) and complications in terms of software implementation (FE formulation at the microstructure level).

The inspection of fracture surfaces with an optical microscope revealed the presence of pores in the only LBW specimen broken at the weld toe and in some HYBRID welded joints. Since fatigue behavior of titanium joints seems to be sensitive to the presence of pores, in order to assess the relative influence of stress concentration produced by pores with respect to local geometry of the weld cord, the numerical model was augmented by including pores.

Results indicate that the presence of pores increases considerably stress concentration: in fact, the stress concentration factor raises to about 2 either in the weld seam region and in the nominal zone. This is in good agreement with the experimental evidence.

Experimental results of fatigue tests carried out on HYBRID welded joints showed that the load carrying capability of these joints is lower than in the case of laser beam welded joints. The geometry of the weld seam is considerably different: in fact, hybrid welding processes cause the heat affected zone and the size of the weld cord to be larger than in the laser welding case. The thicker cord results in lower values of stress concentration factor. However, the stress concentration factor may locally increase because of the higher level of irregularity in the weld seam.

In summary, fatigue behavior of titanium joints depends on welding technique and process parameters. Fatigue resistance of titanium welded joints can be assessed by using the Basquin's law. The characteristic parameters of this law, namely the fatigue resistance and Basquin's exponent, depend on the titanium alloy type (grade 2 and grade 5) and welding process (laser and hybrid laser-MIG). The welding code recently released for structural titanium includes a set of fatigue strength curves for different weld details that however do not account for the type of alloy and the welding process. This limitation is clearly highlighted by the experimental results presented in this research.

The average value found for the Basquin's exponent of welded titanium alloys is about 0.09. Conversely, a value of 0.285 (1/3.5) is suggested in the code. This important difference results in highly conservative predictions when titanium welded joint are designed within the range of fatigue endurance characterized by safety factors larger than 2.

ACKNOWLEDGMENTS

The authors are deeply indebted to their colleagues who participated in the experiments presented in this chapter.

REFERENCES

[1] ASTM B265-10. *Standard Specification for Titanium and Titanium Alloy Strip, Sheet, and Plate*. American Society of Testing & Materials, West Conshohocken, PN (USA), 2010.

[2] AWS D 1.9/D 1.9/M:2007. *Structural Welding Code – Titanium.* American Welding Society, Miami, FL (USA), 2007.

[3] AWS D1.1M:2010. *Structural Welding Code – Steel.* American Welding Society, Miami, FL (USA), 2010.

[4] ASM International. *ASM Handbook Properties and Selection: Nonferrous Alloys and Special – Purpose Materials 10th Edition*, Vol. 2. ASM International, 1991.

[5] BS 7608. *Fatigue Design and Assessment of Steel Structures.* British Standards Institute, London (UK), 1993.

[6] Eurocode 3. *Design of Steel Structures. Part 1-9: Fatigue.* See also UNI-NIEN 1993-1-9. European Committee for Standardization, 2005.

[7] Eurocode 9. *Design of Aluminium Structures. Part 2: Structures Susceptible to Fatigue.* UNI-EN 1999-2, 2002. European Committee for Standardization, 2002.
[8] Recommendations of IIW, XIII-2151-07/XV-1254-07. *Recommendations for Fatigue Design of Welded Joints and Components.* International Institute of Welding, 2007.
[9] *RINA Rules. Testing Procedures for Materials.* Section 2, Part D, Chapter 1. Registro Italiano Navale, Genova (Italy), 2002.
[10] Takao K, Kusukawa K. *Low-cycle fatigue behavior of commercially pure titanium.* Materials Science & Engineering A, 1996, Vol. 213, pp. 81-85.
[11] Torster F, Dos Santos JF, Hutt G, Kocak M. *The use of titanium alloys for dynamic risers: a literature review.* Report No. GKSS 97/E/38. GKSSForschungszentrum Geesthacht GmbH, Geessthacht (Germany), 1997.
[12] Iwata T, Matsuoka K. *Fatigue strength of CP grade 2 titanium fillet welded joint for ship structure.* Welding in the World, 2004, Vol. 48, pp. 40-47.
[13] Bergmann JP. *Mechanical behavior of overlap joints of titanium.* Science and Technology of Welding and Joining, 2005, Vol. 10, pp. 50-60.
[14] Casavola C, Tattoli F, Pappalettere C. *Static and fatigue characterization of titanium alloy welded joints.* Mechanics of Materials, 2009, Vol. 41, pp. 231-243.
[15] Masubuchi K. *Analysis of Welded Structures: Residual Stresses, Distortion, and Their Consequences.* Oxford (UK): Pergamon Press, 1980. ISBN: 978-0080227146.
[16] The Welding Institute. *Residual Stresses and Their Effects,* Abington Hall (UK), 1981. ISBN: 978-0853001416.
[17] Lee YB, Chung CS, Park YK, Kim HK. *Effects of redistributing residual stress on the fatigue behavior of SS330 weldment.* International Journal of Fatigue 1998, Vol. 20, pp. 565-573.
[18] Society for Experimental Mechanics. *Handbook on Residual Stresses.* Bethel, CT (USA), 2005.
[19] Yunlian Q, Ju Q, Quan H, Liying Z. *Electron beam welding, laser beam welding and tungsten arc welding of titanium sheet.* Materials Science and Engineering 2000, Vol. A280, pp. 177-181.
[20] Dattoma V, Pappalettere C. *Local strain for fatigue strength of welded structures.* Journal of Strain Analysis for Engineering Design, 2001, Vol. 36, pp. 605-610.
[21] Casavola C, Nobile R, Pappalettere C. *Fatigue strength by the WEL.FA.RE. local strain method: application to 3-5 mm cruciform and butt welded joints.* In: Proceedings of the 2002 SEM Annual Conference and Exposition on Experimental and Applied Mechanics, Milwaukee, WN (USA), 2002.
[22] Casavola C, Nobile R, Pappalettere C. *Fatigue life predictions by the WEL.FA.RE. method: influence of residual stresses.* In: Proceedings of the 2003 SEM Annual Conference and Exposition on Experimental and Applied Mechanics, Charlotte, NC (USA), 2003.
[23] Casavola C, Pappalettere C. *Industrial application of a new local strain method for fatigue strength evaluation of welded structures.* In: Proceedings of the ICEM12 International Conference on Experimental Mechanics (Edited by C. Pappalettere), Bari (Italy), 2004.
[24] Casavola C, Nobile R, Pappalettere C. *A local strain method for the evaluation of welded joints fatigue resistance: the case of thin main-plates thickness.* Fatigue and Fracture of Engineering Materials and Structures, 2005, Vol. 28, pp. 759-767.

[25] Casavola C, Pappalettere C. *Application of WEL.FA.RE. method on aluminium alloy welded joints.* In: Proceedings of the 2005 SEM Annual Conference and Exposition on Experimental and Applied Mechanics Portland, OR (USA), 2005.

[26] Casavola C, Pappalettere C. *Residual stress on titanium alloy welded joints.* In: Proceedings of the ICEM13 International Conference on Experimental Mechanics, Experimental Analysis of Nano and Engineering Materials and Structures (Edited by E.E. Gdoutos). Dordrecht (The Netherlands): Springer, 2007.

[27] Kalpakjian S, Schmidt SR. *Manufacturing Engineering and Technology. Fifth Edition.* Upper Saddle River, NJ (USA): Prentice Hall, 2006. ISBN: 978-013148965-8.

[28] Schultz H. *Electron Beam Welding.* Cambridge (UK): Woodhead Publishing/The Welding Institute, 1994. ISBN: 978-1855730502.

[29] Caiazzo F, Curcio F, Daurelio G, Memola Capece Minutolo F. *Ti6Al4V sheets lap and butt joints carried out by CO_2 laser: mechanical and morphological characterization.* Journal of Materials Processing Technology, 2004, Vol. 149, pp. 546-552.

[30] Casavola C, Pappalettere C. *Discussion on local approaches for the fatigue design of welded joints.* International Journal of Fatigue, 2009, Vol. 31, pp. 41-49.

[31] Radaj D, Sonsino CM, Fricke W. *Fatigue Assessment of Welded Joints by Local Approaches, 2nd Extended Edition.* Cambridge (UK): Woodhead Publishing, 2006. ISBN: 978-0849384516.

[32] Haibach E. *Die Schwingfestigkeit von Schweibverbindungen aus der Sicht einer örtlichen Beanspruchungsmessung* LBF–Bericht no. FB-77, Lab. Fur Betriebsfestigkt, Fraunhopher Institute, Darmstadt (Germany), 1968.

[33] Pappalettere C, Nobile R. *Fatigue strength of welded joints by the local strain method. Influence of load ratio R and plate thickness.* In: Notch Effects in Fatigue and Fracture (G. Pluvinage and M. Gjonanj Editors), NATO Sciences Series II – Mathematics, Physics and Chemistry, pp. 307-316. Dordrecht (NL): Kluwer Academic Publishers, 2000.

[34] ANSYS Inc. *ANSYS® Version 11.0. Theory and User's Manual.* Canonsburg, PA (USA), 2008.

[35] Radaj D, Sonsino CM, Flade D. *Prediction of service fatigue strength of a welded tubular joint on the basis of the notch strain approach.* International Journal of Fatigue, 1998, Vol. 20, pp. 471-480.

[36] Tovo R, Lazzarin, P. *Relationship between local and structural stress in the evaluation of the weld toe stress distribution.* International Journal of Fatigue, 1999, Vol. 21, pp. 1063-1078.

[37] Hou CY. *Fatigue analysis of welded joints with the aid of real three-dimensional weld toe geometry.* International Journal of Fatigue, 2007, Vol. 29, pp. 772-785.

[38] Peterson RE. *Stress Concentration Factors.* New York (USA): John Wiley & Sons, 1974.

[39] El Haddad MH, Topper TH, Smith KN. *Prediction of non-propagating cracks.* Engineering Fracture Mechanics, 1979, Vol. 11, pp. 573-584.

[40] Anderson TL *Fracture Mechanics: Fundamentals and Applications, Third Edition.* Boca Raton (USA): CRC Press, 2004. ISBN: 978-0849316562.

[41] Ritchie RO, Boyce BL, Campbell JP, Roder O, Thompson AV, Milligan VV. *Thresholds for high-cycle fatigue in a turbine engine Ti-6Al-4V alloy.* International Journal of Fatigue, 1999, Vol. 21, pp. 653-662.

[42] Sheldon JW, Bain KR, Keith Donald J. *Investigation of the effects of the shed-rate, initial K_{max}, and geometric constraint on ΔK_{th} in Ti-6Al-4V at room temperature.* International Journal of Fatigue, 1999, Vol. 21, pp. 733-741.

[43] Casavola C, Pappalettere C. *Residual stresses and fatigue strength of butt welded components.* In: Proceedings of 2004 SEM International Conference and Exposition on Experimental and Applied Mechanics, Costa Mesa, CA (USA), 2004.

[44] Sonsino CM. *Course of S-N curves especially in the high-cycle fatigue regime with regard to component design and safety.* International Journal of Fatigue, 2007, Vol. 29, pp. 2246–2258.

[45] Mohr WC. *The Design Provisions of the New AWS Structural Welding Code – Titanium.* Materials Science Forum, 2008, Vol. 580-582, pp. 93-96.

[46] UK Ministry of Defence. *Fatigue Damage Tolerance Substantiation of Fatigue Life.* Defence Standard 00-970, 1987.

[47] Forgues SA. *Study of material and usage variability for probabilistic analyses and scatter factor determination.* Bombardier/Canadair Report RDA-DSD-123, 1996.

[48] ASTM E837: *Standard method for determining residual stresses by the hole-drilling strain gage method.* Annual Book of ASTM Standards, 2008.

[49] Valentini E, Ferrari M, Ponticelli S. Metodologie di misura. In: *Studi per una proposta di raccomandazione sull'Analisi Sperimentale delle Tensioni Residue con il Metodo del Foro.* Quaderno AIAS 1997, Vol. 3, pp. 53-67. (In Italian)

[50] Beghini M, Bertini L, Raffaelli P. *Numerical analysis of plasticity effects in the hole-drilling residual stress measurement.* Journal of Testing and Evaluation, 1994, Vol. 22, pp. 522-529.

[51] Ajovalasit A, Petrucci G. Influenza dell'eccentricità sulla determinazione delle tensioni residue con il metodo del foro. In: *Studi per una proposta di raccomandazione sull'Analisi Sperimentale delle Tensioni Residue con il Metodo del Foro.* Quaderno AIAS 1997, Vol. 3, pp. 13-19. (In Italian)

[52] Vangi D. Distribuzione non uniforme della tensione nello spessore. In: *Studi per una proposta di raccomandazione sull'Analisi Sperimentale delle Tensioni Residue con il Metodo del Foro.* Quaderno AIAS 1997, Vol. 3, pp. 31-41. (In Italian)

[53] Schajer GS. *Measurements of non uniform residual stress using the hole drilling method. Part I – Stress calculation procedures.* ASME Journal of Engineering Materials and Technology, 1988, Vol. 110, pp. 338-343.

[54] Casavola C, Dattoma V, De Giorgi M, Nobile R, Pappalettere C. Experimental analysis of the residual stresses relaxation of butt-welded joints subjected to cyclic load. In: *Proceedings of 4th International Conference on Fracture and Damage Mechanics (FDM)*, Mallorca (Spain), 2005.

[55] Teng TL, Lin CC. *Effect of welding conditions on residual stresses due to butt welds.* International Journal of Pressure Vessels and Piping, 1998, Vol. 75, pp. 857-864.

[56] Nguyen TN, Wahab MA. *The effect of weld geometry and residual stresses on the fatigue of welded joints under combined loading.* Journal of Materials Processing Technology, 1998, Vol. 77, pp. 201-208.

[57] Chiarelli M, Lanciotti A, Sacchi M. *Fatigue resistance of MAG welded steel elements.* International Journal of Fatigue, 1999, Vol. 21, pp. 1099-1110.

In: Advances in Mechanical Engineering Research, Volume 3 ISBN: 978-61209-243-0
Editor: David E. Malach

Chapter 3

BIRD STRIKE ANALYSIS IN AIRCRAFT ENGINEERING: AN OVERVIEW

Sebastian Heimbs
EADS - European Aeronautic Defence and Space Company, Innovation Works, 81663 Munich, Germany

ABSTRACT

Bird strike is a major threat to aircraft structures, as a collision with a bird during flight can lead to serious structural damage. Computational methods have been used for more than 30 years for the bird-proof design of such structures, being an efficient tool compared to the expensive physical certification tests with real birds. At the velocities of interest, the bird behaves as a soft body and flows in a fluid-like manner over the target structure, with the high deformations of the spreading material being a major challenge for computational simulations based on the finite element method. This chapter gives an overview on the development, characteristics and applications of different soft body impactor modeling methods by an extensive literature survey. Advantages and disadvantages of the most established techniques, which are the Lagrangian, Eulerian or meshless particle modeling methods, are highlighted and further topics like the appropriate choice of impactor geometry or material model are discussed.

1. INTRODUCTION

In January 2009 the bird strike problem became apparent to the general public after the spectacular landing of the Airbus A320 of US Airways flight 1549 on the Hudson River in New York after an engine ingestion of at least two Canada geese [1]. However, the threat of bird strikes is not new and present since the early days of flight one hundred years ago, when mankind began sharing the sky with birds. First records of bird contact were documented by the Wright brothers in the year 1905 [2]. Although exterior aircraft structures are exposed to various threats of foreign object damage (FOD) like hail, runway debris or tire rubber impact, about 90% of all incidences today are reported to be caused by bird strike [3].

Numerous statistics are published each year, illustrating recent numbers on monetary, material and human losses due to bird strike in commercial and military aviation. From 1990 to 2008 almost 90.000 bird strikes on commercial aircraft have been reported to the Federal Aviation Administration (FAA) solely in the USA [4], leading to immense monetary losses due to repair, delay and cancellations. Annual cost values from 614 million to 1.28 billion US$ are reported in [4, 5, 6]. It is estimated that a bird strike event occurs once every 2000 flights [7]. In the period from 1912 to 2008, at least 103 aircraft and 262 lives have been lost in civil aviation due to bird strikes [8, 9]. Further statistics are available on species and altitudes involved in bird strike events, indicating that 72% of all collisions occur near the ground below 500 ft. and 92% under 3000 ft., making the take-off and landing phases especially critical [4]. Therefore, also military flight missions at low altitudes are frequently exposed to bird strike incidences.

A collision with a bird during flight can lead to serious damage to the aircraft. All forward facing components are concerned, i.e. the engine fan blades and inlet, the windshield, window frame, radome and forward fuselage skin as well as the leading edges of the wings and empennage (Fig. 1). For helicopters the windshield, forward fuselage structure and rotor blades are especially vulnerable against bird strike.

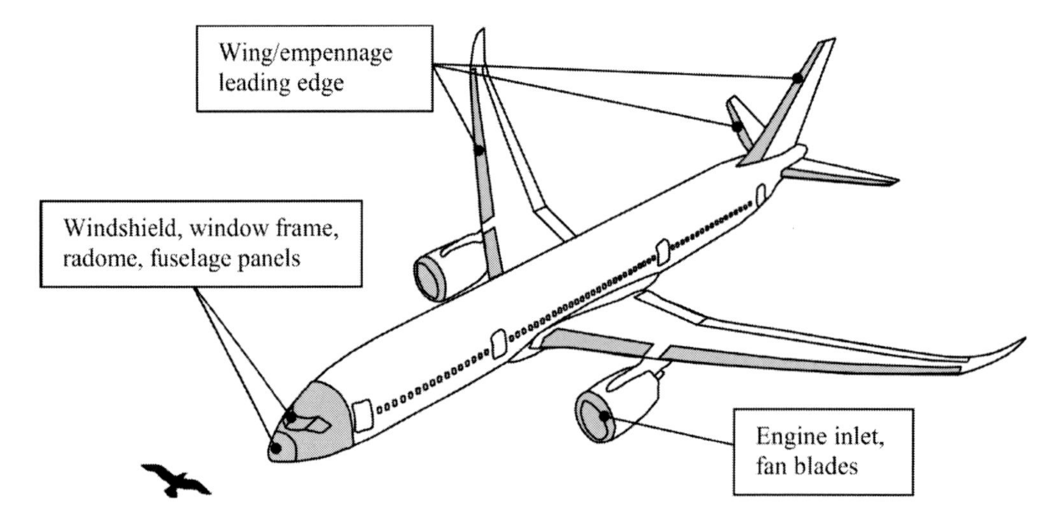

Figure 1. Illustration of aircraft components exposed to the risk of bird strike.

Consequently, the aviation authorities require that all forward facing components need to prove a certain level of bird strike resistance in certification tests before they are allowed for operational use. These requirements are compiled in the Federal Aviation Regulations (FAR) and in the European Joint Aviation Regulations (JAR):

- Windshields or military aircraft canopies, which were first made from brittle glass and later from more ductile PC, PMMA or stretched acrylic plastics in a monolithic or laminated configuration [10, 11, 12], are a key component to ensure safety during flight. Besides air tightness and good visibility they are required to withstand the impact of a 4 lb (1.8 kg) bird at cruise speed without penetration (FAR/JAR 25.775).

- For wing leading edges the certification criteria require that even in case of penetration of the leading edge skin no critical damage may be introduced to the front spar elements or the wing tank, so that a continued safe flight and landing after impact are assured. This has to be proven for 4 lb (1.8 kg) birds impacting the wing and 8 lb (3.6 kg) birds impacting the empennage leading edge at operational speed (FAR/JAR 25.571, 25.631).
- All gas turbine engines and particularly turbofan engines must be durable in order to withstand ingested birds. For small and medium birds the certification regulation requires the demonstration of structural integrity and a continued thrust level of 75% power. For large 4 lb (1.8 kg) birds it has to be proven that the engine can be shut down in a controlled manner without fire. This must be shown in a certification test on an entire engine, although manufacturers usually perform bird strike tests on single or multiple fan blades for design studies. The engine inlet also has to withstand a 4 lb (1.8 kg) bird impact without damage of critical components (FAR/JAR 25.571, 33.76).

An overview on international certification regulations for military aircraft that are very similar can be found in [13].

In past years it was common practice for bird-proof design of aircraft components to build and test, then redesign and test again [14]. One example of this procedure is documented for the development of the bird-proof Dash 8 wing leading edge [15]. Without doubt, this is not only a very time-consuming, but also cost-consuming practice. Therefore, numerical methods were developed and applied since the late 1970s for the purpose of rapid and improved design optimization, ensuring that the very first full-scale bird strike certification test is successful [16, 17]. The definition of a suitable bird model is often the main problem in the numerical simulation of bird strike. Starting with relatively simple nonlinear calculations and a pressure load applied to the target structure in the 1970s, complex fluid-structure interactions are treated today with explicit simulation codes and high performance computing (HPC). Besides load cases like water ditching or crashworthiness evaluations in drop test simulations, bird strike simulations are the major load case for the aircraft industry treated with explicit finite element (FE) analysis codes today.

Therefore, many studies and investigations on bird strike modeling have been conducted in the past and a lot of papers on different modeling methods have been published in the technical literature. Most interestingly, this evolution from simple to complex and accurate methods did not lead to the establishment of one generally accepted bird impactor modeling approach. Instead, there are still at least three techniques today, which are widely used, each having its own advantages and disadvantages. Therefore, it can be difficult for the analyst to have an overview and to decide for an appropriate modeling method.

The intention of the current chapter is to give an overview on computational methods for bird strike simulation by an extensive and so far unique literature review on the topic of soft body bird impactor modeling. Although a lot of in-house bird strike research work is conducted in aerospace companies documented in internal reports, the focus of this literature review is on generally accessible papers and publications. More than 170 papers have been evaluated in a meticulous study, giving a considerable impression of the progress in bird strike modeling over the last 30 years, covering the time period from 1979 to 2010. Though, a guarantee on completeness is by no means given. After some introductory words on the hydrodynamic basics for bird impactor modeling, typical substitute bird materials used and available test data for model validations, the focus will be on the successive development of different numerical methods for the bird modeling. The three most established methods (Lagrangian, Eulerian and meshless particle modeling) will be introduced in more detail, highlighting their advantages, disadvantages and potential applications. Additional topics like the appropriate bird impactor geometry, material model and contact algorithm are addressed.

2. Theoretical Background

According to [18], the projectile response during an impact can be divided into five categories as a function of the impact velocity: elastic, plastic, hydrodynamic, sonic or explosive. During an elastic impact, the internal stresses in the projectile are below the material strength so that it will rebound. With increasing impact velocity, plastic response of the impactor begins but the material strength is still sufficient to prevent a fluid-like behavior. A further increase of impact velocity causes internal stresses to exceed the projectile's strength and fluid-like flow occurs. At this impact velocity, the material density and not the material strength determines the response of the impactor. This flow behavior of real birds can typically be observed in high-speed films of impact tests. Therefore, the bird impactor is treated as a so-called 'soft body' at the velocities of interest, since the stresses that develop within the bird are significantly higher than its own strength. The load is spread over a relatively large area during the impact.

Pioneering work on the characterization of bird impact loads on flat targets has been performed by Wilbeck and Barber [19, 20, 21, 22]. Besides extensive experimental studies to assess the flow behavior and pressure loads, the basics of the hydrodynamic theory of a soft body impacting a flat target were derived, which are shortly summarized here.

The impact behavior consists of four main phases: a) initial shock at contact, b) impact shock decay, c) steady flow and d) pressure decay (Fig. 2) [18, 23]. If the impactor with an initial velocity u_0 hits a surface, the material at the contact point is instantaneously brought to rest and a shock wave with the velocity u_s is generated with the wave front being parallel to the surface and the running direction being perpendicular to the surface, propagating up the impactor body. A significant pressure gradient develops at the outer surface because of the shock-load on the one side and the free surface on the other side. This leads to an outward acceleration of the material particles and a release wave is formed, which propagates inwards towards the centre of impact, interacting with the shock wave. This release wave causes a significant decrease in the pressure at the impact point. Therefore, the high initial pressure peak at the centre of impact lasts less than a microsecond (Fig. 3) [24, 25]. After several reflections of the release waves, the material flows steadily, leading to a constant pressure and velocity in the impactor. The existence of steady flow depends on the length/diameter ratio of the impactor. Since for a very short impactor the impact event will be over before steady flow can establish, a length/diameter ratio of 2 is recommended. It should be noted that the velocities of the shock and release waves are much greater than the initial velocity of the impactor. During the phase of steady flow, the shock wave is constantly weakened by the release waves, until it disappears. A typical pressure curve for such a soft body impact on a rigid target plate is depicted in Fig. 3.

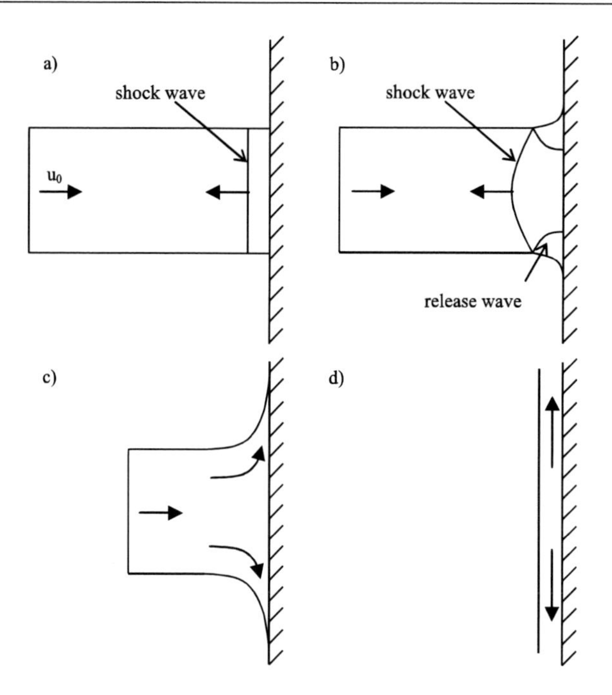

Figure 2. Illustration of shock and release waves in soft body impacting a rigid wall.

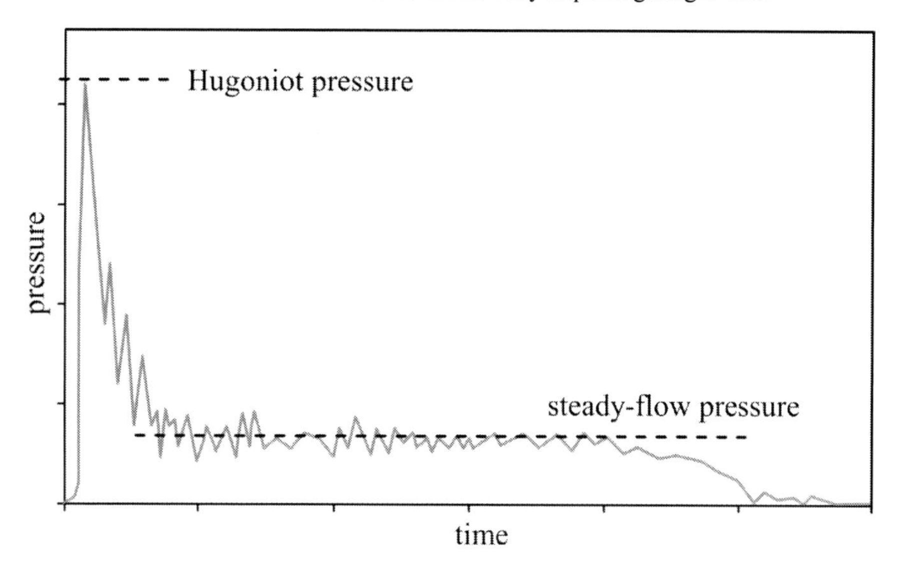

Figure 3. Typical pressure curve for normal soft body impact on a rigid wall.

The initial pressure peak in the contact point in a perpendicular impact is typically referred to as the Hugoniot pressure P_H:

$$P_H = \rho_0 u_0 u_s \,, \tag{1}$$

with ρ_0 as the initial density of the impactor.

The steady-flow pressure P_S can be estimated by the Bernoulli relationship:

$$P_S = \frac{1}{2}\rho_0 u_0^2 \, . \tag{2}$$

The total duration t_D of the impact can be approximated by the time needed for the impactor to flow through its own length L:

$$t_D = \frac{L}{u_0} \, . \tag{3}$$

3. Substitute Bird Materials

For bird strike certification tests on aircraft components real birds have to be used, typically dead chickens. However, the use of real birds is not ideal due to the large scatter between individual tests. As only the mass is defined by the certification regulation, the species may vary, having different body densities. This can have a significant effect on the impact load. The irregular shape of the bird may also pose difficulties in striking the desired target point on the structure [26]. For these and further reasons like hygiene, many companies use artificial birds or substitute birds for pre-certification impact testing, leading to advantages in convenience, cost and reproducibility [27]. The substitute bird shall not copy the real bird itself with its flesh and bones, but it should reproduce the same pressure loading during impact as a real bird. The substitute bird geometry and material have to be selected appropriately for this purpose.

Typical substitute artificial birds have a simplified regular geometry like a cylinder, a cylinder with hemispherical ends, an ellipsoid or a sphere, representing the torso of the bird. The influence of the impactor geometry is discussed later.

Several impact tests were conducted with wax, foam, emulsion, beef, rubber, neoprene and gelatin as substitute materials for birds [14, 21, 28]. It could be concluded from these studies that soft material substitutes that have the specific gravity of water produced loading profiles similar to those of real birds. Gelatin, or porous gelatin, with the specific gravity of water, reproduced the behavior of a real bird with a high degree of accuracy. The corresponding pressure curves during an impact on a rigid plate correlate to the real bird's pressure curve. For this reason, gelatin is today the most frequently used substitute bird material in both experimental tests and numerical simulations [14].

4. Test Data for Bird Model Validation

Once an appropriate numerical bird impactor model is developed, it has to be validated against experimental data before using it for impact simulations on complex aircraft components.

For this purpose, it is common practice to simulate the impact on a rigid, flat plate and to qualitatively assess the flow behavior for comparisons with high speed films and to quantitatively evaluate the pressure curve for comparisons against data from the pressure transducer. The physical tests are typically performed with a high-powered gas cannon, which is used for bird strike tests since the mid 1950s. The component is mounted on a test rig and the bird is fired at a realistic operational velocity on the target structure. Inside the cannon, the real bird or substitute bird is placed in a cylindrical, open-ended carrier, a so-called sabot, which is accelerated by gas pressure to the required velocity and separates from the projectile before impact [14].

Because the availability of test data is very limited, the majority of all studies in the literature use the experimental study of Barber and Wilbeck [19, 21, 22] from the late 1970s for validations. They conducted impact tests on a rigid plate with pressure transducers mounted on the surface, using different impactors like real birds, beef, neoprene, rubber, gelatin and porous gelatin. The impactors of different weights, except the real birds, had a cylindrical shape and velocities of 100–300 m/s. The pressure-time-curves, measured by the centre pressure transducer with a sampling frequency of 300 kHz, were taken as reference curves for numerous model validations. However, uncertainties of these 30 years old test data arise from the fact that the piezoelectric quartz pressure transducers used in this study were not designed for transient impact loads as they had no adequate acceleration compensation with noise occurring in the curves. Furthermore, their resonance frequency was about 300 kHz, leading to resonance-induced overshots in some of the curves. The accuracy of the measurements was also limited by the finite frequency response of the transducers, which prevented measurements of rise times of less than 5 μs. One additional point is the fact that the distance between the gas gun and the pressure plate is not known and no information is given, if the geometrical consistency of the fluid-like impactor was maintained at the time of impact or if the impactor was already distorted during the time of flying, as it was often observed in later studies [17]. Lavoie et al. [29, 30] remind that the experimental pressure data by Wilbeck should only be seen as a reference for the general behavior rather than as a tool of evaluation.

More recent data were generated by the GARTEUR bird strike group (Group for Aeronautical Research and Technology in Europe), which were used for validations in [3, 31, 32, 33], but those results are not available to the public.

In order to provide new test data, Lavoie et al. [34] performed new tests lately. A 1 kg cylindrical impactor with hemispherical ends was used for impact tests at 0° and 30°. However, the authors put those experimental data into question, since the pressure curves are about 3–4 times higher than theoretically expected. Furthermore, the transferred energy is higher than the initial energy. Hence, values of the tests were corrected by normalizing them. No final explanation of this error could be given, making the use of these data for model validations questionable.

Some experimental data of bird strike tests on wing leading edges or windshields can be found in the literature, but they typically only apply for the specific configuration of the respective study and cannot be applied for general bird impactor model validation purposes.

Besides the pure validation of bird impactor models in impact simulations on rigid plates, it is proposed in [35] that additional reference tests on a test rig with a bird splitter should be used for the validation of the numerical model, if the bird splits during the simulation, e.g. during the impact on a sharp leading edge of military aircraft. Moffat and Cleghorn [36] and Kamoulakos [37] advise to include impact simulations on deformable metallic plates with plastic deformations to the validation, as documented in [38], to calibrate the impactor model for realistic impact damages. Soft body impact tests on fiber-reinforced composite plates are documented in [39, 40, 41, 42, 43].

5. Bird Modeling Methods

Impact simulations of a soft body that is highly deformed during the analysis are a major challenge for FE codes. The following section summarizes the development of different numerical approaches to cope with this challenge and characterizes the modeling methods with their advantages and disadvantages.

5.1. First Bird Strike Simulations: Bird as Pressure Function

While the first theoretical bird strike investigations were based on pure analytical calculations [44], the finite element method was adopted as a bird strike analysis tool in the late 1970s with pioneering work conducted by the U.S. Air Force research laboratories [10]. However, the requirements of complex bird strike analyses for existing FE codes at the time of 1975 were too demanding, because of the large deflections, high nonlinearity of the plastic target materials typically used, three-dimensionality and the high loading rate [10]. Linear FE codes like the program IMPACT, developed in 1977, were not able to deliver acceptable numerical predictions, with deviations of 50% to experimentally measured strains. In 1978 the nonlinear FE code MAGNA was developed and applied for bird strike simulations on windshields [10], accounting for very severe geometrical and material nonlinearities. For the bird impact load a uniform pressure vs. time curve was applied in normal direction, which represents the nearly constant steady-flow pressure observed in bird impact tests on rigid plates, neglecting the initial pressure peak and tangential loads. This pressure was applied as a uniform force over a specified 'bird impact footprint' area [45]. The same approach was used in [46] for bird impact analyses on fan blades with the program MARC. In [47, 49, 48], [50, 51], [52] and [53] a bilinear or stepwise instead of a uniform pressure-time-curve was used to cover the initial pressure peak for bird strike simulations with the codes NONSAP, DYNA3D, BASIS and BINA, respectively. Reference [54] used a trilinear pressure function to approximate the pressure peak and steady-flow pressure and scaled this pressure curve as a function of the radius to the central point to cover the spreading. References [28] and [55] take the spreading of the bird into account by increasing the oval 'footprint' area with time.

In contrast to pure bird strike loads on rigid plates, it was early understood that interactive effects must be included in the solution of deformable targets, i.e. the applied loads are a function of the structural response (coupled fluid-structure interaction) [12, 45, 46, 49, 56]. Typically, subroutines were used to update the bird impact pressure continually during the solution. McCarty used this approach within the code MAGNA for bird strike analyses on windshields of numerous aircraft like the F-16 [45], TF-15 [12], T-38 [57], T-46A [58] and the Space Shuttle orbiter [59, 60], Leski et al. [54] for the Russian fighter Su-22, Wen [61] for the Chinese Y-12 aircraft and Boroughs [62] for the Learjet 45.

5.2. Lagrangian Bird Impactor Model

First attempts to actually model the soft body impactor with Lagrangian elements instead of applying pressure loads were documented in 1984 by Brockman [63, 64, 65] and West

[66], using eight-node solid elements, explicit time integration and rezoning or mesh regeneration to avoid excessive mesh distortion. A contact algorithm covers the fluid-structure interaction in this case. For the transient bird strike problem the conditionally stable explicit central-difference time integration is clearly seen to be more appropriate than the implicit integration procedure. Stability requires the use of time steps shorter than the wave propagation time in the smallest element of the model [67].

5.2.1 Characteristics of Lagrangian Modeling

The Lagrangian modeling method is the standard approach for most structural finite element analyses. The nodes of the Lagrangian mesh are associated to the material and therefore each node of the mesh follows the material under motion and deformation (Fig. 4a). The boundary of the body is always clearly defined, since the boundary nodes remain on the material boundary, leading to a simplified definition of boundary conditions. This approach is typically used for solid materials.

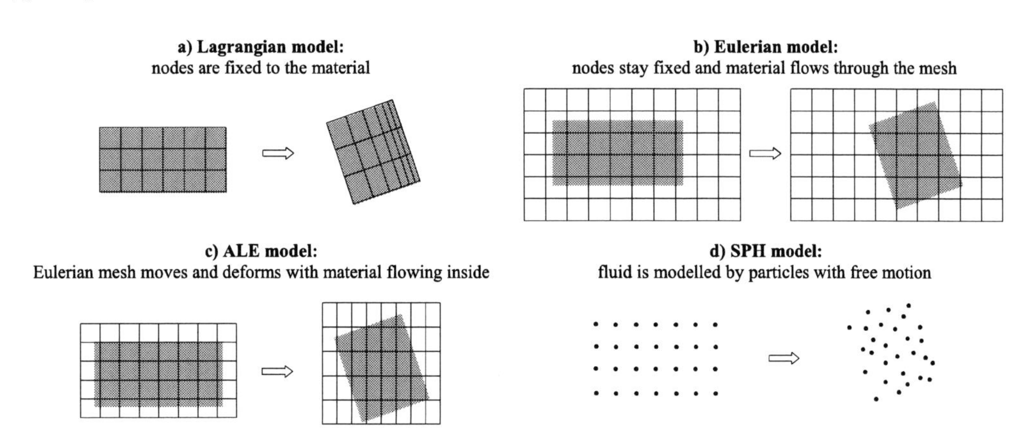

Figure 4. Different finite element modeling methods for soft body projectiles.

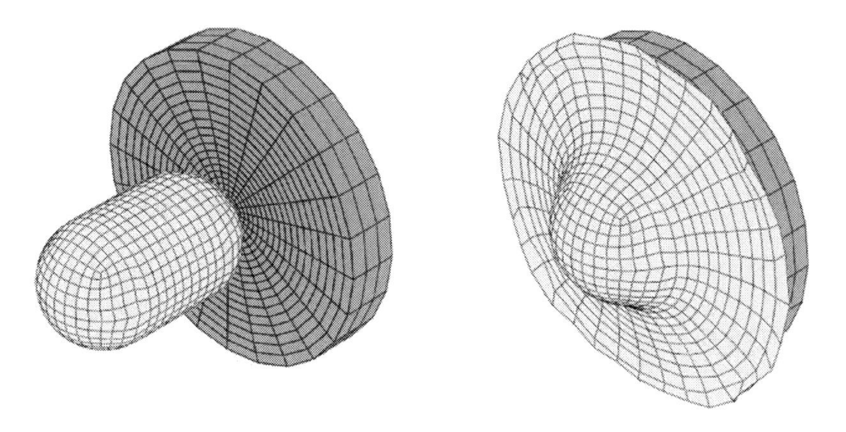

Figure 5. Bird strike simulation on rigid plate with Lagrangian impactor model.

The major problem of Lagrangian bird impactor models are the severe mesh deformations. Large distortions of the elements may lead to inaccurate results, severe hourglassing and even error termination due to negative volume elements [5, 68, 69, 70, 71,

72]. In [73] it is highlighted that large distortion of elements may introduce an artificial stiffening effect into the soft body finite element model, negatively affecting the impact pressure curve. Furthermore, the time step in explicit calculations drastically decreases, when the solid elements are compressed to such an extent, since the time step depends on the shortest element length in the model. The element deformation in a Lagrangian bird model is shown in Fig. 5.

Several choices exist for dealing with the mesh distortion problem, including adaptive remeshing or element erosion. Adaptive remeshing involves remeshing the region of severe mesh tangling. According to [14], this mapping procedure often increases the numerical errors associated with the approximation and may be computationally expensive. In [74] elements are deleted from the calculation, when they reach a defined degree of flattening, which is referred to as element erosion. Zhu et al. [75] used a shear failure criterion for element deletion. The deletion of highly distorted elements prevents numerical problems, and was often adopted in other studies, with the argument that most of the momentum of those elements is probably already transferred to the structure [70, 76]. In contrast, Lavoie et al. [29] report that especially for finely meshed Lagrangian impactors with element erosion a large part of the mass is eroded during the simulation, which makes it very difficult to obtain accurate results. However, the mass can be retained after element deletion by lumping it at the free nodes still interacting with the structure [77, 78, 79]. With this approach of element erosion even the splitting of a bird can be modeled with Lagrangian elements [35, 77, 80, 81]. However, Castelletti and Anghileri [69] remind that the numerical-experimental correlation gets worse, when a failure criterion is introduced, compared to a good correlation without failure criterion. Another typical problem arising from element erosion is the artificial oscillations in the contact forces due to the discretised nature of the simulated contact algorithms, especially for coarse meshes. Once the frontal elements are deleted, the contact force will decrease dramatically until the impactor comes into contact again with the target, and this introduces artificial noise into the contact forces [31]. This can only be reduced by using very fine meshes [3]. A severe influence of the mesh density in the Lagrangian impactor was also identified in [29, 79].

Georgiadis et al. [5] conclude that the Lagrangian approach remains an impractical way to model bird strike.

5.2.2 Application of Lagrangian Model for Bird Strike Simulations

Lawson and Tuley [82], Schuette [83] and Niering [84] adopted this approach with Lagrangian elements for the impactor in the late 1980s and early 1990s for explicit bird strike simulations on engine fan blades with DYNA3D. The results of Niering [84] indicated that internal friction in the bird material should be accounted for, otherwise the pressure maxima occur at the rim of the impact area and not in the impact centre.

The Lagrangian method became the standard approach for bird impactor modeling for the next decade and beyond. It was used by several researchers for impact simulations on windshields [11, 75, 78, 85, 86, 87, 88, 89, 90], engine fan blades [3, 31, 32, 33, 76, 91, 92, 93, 94, 95, 96], helicopter blades [80], aircraft radomes [13, 24, 97], the forward bulkhead behind the radome [98], fuselage panels [23, 99], composite plates [40, 42] and wing/tail plane leading edges [26, 67, 77, 78, 97, 100, 101, 102, 103, 104, 105, 106, 107, 108, 109, 110, 111, 112].

5.3. Eulerian Bird Impactor Model

The major limitation of the Lagrangian model with respect to the flowing behavior is the excessive mesh distortion and reduced time step. A promising alternative is the Eulerian modeling technique, which is mostly applied to the simulation of fluid behavior.

5.3.1 Characteristics of Eulerian Modeling

In the Eulerian formulation, the mesh remains fixed in space and the material under study flows through the mesh (Fig. 4b) [70]. Since the mesh does not move, mesh deformations do not occur and the explicit time step is not influenced. Stability problems due to excessive element deformation do not occur. This approach is typically used for fluid materials and flow processes. Each element has a certain volume fraction of different materials, those can be for example a fluid material and void, or even more materials. This means that each element may be partially filled with the fluid material. The disadvantage of this approach is that the body's boundary is not exactly defined, and depends on the mesh size. Visualization tools in the post-processing software can be used to estimate the outer boundaries.

Since in a bird strike simulation typically only the impactor is modeled as a fluid-like body with Eulerian elements and the target as a solid structure with Lagrangian elements, a coupled Eulerian-Lagrangian approach is used for this fluid-structure interaction problem. Since the mesh in the classical Eulerian technique is fixed in space, the computational domain should cover not only the region where the material currently exists, but also additional void space to represent the region where material may exist at a later time of interest (Fig. 6). Therefore, the computational domain for structural analysis with the Eulerian technique is much larger than with the Lagrangian approach [14]. In general, the computational cost of the Eulerian model is relatively high, due to the high number of elements and the cost-intensive calculation of element volume fractions and interactions [114]. Typically, the element size of the Eulerian mesh has to be defined very small in order to achieve accurate results [73]. The Eulerian formulation is also not free from numerical problems. There are well-known dissipation and dispersion problems associated with the flux of mass between elements (numerical leakage) [87, 91, 113, 114, 115, 116]. McCallum and Constantinou [117] conclude that the total energy decreases by approximately 6% from the starting value, which is associated with a loss of energy in the contact interface during the bird strike simulation.

In contrast to the classical Eulerian modeling with a fixed Eulerian mesh, the Arbitrary Lagrangian-Eulerian (ALE) formulation was adopted to make the simulation more efficient. The ALE method is basically similar to the classical Eulerian method, but the surrounding Eulerian box can move and stretch if needed and is not fixed in space (Fig. 4c) [73]. Depending on the current position of the bird material in space, the geometry and position of the Eulerian background mesh is currently updated. Since the background mesh can move in the same direction as the projectile, the number of elements for the computational domain can significantly be reduced, leading to computational time savings, especially when motion of the material covers a wide region in space [69, 73].

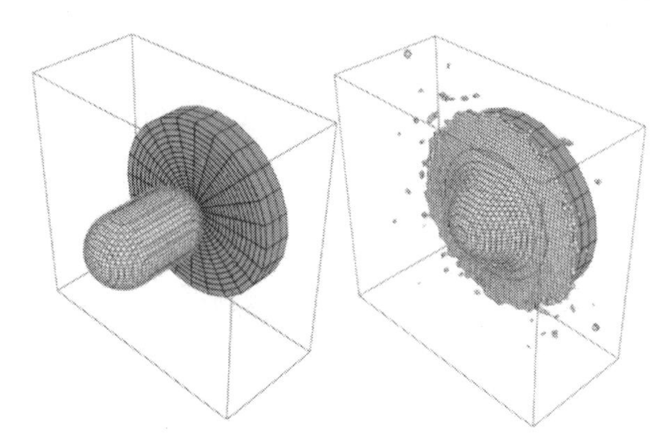

Figure 6. Bird strike simulation on rigid plate with Eulerian impactor model.

However, it has to be kept in mind that through the spreading of the bird material the lateral expansion of the Eulerian box is significant (Fig. 7). As the number of Eulerian elements is constant, they are also elongated significantly, and as stated before, the accuracy of the results is strongly mesh dependent and requires fine meshes. Therefore, the accuracy of the ALE model for severe deformations of the Eulerian box may be questionable [69, 116]. This mainly concerns the later part of the steady-flow regime, though.

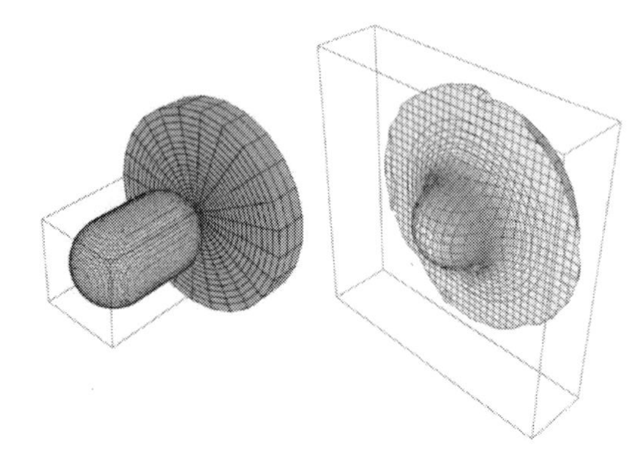

Figure 7. Bird strike simulation on rigid plate with ALE impactor model.

5.3.2 Application of Eulerian Model for Bird Strike Simulations

The Eulerian approach was first used for bird strike simulations in the late 1990s in [23, 118] in the DYTRAN code. The authors compared the results against PAM-CRASH results from a pure Lagrangian simulation for the load cases of impacts on a fuselage panel and an engine nacelle with quite similar results, but without error termination, which typically occurs in Lagrangian simulations due to severe element distortion. Also in [36] and [119] the Eulerian approach was used in DYTRAN for bird strike simulations (on rigid and deformable plates and on a radome, respectively).

A classical Eulerian modeling approach in LS-DYNA with a fixed Eulerian mesh was compared in [113] to the new ALE approach with a moving Eulerian mesh. By translating, rotating and deforming the multi-material mesh in a controlled way, the mass flux between elements could be minimized, which occurred to a higher extent in the fixed Eulerian mesh.

The number of elements could be kept smaller than in a classical Eulerian model, increasing the efficiency.

In [80] the Lagrangian bird model was compared to a Eulerian model in DYTRAN, with the Eulerian results correlating better to test data. The Lagrangian model in general produced much higher predicted impact loads, about two times higher than in the test data.

The study in [99] compared Lagrangian and Eulerian bird impactors in RADIOSS. Though both approaches correlated well with experimental pressures, the calculation time with the Lagrangian model was much higher due to the decreasing time step. In the comparison study of [76], a tensile failure criterion for the Lagrangian impactor was introduced, leading to significantly lower CPU costs than the Eulerian model and was therefore preferred, since the results were similar.

In [120] a Lagrangian model is compared to a Eulerian model in LS-DYNA impacting a tail plane leading edge. The results differ to a certain extent, as the Eulerian loading is less localized and more spread, which is attributed to a too stiff representation of the Lagrangian bird due to high element distortions.

Further examples for Eulerian bird strike simulations are found in [115, 121, 122, 123] for fan blade impact and in [124] for impact on fuselage panels, in [125] on a cockpit bulkhead plate, in [71, 126, 127] on a composite plate, in [128, 129] on a windshield, in [127, 130, 131] on a leading edge and in [127] on a rotor spinner.

An impression of a Eulerian bird strike simulation on a composite wing leading edge is given in Fig. 8 and Fig. 9, taken from a recent study of the author of this chapter. This simulation was performed with ABAQUS and shows the potential of the Eulerian impactor model when it comes to splitting and secondary impact after penetration. It was not possible to obtain results of similar quality with a Lagrangian bird model.

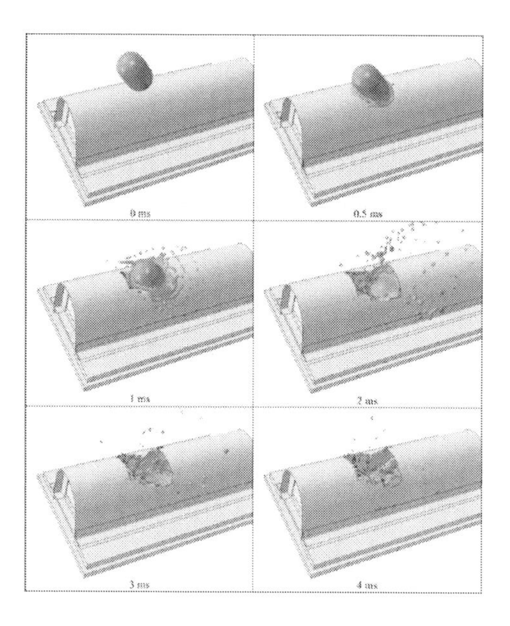

Figure 8. Bird strike simulation on composite wing leading edge with Eulerian impactor model (top view).

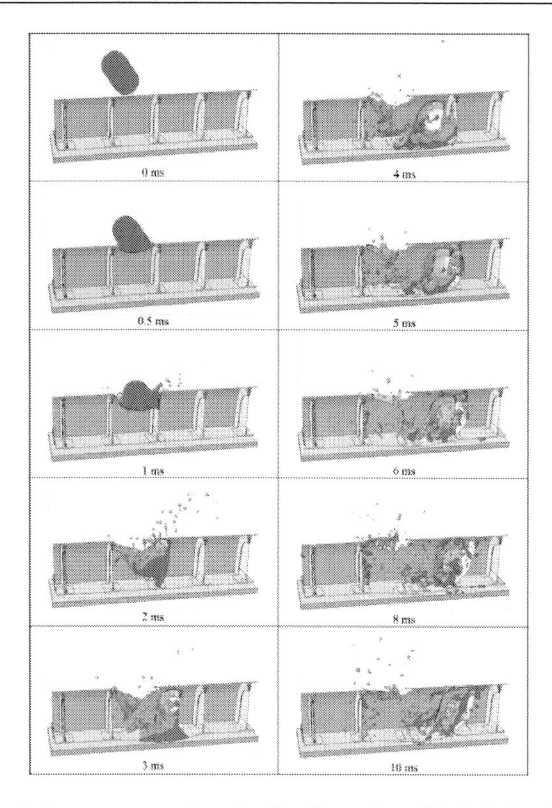

Figure 9. Bird strike simulation on composite wing leading edge with Eulerian impactor model (cross-sectional view).

5.4. Sph Bird Impactor Model

As a contrast to the Lagrangian and Eulerian soft body impactor modeling approaches, where a regular finite element mesh is used, different meshless particle methods were developed over the years, aiming at the independence of mesh distortion problems and computational efficiency.

Anghileri et al. [132, 133] investigated a simplification of the bird modeling in the early 2000s by describing the impactor using only nodes with added lumped mass and initial velocity. This approach is inspired by the discrete element method (DEM). Contact definitions apply the load onto the impacted structure. The advantage, compared to a Lagrangian impactor model, is the reduction of the CPU time by the factor 10 with no effect on the time step size and the ability to easily cover severe deformations and splitting of the impactor. The disadvantage of this method is the lack of internal interaction of the nodal masses leading to the lack of dissipation mechanisms and hence to an unrealistic bird behavior [69]. Reference [77] states that this lumped mass model exerts high frequency force peaks, having a strong negative influence on local structural response.

Further applications of particle methods are documented in [16] and [18]. Spherical fluid elements with a centrally located node were developed in the code WHAM for the impactor modeling in a fan blade impact study. The spherical elements are assembled in a closest-packed formation without structural connection, producing the soft body impactor shape.

Contact algorithms avoid penetration of the elements and generate reaction forces. Another particle method was proposed in the early 2000s in [17] based on the DEM. The gelatin impactor is represented by spherical rigid particles in conjunction with a viscoelastic-plastic penalty form of the Hertzian contact law. The parameters of this interaction law therefore control the global fluid-like behavior of the projectile. Promising results were obtained, although the spread of the bird was slightly underpredicted.

Another particle method was initially developed and applied for astrophysical problems in the 1970s: the SPH (Smoothed Particle Hydrodynamics) method [134, 135, 136, 137]. It caught attraction for the splashing simulation of fluids and gained increased attention for bird strike modeling in recent years.

5.4.1 Characteristics of SPH Modeling

The SPH method is a meshless Lagrangian technique, based on interpolation theory and smoothing kernel functions [138]. The fluid is represented as a set of discrete interacting particles (Fig. 4d), which are independent from each other, being able to cover large deformations without the problem of mesh distortion. Each particle has a mass, velocity and material law assigned to it, which is not localized but smoothed in space by a smoothing kernel function, typically based on a B-spline approximation [138, 139, 140], defining the range of influence of the particle. Field variables of an individual particle are computed through interpolation of the neighboring particles, whereas a particle is considered a neighbor when it is located within the smoothing length of another particle [14]. As it is based on a Lagrangian technique, it can easily be linked to conventional Lagrangian finite element models, avoiding possible material interface problems associated with Eulerian codes [139].

Compared to the conventional solid Lagrangian mesh, where the explicit time step decreases significantly with element deformation, the time step is constant for the SPH model. However, a sufficient particle density is required in order to achieve accurate results, which may necessitate high memory resources and typically is a compromise between accuracy and required CPU time [68, 69]. The influence of the SPH mesh density was analyzed in detail in [29, 70, 141], with the conclusion of a high influence on the soft body pressure curve in impact simulations on a flat plate (Fig. 10). Rössler [142] could also show that the time step has an influence on the bird's flow behavior and therefore on the pressure curve.

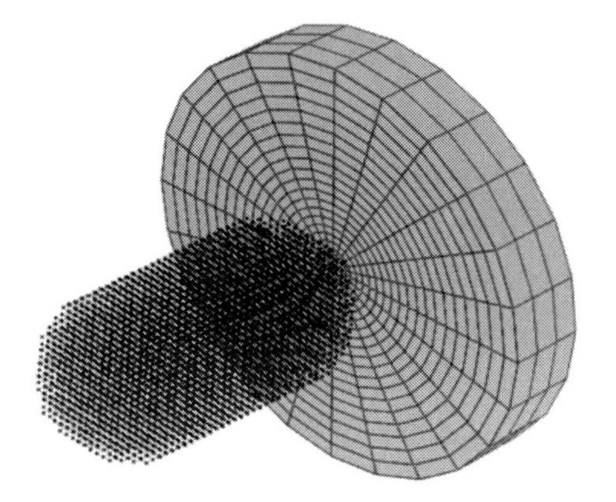

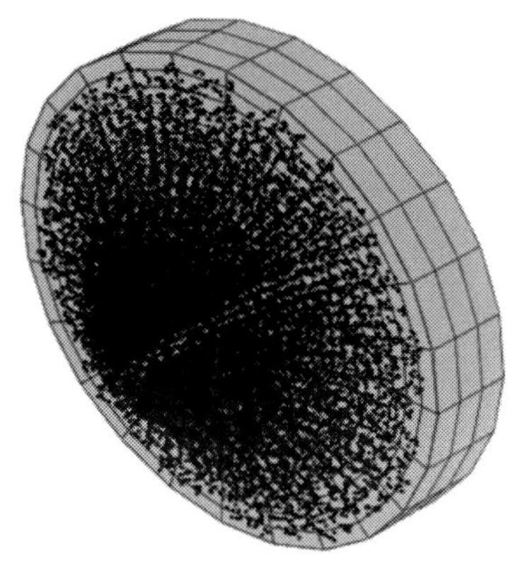

Figure 10. Bird strike simulation on rigid plate with SPH impactor model.

In general, compared to the Eulerian model, the SPH method requires fewer elements, avoids the material interface problems associated with it and normally has a shorter solution time [29]. The numerical robustness, compared to the conventional Lagrangian mesh with its mesh distortion problems, is very high.

A disadvantage of the SPH method, as discussed in [69], is the lack of sharp boundaries, which makes the application of boundary conditions difficult and may also affect the fluid-structure interaction. The so-called tension instability is identified as another weakness of the SPH method, which is referred to as the numerical collapse of the continuum under tension. This, however, should have a negligible influence in case of bird strike simulations.

5.4.2 Application of SPH Model for Bird Strike Simulations

The first adoption of the SPH method for bird strike simulations is documented in the late 1990s and early 2000s [114, 143, 144], where this approach was implemented in the code PLEXUS and used for fan blade impact studies. The slicing of the bird could be represented successfully. In a Lagrangian model like in [91], the slices had to be predefined. In [145] not only the impactor but also the target plate was modeled with the SPH method, which reduced the global time step compared to a simulation with a Lagrangian plate. References [68, 108, 146, 147, 148, 149] and [41, 138, 139, 140, 142, 150, 151, 152, 153, 154, 155, 156, 157, 158, 159, 160, 161] used the SPH approach in LS-DYNA and PAM-CRASH for bird strike investigations on leading edges and highlight the increased stability, good potential of bird splitting and reduced cost of the simulations compared to Lagrangian bird impactors. SPH bird impactors for impact on fan blades are documented in [95, 144, 162, 163, 164], for impact on engine nacelles in [165], on a windscreen in [166], on composite plates in [142, 154], on wing flaps in [5], on an aircraft radome in [167] and on a helicopter rotor blade in [168].

5.5. Comparison of Different Bird Modeling Methods

A large number of publications exist, where the most established modeling approaches were compared for specific applications and compared to experimental data.

A benchmark comparison study was performed in [69, 116], comparing five modeling approaches in LS-DYNA: Lagrangian, classical Eulerian, ALE, SPH and the DEM-based nodal mass model. The Lagrangian impactor model showed reliable results in absence of large distortions, otherwise it led to high computational cost and early error terminations. The ALE approach led to high costs compared to the meshless methods and questionable results due to severe deformation of the moving Eulerian box. Also for the classical Eulerian model the numerical-experimental correlation was not satisfying. The lumped mass model provided poor results due to lack of internal interaction. The SPH method is the recommended approach in this study, due to high stability, low cost and good correlation with experimental observations.

In an extensive study in [70, 141, 169], Lagrangian, Eulerian and SPH impactors were compared in LS-DYNA in an impact study on a flat plate, achieving consistent results within a range of 10% of experimental reference data with slightly higher forces for the Eulerian model. The Lagrangian method with element deletion was found to be computationally more efficient than the SPH or Eulerian technique. But still all three methods were recommended by the authors. A similar study was performed by Chuan [170], adopting the Lagrangian, Eulerian and SPH approaches in LS-DYNA for impact simulations on flat plates. All models led to very similar qualitative and quantitative results, although all being slightly higher than the experimental reference pressure curves. The higher computational cost of the Eulerian approach was highlighted as well as the problems of bird splitting for the Lagrangian bird. Lavoie et al. [29, 30] performed a similar comparison study in LS-DYNA for the Lagrangian, ALE and SPH bird impacting a rigid plate. The Lagrangian bird was concluded to be worst due to the problem of element distortion, having a very negative influence on the pressure curve. Although element erosion was used and a significant amount of the impactor mass was deleted, the computational time was 30 times higher than with the ALE or SPH model. Both ALE and SPH results were found to be satisfying, with the SPH model being slightly faster and easier to implement with fewer parameters. Hachenberg [81] also compared the three methods Lagrangian, Eulerian and SPH in LS-DYNA for bird strike simulations on rigid plates and curved leading edges with respect to qualitative deformation and splitting, contact forces and CPU time. Due to mesh distortion problems, the Lagrangian models often led to the highest CPU time and too low contact forces, while the Eulerian model and SPH led to consistent results, though being strongly mesh dependent. Jenq et al. [73] used the Lagrangian, classical Eulerian and ALE approach in LS-DYNA for bird strike simulations on a rigid plate. The Lagrangian impactor was found to be worst due to severe mesh distortion, having an unrealistic pressure curve. The classical Eulerian and ALE approaches both led to good results with the ALE model being preferred due to slightly better results and lower CPU cost.

McCallum and Constantinou [117] compared the Eulerian and SPH approach in LS-DYNA for impact simulations on a flat plate leading to close agreement in the results. Similar qualitative results were obtained in the comparison study of [115] with Eulerian and SPH models in LS-DYNA for impact simulations on a turbine engine. However, a certain loss of accuracy in the Eulerian simulation due to dispersion and numerical leakage is highlighted. Also Tho and Smith [127] compared ALE and SPH in LS-DYNA for impact simulations on rigid plates. Although both correlate well with experimental test data, the ALE model was preferred. In following case studies on leading edges both approaches again performed in a very similar and satisfying way in terms of bird spreading and amount of damage. Similar conclusions were gained in [34], where ALE and SPH models impacting a rigid plate were compared in LS-DYNA. Both approaches were found to be useful. The deformations were slightly better with the SPH approach but the pressure curves were somewhat better with the ALE model.

Ryabov et al. [171] compared a Lagrangian, Eulerian and SPH approach for fan blade impact simulations. The Eulerian approach was not recommended because of high CPU cost and inaccurate results. The Lagrangian and SPH models both led to similar results with the Lagrangian approach being cheaper and therefore recommended. In a similar study by Shmotin et al. [95] with Lagrangian and SPH bird models in fan blade impact simulations with LS-DYNA the SPH model was found to correlate better to experimental results than the Lagrangian model.

Guida et al. [72] analyzed Lagrangian and Eulerian bird models in DYTRAN for the application of a bird strike on a wing leading edge and found the Lagrangian model to be 10 times faster and by far more accurate with respect to force peaks and extent of damage in correlation with experiments. In contrast, the impact simulations of Smojver and Ivancevic

[131] on an aircraft wing flap with ABAQUS showed that the Eulerian bird model is more stable than the Lagrangian bird and the results appear to be physically more realistic.

5.6. Summary on Bird Modeling Methods

From this literature study it can be concluded that the three most established bird modeling approaches (Lagrangian, Eulerian and SPH) have their advantages and disadvantages and it often depends on the specific application, software package and parameter setup, which approach is best suited for the individual problem. However, a global trend is visible, that the Eulerian and SPH approaches are preferred today compared to the Lagrangian modeling approach with its problem of element distortion. In most studies more than one approach for the bird modeling was used, in order to identify the best approach for the given problem. An overview of the advantages and disadvantages of these three modeling methods is given in Tab. 1.

Table 1. Overview of advantages and disadvantages of bird modeling methods.

	Lagrangian model	Eulerian model	SPH model
Advantage	+ simple model generation + low CPU time (before mesh distortion) + impactor boundary clearly defined	+ no mesh distortion, constant time step + numerically stable simulations + complex bird splitting can be simulated + good representation of splashing behavior + Eulerian mesh motion (ALE) reduces computational cost	+ no mesh distortion, constant time step + numerically stable simulations + complex bird splitting can be simulated + good representation of splashing behavior + lower computational cost than Eulerian model
Disadvantage	- severe mesh distortion leads to reduced time step, artificial stiffening, loss of accuracy and error termination - hourglass problems - element erosion may remove mass from the simulation - splashing behavior difficult to represent with Lagrangian elements - bird splitting difficult to cover	- model generation and results visualization more complex - no clear outer boundary - numerical leakage: total energy reduces with time - fine Eulerian mesh necessary in impact zone: expensive - relatively high computational time	- model generation more complex - no tensile behavior - no clear outer boundary - higher CPU time than Lagrangian model (before mesh distortion)

6. Impactor Geometry for Bird Strike Simulations

Although substitute bird impactors have been used for many years in bird strike tests and simulations by a variety of organizations there is no standardized artificial bird shape. The geometries are typically chosen to reflect the principal mass and shape of the torso of a real bird [117]. The four most established substitute bird geometries are the cylinder, the cylinder with hemispherical ends, the ellipsoid and the sphere (Fig. 11). The use of such a simple geometry is also beneficial for an easy manufacturing. While the cylindrical shape of the impactor was preferred in early studies, the cylinder with hemispherical ends is dominating today as the geometry for the substitute gelatin bird. This results from the fact that the first experimental studies by Barber and Wilbeck [19, 21] were performed with cylindrical projectiles and these tests were often taken as the basis for the numerical model development.

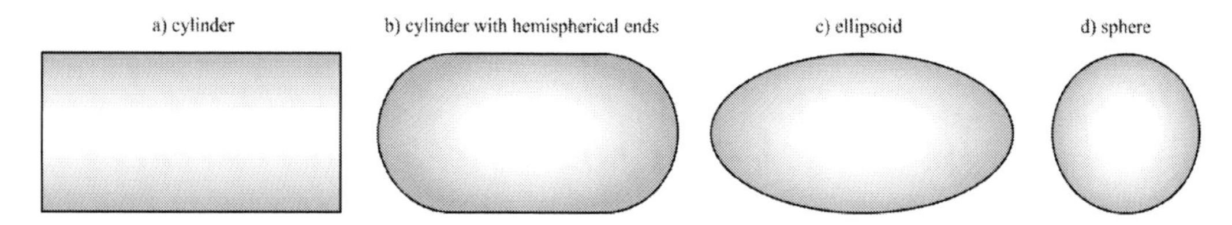

Figure 11. Different substitute bird impactor geometries.

Numerous studies investigated the influence of the projectile geometry. Nizampatnam [14] investigated the influence of the four typical projectile shapes in Fig. 11 on the shock and steady-flow pressure. Although all geometries showed rather similar results, the cylinder with hemispherical ends in both cases was closest to experimental data. In [79] both a cylindrical impactor and a cylinder with hemispherical ends were compared, with the latter geometry considered as closer to a real bird. These two geometries were also compared in [172, 173] in an impact study on a windshield. Again, a better correlation to actual bird strike test data were found for the cylinder with hemispherical ends, which is also stated in [68]. In older studies [23, 97] the ellipsoid was preferred to the right cylinder since sharp or regular shapes such as a cylinder produced unrealistic impact pressure profiles. In the study in [3, 32] the three geometries right cylinder, cylinder with hemispherical ends and ellipsoid were compared in a numerical study. A strong influence especially on the shock pressure was observed, which is highest for the straight-ended cylinder due to the largest instantaneous contact area. It was 43% higher than with the hemispherical-ended cylinder, which in turn is 30% higher than with the ellipsoidal projectile. The length-to-diameter aspect ratio of the bird (1.5:1, 2:1, 2.5:1) was found to have little influence on the results. Increasing the bird mass from 4 to 6 and 8 lb also slightly increased the pressure peak.

In contrast to these simplified bird geometries, the realistic shape of a Canadian goose was represented by a multi-material modeling approach based on biometric data in the study of McCallum and Constantinou [117]. The long neck, torso and wings were modeled with different densities using the SPH approach in LS-DYNA, although still sharing the same material law. In a direct comparison to the cylinder with hemispherical ends, also modeled as an SPH impactor, significant differences in the load curve were identified. Resulting from the initial neck impact, a noticeable plate displacement was obtained before the main torso impacted the target. Therefore, the authors concluded that the target panel is under a state of pre-stress when the torso impacts, which may have a significance on the final level of damage predicted for the structure [174]. The study was yet extended by Nizampatnam [14] for an even more realistic bird geometry modeling with SPH in LS-DYNA. The head, neck, torso, bones, lungs and wings with different densities and different equations of state were included in the model. The individual peaks in the pressure-time diagram of head, torso etc. could clearly be seen, also leading to a slightly different loading scenario than with the simplified impactor. However, no experimental data of a real bird with a forward neck and wings extracted as simulated are available for comparison. Still, such studies legitimate the question if the chickens that are typically used in bird strike certification tests are appropriate at all to represent the bird strike incident during flight with a real bird with forward facing head and extracted wings [27].

7. Bird Impactor Material Models

Just like the heterogeneity of projectile shapes, numerous different approaches for the bird impactor material modeling can be found in the various studies in the literature.

Generally spoken, real birds are mostly composed of water. Therefore, a water-like hydrodynamic response can be considered as a valid approximation for a constitutive model for bird strike analyses [79]. Furthermore, the anatomic structure of birds includes several internal cavities like pneumatic bones, lungs and peculiar air sacs, reducing the bird average density. In order to cover the effects of these cavities in the numerical model, a homogenized bird material with an average density between 900 and 950 kg/m^3 after elimination of the feathers can be estimated [79, 173]. This corresponds to the void content of 10-15% in the porous gelatin of an artificial bird, which is typically used for bird strike tests. Nizampatnam [14] raises the point that a higher porosity of 30-40% gave better overall agreement to experimental results than the typically used 10-15%.

Several authors tried to model the bird with a simple elasto-plastic material law with a defined failure strain [75, 81, 107], and some of them highlight the limitations of this simplified approach [88, 103]. It is observed that no fluid-like flow response can be achieved with such an elasto-plastic material law, only if the shear modulus G is set very low [74]. In [11] even a rubber-like hyperelastic Mooney-Rivlin material is applied to simulate the bird behavior during impact.

It is much more common to use an equation of state (EOS) for the constitutive modeling of the bird impactor, defining the pressure-volume relationship with parameters of water at room temperature. In most studies [3, 7, 30, 33, 42, 71, 73, 125, 126, 127, 130, 138, 145, 148, 150, 158, 170, 173, 175] the polynomial form of the EOS was used, where the pressure p is calculated by the following equation:

$$p = C_0 + C_1\mu + C_2\mu^2 + C_3\mu^3 , \tag{4}$$

where C_0, C_1, C_2 and C_3 are material constants and μ is a dimensionless parameter based on the ratio of current density ρ to initial density ρ_0:

$$\mu = \frac{\rho}{\rho_0} - 1 . \tag{5}$$

In other studies [139, 153, 166] the simpler Murnaghan EOS was used:

$$p = p_0 + B\left[\left(\frac{\rho}{\rho_0}\right)^{\gamma} - 1\right], \tag{6}$$

where p_0 is a reference pressure and B and γ are material constants. A further approach, adopted in [94, 95, 116, 162], is the Grüneisen EOS:

$$p = \frac{\rho_0 C^2 \mu \left[1 + \left(1 - \frac{\gamma_0}{2} \right) \mu - \frac{a}{2} \mu^2 \right]}{\left[1 - (s_1 - 1)\mu - s_2 \frac{\mu^2}{\mu + 1} - s_3 \frac{\mu^3}{\mu + 1} \right]^2} + (\gamma_0 + a\mu)E, \quad (7)$$

where E is the internal energy, γ_0 is the Grüneisen parameter and a, C, s_1, s_2 and s_3 are constants. However, in [14] it is recalled that the Grüneisen EOS is only valid for solid materials that remain in the solid state throughout the impact event and should therefore be used with care for bird strike simulations.

Which EOS is used often depends on the individual software code that is utilized, as most commercial codes are limited to one or few of the abovementioned formulations. All EOS have in common that certain material constants have to be defined that typically cannot be measured directly. A regular technique is to use simplified analytical approximations or parameter calibrations in reference simulations with comparisons to experimental data for the parameter identification. This was e.g. performed in [116] or [139] using an optimization software to determine the parameters in conjunction with bird strike tests on instrumented plates.

8. Contact Calculation in Bird Strike Simulations

The fluid-structure interaction in a soft body impact simulation like a bird strike is generally based on a contact algorithm, which prevents penetrations and calculates reaction forces. From computational point of view, the contact algorithm plays a significant role in the bird strike simulation, since it has to cope with large deformations and splitting of the projectile, sliding of the bird material over the target surface and the creation of multiple contact interfaces due to possible fracture and penetration of the structure [149]. An appropriate contact formulation is of great importance as it significantly influences the pressure curve during impact simulations on a flat pressure plate [70]. Most contact algorithms in explicit simulations are based on the penalty formulation, which means that a certain penetration of the slave node (impactor) into the master surface (target plate) occurs, before the respective reaction forces as a function of the contact stiffness and penetration depth are generated, pressing out the penetrating node. During the flow regime of the projectile, significant oscillations in the contact force can occur due to this penalty contact algorithm that are often dependent in their frequency and peaks on the penalty stiffness scale factor, which has to be selected with care for this contact pair with highly different stiffnesses. Ryabov et al. [171] investigated numerous different contact types in Lagrangian bird strike simulations in LS-DYNA and found them to have a strong influence on the results. However, in most references, no detailed information on the contact type that was used is given.

Shmotin et al. [95] investigated the influence of the friction coefficient in the contact calculation between bird and metallic target structure with values from 0 to 1. Best results compared to experimental results were obtained with zero friction.

A further important point is the contact area, where the pressure is calculated for comparisons with experimental pressure curves. In the reference test, the pressure transducer has a finite size, which should be similar to the nodal area in the model, where the nodal contact forces are evaluated and converted into a pressure. In almost no paper information can be found on this topic. Only in a few studies [41, 140, 142, 150, 154] the influence of selected areas of pressure evaluation is analyzed. It could be shown that this area has a great influence on the peak pressure, which is very high for a small area and consequently lower with increasing area, as the stress peak in the impact centre is smeared over a larger area.

Conclusion

During a bird strike on aircraft structures at the velocities of interest, the bird behaves as a soft body and flows in a fluid-like manner over the target structure, with the high deformations of the spreading material being a major challenge for finite element simulations. In an extensive literature survey the progress in bird strike modeling over the last 30 years was assessed. The most interesting conclusion is that no homogeneity or uniformity in terms of a generally accepted modeling approach exists, neither with respect to the discretisation method nor the material model, contact formulation, impactor geometry, size or mass. This makes such a literature overview especially valuable, as the global picture in terms of appropriate modeling techniques for bird strike simulations is sometimes difficult to paint due to the high number of different options.

First simple simulation models in the 1970s and 1980s mainly used nodal pressure loads to represent the bird impact load. They were displaced by Lagrangian impactor models in the following years. However, the main problem in using a Lagrangian impactor model is the mesh distortion due to the large spreading of the material, which causes time step and stability problems and the difficulty in modeling splitting of the bird, which is very important to correctly cover secondary impact loadings of bird fragments after structural penetration. Therefore, more suitable methods for the simulation of fluid-like flow behavior like the Eulerian or meshless SPH technique were adopted in recent years. However, none of these approaches is free of disadvantages, which is the main reason why none of the methods has established as the standard modeling technique until today. It often depends on the specific application, software package and parameter setup, which approach is best suited for the individual problem. The best solution for one case may just be the second best option for another case. It is therefore common practice to adopt different modeling methods and to assess in validation studies, which approach is most beneficial for the current problem. Generally speaking, a trend is visible towards the utilization of Eulerian or SPH models with an equation of state for the impactor material using the properties of a water-air-mixture at room temperature. Very good correlation to experimental results could be obtained with these models.

Although the certification of bird-proof aircraft components today still depends on real physical tests, proposals are increasingly put forward to use more simulation techniques instead of experiments during certification in certain well-defined scenarios. In this context, it would be desirable to increase the uniformity, and herewith comparability, of bird strike analyses e.g. in terms of projectile geometry, mass and composition, which was encouraged by many authorities before but has not been achieved yet. Also the general availability of reliable up-to-date test data for common bird impactor model validations is desirable for future bird strike analyses.

REFERENCES

[1] R.A. Dolbeer, *Birds and aircraft - fighting for airspace in ever more crowded skies*, Human-Wildlife Conflicts 3, 2 (2009) 165-166.

[2] B. MacCinnon, *Sharing the skies: an aviation industry guide to the management of wildlife hazards,* Report TP 13549, Transport Canada, 2nd ed., 2004.

[3] S.A. Meguid, R.H. Mao, T.Y. Ng, *FE analysis of geometry effects of an artificial bird striking an aeroengine fan blade,* International Journal of Impact Engineering 35, 6 (2008) 487-498.

[4] R.A. Dolbeer, S.E. Wright, J. Weller, M.J. Begier, *Wildlife strikes to civil aircraft in the United States 1990-2008*, FAA National Wildlife Strike Database, Serial Report Number 15, September, 2009.

[5] S. Georgiadis, A.J. Gunnion, R.S. Thomson, B.K. Cartwright, *Bird-strike simulation for certification of the Boeing 787 composite moveable trailing edge*, Composite Structures 86, 1-3 (2008) 258-268.

[6] J.R. Allan, A.P. Orosz, *The costs of birdstrikes to commercial aviation,* Proceedings of the 2001 Bird Strike Committee-USA/Canada 3rd Joint Annual Meeting, Calgary, Canada, 2001.

[7] A.I. Khan, R.K. Kapania, E.R. Johnson, *A review of soft body impact on Composite Structure,* 51st AIAA/ASME/ASCE/AHS/ASC Structures, Structural Dynamics, and Materials Conference, Orlando, FL, April 12-15, 2010.

[8] J. Thorpe, *Fatalities and destroyed civil aircraft due to bird strikes 1912 – 2002,* International Bird Strike Committee, 26th Meeting, Warsaw, Poland, May 2003.

[9] J. Thorpe, *Update on fatalities and destroyed civil aircraft due to bird strikes with appendix for 2006 to 2008*, International Bird Strike Committee, 28th Meeting, Brasilia, Brazil, November 2008.

[10] R.E. McCarty, *Computer analysis of bird-resistant aircraft transparencies*, Proceedings of the 17th Annual SAFE Symposium, Las Vegas, NV, December 2-6, 1979, pp. 93-97.

[11] Y.N. Gong, S.Q. Xu, *Bird impact analysis of aircraft windshield transparency*, Chinese Journal of Aeronautics 5, 2 (1992) 106-112.

[12] R.E. McCarty, *MAGNA computer simulation of bird impact on the TF-15 aircraft canopy,* in: S.A. Morolo (Ed.), Proceedings of the 14th Conference on Aerospace Transparent Materials and Enclosures, Scottsdale, AZ, July 11-14, 1983, pp. 974-1008.

[13] H. Odebrecht, *Vogelschlag-Probleme an fliegendem Gerät - dargestellt am Beispiel Militärflugzeug*, Deutscher Luft- und Raumfahrtkongress 2001, Hamburg, Germany, September 17-20, 2001.

[14] L.S. Nizampatnam, *Models and methods for bird strike load predictions*, PhD thesis, Wichita State University, 2007.

[15] L.K. John, *Wing leading edge design with composites to meet bird strike requirements*, in: A.B. Strong (Ed.), Composites in Manufacturing - Case Studies, SME, Dearborn, 1991, pp. 3-18.

[16] H.C. Teichman, R.N. Tadros, *Analytical and experimental simulation of fan blade behavior and damage under bird impact*, Journal of Engineering for Gas Turbines and Power 113, 4 (1991) 582-594.

[17] N. Petrinic, R. Duffin, *Discrete element modeling of soft body impact against rigid targets,* 3rd B2000 Users Workshop, Enschede, The Netherlands, November 27-28, 2000.

[18] N.F. Martin, *Nonlinear finite-element analysis to predict fan-blade damage due to soft-body impact*, Journal of Propulsion and Power 6, 4 (1990) 445-450.

[19] J.P. Barber, H.R. Taylor, J.S. Wilbeck, *Characterization of bird impacts on a rigid plate: part 1.* Technical Report AFFDL-TR-75-5, Air Force Flight Dynamics Laboratory, 1975.

[20] J.P. Barber, H.R. Taylor, J.S. Wilbeck, *Bird impact forces and pressures on rigid and compliant targets,* Technical Report AFFDL-TR-77-60, Air Force Flight Dynamics Laboratory, 1978.

[21] J.S. Wilbeck, *Impact behavior of low strength projectiles,* Technical Report AFML-TR-77-134, Wright-Patterson Air Force Base, 1978.

[22] J.S. Wilbeck, J.P. Barber, *Bird impact loading*, The Shock and Vibration Bulletin 48, 2 (1978) 115-122.

[23] M. Anghileri, G. Sala, *Theoretical assessment, numerical simulation and comparison with tests of birdstrike on deformable structures*, Proceedings of the 20th ICAS Congress, Sorrento, Italy, September 8-13, 1996, pp. 665-674.

[24] L. Iannucci, *Bird strike on composite panels,* DYNA3D User Conference 1992, Manchester, UK, September 22-23, 1992.

[25] L. Iannucci, *Bird-strike impact modeling*, in: I. Mech. E. Seminar, Foreign Object Impact and Energy Absorbing Structures, Inst. of Mech. Engineers, London, 1998.

[26] M. Guida, F. Marulo, M. Meo, M. Riccio, S. Russo, *Fiber metal laminate for bird impact conditions - numerical and experimental analysis,* Proceedings of the 6th International Conference on Composite Science and Technology, ICCST/6, Durban, South Africa, January 22-24, 2007.

[27] R. Budgey, *The development of a substitute artificial bird by the International Birdstrike Research Group for use in aircraft component testing,* International Bird Strike Committee ISBC25/WP-IE3, Amsterdam, The Netherlands, 2000.

[28] T.V. Baughn, L.W. Graham, *Simulation of a birdstrike impact on aircraft canopy material,* Journal of Aircraft 25, 7 (1988) 659-664.

[29] M.A. Lavoie, A. Gakwaya, M. Nejad Ensan, D.G. Zimcik, *Validation of available approaches for numerical bird strike modeling tools,* International Review of Mechanical Engineering 1, 4 (2007) 380-389.

[30] M.A. Lavoie, A. Gakwaya, M. Nejad Ensan, D.G. Zimcik, *Review of existing numerical methods and validation procedure available for bird strike modeling,* International Conference on Computational & Experimental Engineering and Sciences 2, 4 (2007) 111-118.

[31] R.H. Mao, S.A. Meguid, T.Y. Ng, *Finite element modeling of a bird striking an engine fan blade,* Journal of Aircraft 44, 2 (2007) 583-596.

[32] R.H. Mao, S.A. Meguid, T.Y. Ng, *Transient three dimensional finite element analysis of a bird striking a fan blade,* International Journal of Mechanics and Materials in Design 4, 1 (2008) 79-96.

[33] R.H. Mao, S.A. Meguid, T.Y. Ng, *Effects of incidence angle in bird strike on integrity of aero-engine fan blade,* International Journal of Crashworthiness 14, 4 (2009) 295-308.

[34] M.A. Lavoie, A. Gakwaya, M. Nejad Ensan, D.G. Zimcik, D. Nandlall, *Bird's substitute tests results and evaluation of available numerical methods,* International Journal of Impact Engineering 36, 10-11 (2009) 1276-1287.

[35] C.H. Edge, J. Degrieck, *Derivation of a dummy bird for analysis and test of airframe structures,* Proceedings of the 1999 Bird Strike Committee-USA/Canada, First Joint Annual Meeting, Vancouver, Canada, 1999.

[36] T.J. Moffat, W.L. Cleghorn, *Prediction of bird impact pressures and damage using MSC/Dytran*, Proceedings of ASME Turbo Expo 2001, New Orleans, LO, June 4-7, 2001.

[37] A. Kamoulakos, *Bird impactor model,* CRAHVI Technical Workshop, Bristol, UK, March 17, 2004.

[38] R.S. Bertke, J.P. Barber, *Impact damage on titanium leading edges from small soft-body objects,* Report AFML-TR-79-4019, Wright Patterson Air Force Base, Ohio, 1979.

[39] H. Morita, B.R. Tittmann, *Ultrasonic characterization of soft body impact damage on CF/PEEK laminates with gelatin projectiles*, in: D.O. Thompson, D.E. Chimenti (Eds)., Review of Progress in Quantitative Nondestructive Evaluation, Vol. 16, Plenum Press, New York, 1997, pp. 1885-1892.

[40] J. Cheng, G.D. Roberts, W.K. Binienda, *Finite element simulation of soft projectiles impacting composite targets,* 35th International SAMPE Technical Conference, Dayton, OH, September 28 - October 2, 2003.

[41] J. Tischler, *Numerische Simulation des Vogelschlag-Impakts auf eine CFK-Flügelvorderkante mit Hilfe der Finite Elemente Methode*, Diploma thesis, Universität der Bundeswehr München, 2003.

[42] R.L. Azevedo, M. Alves, *Numerical simulation of bird strike impact against balanced fiberglass/epoxy composite plates,* Proceedings of the 19th International Congress of Mechanical Engineering, Brasilia, Brazil, November 5-9, 2007.

[43] J.P Hou, C. Ruiz, *Soft body impact on laminated composite materials*, Composites Part A 38, 2 (2007) 505-515.

[44] R.W. Cornell, *Elementary three-dimensional interactive rotor blade impact analysis,* Journal of Engineering for Power 98, 4 (1976) 480-486.

[45] R.E. McCarty, *Finite element analysis of F-16 aircraft canopy dynamic response to bird impact loading,* Proceedings of the 21st AIAA/ASME/ASCE/AHS Structures, Structural Dynamics and Material Conference, Seattle, WA, May 12-14, 1980, pp. 841-852.

[46] J.J Engblom, *Coupled fluid/structure response predictions for soft body impact of airfoil configurations*, in J.R. Vinson (Ed.), Emerging Technologies in Aerospace

Structures, Design, Structural Dynamics and Materials, ASME Century 2 Publication, 1980, pp. 209-223.

[47] R.P. Nimmer, *Nonlinear transient response of turbofan blades due to foreign object impact,* Proceedings of the 2nd ADINA Conference, Cambridge, MA, August 1, 1979.

[48] A.F. Storace, R.P. Nimmer, R. Ravenhall, *Analytical and experimental investigation of bird impact on fan and compressor blading*, Journal of Aircraft 21, 7 (1984) 520-527.

[49] R.P. Nimmer, L. Boehman, *Transient nonlinear response analysis of soft bodied impact on flat plates including interactive load determination,* 22nd AIAA/ASME/ASCE/AHS Structures, Structural Dynamics and Materials Conference, Atlanta, GA, April 6-8, 1981.

[50] S.G. Zang, C.H. Wu, R.Y. Wang, J.R. Ma, *Bird impact dynamic response analysis for windshield,* Journal of Aeronautical Materials 20, 4 (2000) 41-45.

[51] C.H. Wu, J.L. Yang, S.G. Zang, J.R. Ma, *Study of bird impact loading model*, Journal of Beijing University of Aeronautics and Astronautics 27, 3 (2001) 332-335.

[52] A. Samuelson, L. Sörnäs, *Failure analysis of aircraft windshields subjected to bird impact,* 15th ICAS Congress, London, UK, September 7-12, 1986, pp. 724-729.

[53] Q. Zhang, Z. Xu, *A study of dynamic response for bird impact on arc windshields of aircrafts,* Acta Aeronautica et Astronautica Sinica 12, 2 (1991) B100-105.

[54] A. Leski, R. Baraniecki, J. Malachowski, *Numerical simulation to study the influence of the thickness of canopy at a bird strike*, Proceedings of the 7th International Design Conference, DESIGN 2002, Dubrovnik, Croatia, May 14-17, 2002, pp. 667-672.

[55] A. Alexander, *Interactive multi-mode blade impact analysis*, ASME Gas Turbine Conference and Products Show, Houston, TX, March 9-12, 1981.

[56] R.E. McCarty, *Aircraft transparency bird impact analysis using the MAGNA computer program,* Conference and Training Workshop on Wildlife Hazards to Aircraft, Charleston, SC, May 22-24, 1984, pp. 85-94.

[57] R.E. McCarty, D.E. Trudan, A.D. Davis, *Nonlinear dynamic finite element analysis for the bird impact response of a preprototype T-38 aircraft windshield system,* Proceedings of the 15th Conference on Aerospace Transparent Materials and Enclosures, Monterey, CA, January 16-20, 1989, pp. 1447-1485.

[58] R.E. McCarty, P. Landry, *Analytic assessment of bird impact resistant T-46A aircraft windshield system designs using MAGNA,* Proceedings of the 15th Conference on Aerospace Transparent Materials and Enclosures, Monterey, CA, January 16-20, 1989, pp. 1379-1424.

[59] K.S. Edelstein, R.E. McCarty, *Space shuttle orbiter windshield bird impact analysis,* 16th ICAS Congress, Jerusalem, Israel, August 28-September 2, 1988, pp. 1267-1274.

[60] R.E. McCarty, R.A. Smith, *Assessment of bird impact protection provided by the space shuttle orbiter windshield system using the MAGNA computer program*, Proceedings of the 15th Conference on Aerospace Transparent Materials and Enclosures, Monterey, CA, January 16-20, 1989, pp. 1327-1359.

[61] J. Wen, *The nonlinear dynamic response analysis of the front windshield of Y12 under bird-impact loads,* Acta Aeronautica et Astronautica Sinica 11, 11 (1990) A573-A577.

[62] R.R. Boroughs, *High speed bird impact analysis of the Learjet 45 windshield using DYNA3D,* 39th AIAA/ASME/ASCE/AHS/ASC Structures, Structural Dynamics, and Materials Conference and Exhibition, Long Beach, CA, April 20-23, 1998.

[63] R.A. Brockman, *Current problems and progress in transparency impact analysis*, Proceedings of the 14th Conference on Aerospace Transparent Materials and Enclosures, Scottsdale, AZ, July 11-14, 1983, pp. 1058-1082.

[64] R.A. Brockman, *Finite element analysis of soft-body impact,* Report AFWAL-TR-84-3035, Wright-Patterson Air Force Base, Ohio, 1984.

[65] R.A. Brockman, *Computational methods for soft-body impact problems*, in: S. Atluri, G. Yagawa (Eds.), Computational Mechanics 1988, Vol. 1, Springer, 1988.

[66] B.S. West, *Development of bird impact resistant crew enclosures for aircraft*, in: J. Morton (Ed.), Structural Impact and Crashworthiness, Conference Proceedings, 1984, pp. 696-709.

[67] F. Martinez, G. Rico, A. Franganillo, H. Climent, *Bird impact on an aircraft leading edge*, ABAQUS Users' Conference, Paris, France, May 31-June 2, 1995.

[68] P. Starke, G. Lemmen, K. Drechsler, *Anwendung von FE-Simulationsmethoden bei Vogelschlag.* Deutscher Luft- und Raumfahrtkongress, Stuttgart, September 23-26, 2002.

[69] L.M.L. Castelletti, M. Anghileri, *Multiple birdstrike analysis - a survey of feasible techniques,* 30th European Rotorcraft Forum, Marseilles, France, September 14-16, 2003, pp. 495-505.

[70] C.A. Huertas, *Robust bird-strike modeling using LS-DYNA*, MSc thesis, University of Puerto Rico at Mayagüez, 2006.

[71] J. Cheng, W.K. Binienda, *Simulation of soft projectiles impacting composite targets using an arbitrary Lagrangian-Eulerian formulation*, Journal of Aircraft 43, 6 (2006) 1726-1731.

[72] M. Guida, F. Marulo, M. Meo, M. Riccio, *Evaluation and validation of multi-physics FE methods to simulate bird strike on a wing leading edge*, Proceedings of the 13th European Conference on Composite Materials, ECCM/13, Stockholm, Sweden, June 2-5, 2008.

[73] S.T. Jenq, F.B. Hsiao, I.C. Lin, D.G. Zimcik, M. Nejad Ensan, *Simulation of a rigid plate hit by a cylindrical hemi-spherical tip-ended soft impactor,* Computational Materials Science 39, 3 (2007) 518-526.

[74] F. Stoll, R.A. Brockman, *Finite element simulation of high-speed soft-body impacts,* 38th AIAA/ASME/ASCE/AHS/ASC Structures, Structural Dynamics and Materials Conference, Kissimmee, FL, April 7-10, 1997, pp. 334-344.

[75] S. Zhu, M. Tong, Y. Wang, *Experiment and numerical simulation of a full-scale aircraft windshield subjected to bird impact,* 50th AIAA/ASME/ASCE/AHS/ASC Structures, Structural Dynamics, and Materials Conference, Palm Springs, CA, May 4-7, 2009.

[76] K. Shimamura, T. Shibue, D.J. Grosch, *Numerical simulation of bird strike damage on jet engine fan blade,* ASME Pressure Vessels and Piping Division, Emerging Technology in Fluids, Structures, and Fluid-Structure Interactions, 2004, Vol. 485, Part 1, pp. 161-166.

[77] A. Airoldi, D. Tagliapietra, *Bird impact simulation against a hybrid composite and metallic vertical stabilizer,* 42nd AIAA/ASME/ASCE/AHS/ASC Structures, Structural Dynamics and Materials Conference, Seattle, WA, April 16-19, 2001.

[78] A. Airoldi, B. Cacchione, *Numerical analyses of bird impact on aircraft structures undergoing large deformations and localized failures,* in: M. Alves, N. Jones (Eds.),

Impact Loading of Lightweight Structures, WIT Transaction on Engineering Sciences, Vol. 49, 2005.

[79] A. Airoldi, B. Cacchione, *Modeling of impact forces and pressures in Lagrangian bird strike analyses*, International Journal of Impact Engineering 32, 10 (2006) 1651-1677.

[80] A. Dobyns, F. Federici, R. Young, *Bird strike analysis and test of a spinning S-92 tail rotor,* Proceedings of the 57th AHS International Annual Forum, Washington, DC, May 9-11, 2001.

[81] D. Hachenberg, O. Graf, T. Leopold, *Comparison of different approaches for bird strike simulation*, 2nd EADS Workshop on Crash and Impact Simulation, Ottobrunn, Germany, December 11, 2003.

[82] M. Lawson, R. Tuley, *Supercomputer simulation of a birdstrike on a turbofan aero engine*. Finite Element News 3 (1987) 10-11.

[83] W. Schuette, *Blade behavior during birdstrike, Science and Engineering on Supercomputers,* Proceedings of the Fifth International Conference, London, UK, October 22-24, 1990, pp. 145-157.

[84] E. Niering, *Simulation of bird strikes on turbine engines*, Journal of Engineering for Gas Turbines and Power 112, 4 (1990) 573-578.

[85] R.A. Brockman, T.W. Held, *Explicit finite element method for transparency impact analysis,* Report WL-TR-91-3006, Wright Laboratory, Ohio, 1991.

[86] Y.N. Gong, C. Qian, *New model of bird impact response analysis and its engineering solution*, Proceedings of the 18th ICAS Congress, Beijing, China, September 20-25, 1992, pp. 1476-1481.

[87] A. Wang, X. Qiao, L. Li, *Finite element method numerical simulation of bird striking multilayer windshield,* Acta Aeronautica et Astronautica Sinica 19, 4 (1998) 446-451.

[88] J. Yang, X. Cai, C. Wu, *Experimental and FEM study of windshield subjected to high speed bird impact,* Acta Mechanica Sinica 19, 6 (2003) 543-550.

[89] J. Bai, Q. Sun, *On the integrated design technique of windshield against bird-strike,* Mechanics and Practice 27, 1 (2005) 14-18.

[90] F.S. Wang, Z.F. Yue, *Numerical simulation of damage and failure in aircraft windshield structure against bird strike*, Materials & Design 31, 2 (2010) 687-695.

[91] J.D. Frischbier, *Bird strike investigations in the development process of a transonic fan blisk,* International Gas Turbine & Aeroengine Congress & Exhibition, Orlando, FL, June 2-5, 1997.

[92] W. Chen, Y.P. Guan, D.P. Gao, *Numerical simulation of the transient response of blade due to bird impact,* Acta Aeronautica et Astronautica Sinica 24, 6 (2003) 531-533.

[93] J. Frischbier, *Vogelschlag in Flugtriebwerken - eine impulsartige Fluid-Struktur-Wechselwirkung in der Triebwerksauslegung.* VDI-Berichte Nr. 1682, VDI-Verlag, 2002.

[94] Y. Guan, Z. Zhao, W. Chen, D. Gao, *Foreign object damage to fan rotor blades of aeroengine part II: numerical simulation of bird impact*, Chinese Journal of Aeronautics 21, 4 (2008) 328-334.

[95] Y.N. Shmotin, P.V. Chupin, D.V. Gabov, A.A. Ryabov, V.I. Romanov, S.S. Kukanov, *Bird strike analysis of aircraft engine fan,* 7th European LS-DYNA Users Conference, Salzburg, Austria, May 14-15, 2009.

[96] T.J. Vasko, *Fan blade bird-strike analysis and design*, 6th International LS-DYNA Users Conference, Dearborn, MI, April 9-11, 2000.

[97] S. Kari, J. Gabrys, D. Lincks, *Birdstrike analysis of radome and wing leading edge using LS-DYNA,* 5th International LS-DYNA Users Conference, Southfield, MI, September 21-22, 1998.

[98] B. Dopker, M. Albi, *When bird and plane collide*, Catia Solutions Magazine 3, 6 (1999) 2-5.

[99] B. Langrand, A.S. Bayart, Y. Chauveau, E. Deletombe, *Assessment of multi-physics FE methods for bird strike modeling – application to a metallic riveted airframe,* International Journal of Crashworthiness 7, 4 (2002) 415-428.

[100] R.D. Furtado, *Finite element analysis of birdstrike on aircraft structures,* MSc thesis, Wichita State University, 2000.

[101] L. Iannucci, R. Dechaene, M. Willows, J. Degrieck, *A failure model for the analysis of thin woven glass composite structures under impact loadings*, Computers & Structures 79, 8 (2001) 785-799.

[102] F. Rueda, F. Beltran, C. Maderuelo, H. Climent, *Birdstrike analysis of the wing slats of EF-2000*, in: N. Jones, C.A. Brebbia, A.M. Rajendran (Eds.), Structures under Shock and Impact VII, WIT Press, Southampton, 2002.

[103] T. Kersten, J. Mendler, *Numerical bird strike simulation on aircraft structures*, Vogel und Luftverkehr 22, 2 (2002) 5-15.

[104] E. Kirtil, D. Pestal, A. Kollofrath, N. Gänsicke, J. Mendler, *Simulating the impact behavior of composite aircraft structures,* ABAQUS Users' Conference, Munich, Germany, June 4-6, 2003, pp. 287-301.

[105] D. Berry, *Abaqus simulates, analyses composites for aerospace*, JEC Composites 32 (2007) 45-47.

[106] M. Guida, F. Marulo, M. Meo, M. Riccio, *Analysis of bird impact on a composite tailplane leading edge,* Applied Composite Materials 15, 4-6 (2008) 241-257.

[107] Y. Zhang, Y. Li, *Analysis of the anti-bird impact performance of typical beam-edge structure based on ANSYS/LS-DYNA*, Advanced Materials Research 33-37 (2008) 395-400.

[108] M. Guida, A. Grimaldi, F. Marulo, M. Meo, G. Olivares, *Bird impact on leading edge wing with SPH formulation,* 17th International Conference on Composite Materials, ICCM-17, Edinburgh, UK, July 27-31, 2009.

[109] I. Smojver, D. Ivancecic, *Numerical modeling of bird strike damage in airplane flap structure*, 15th International Conference on Composite Structures, ICCS-15, Porto, Portugal, June 15-17, 2009.

[110] I. Smojver, D. Ivancevic, *Numerical simulation of bird strike damage prediction in airplane flap structure,* Composite Structures 92, 9 (2010) 2016-2026.

[111] M. Guida, F. Marulo, M. Meo, M. Riccio, *Transient three dimensional finite element analysis of a birdstrike on a sandwich fin leading edge,* 7th International Conference on Composite Science and Technology, ICCST-7, Sharjah, United Arab Emirates, January 20-22, 2009.

[112] I. Smojver, D. Ivancevic, *Numerical modeling of impact damage in multi-material airplane structure,* 17th International Conference on Composite Materials, ICCM-17, Edinburgh, UK, July 27-31, 2009.

[113] L. Olovsson, M. Souli, *ALE and fluid-structure interaction capabilities in LS-DYNA,* 6th International LS-DYNA Users Conference, Dearborn, MI, 2000.

[114] S. Audic, M. Berthillier, J. Bonini, H. Bung, A. Combescure, *Prediction of bird impact in hollow fan blades*, 36th AIAA/ASME/SAE/ASEE Joint Propulsion Conference and Exhibit, Huntsville, AL, 16-19 July, 2000.

[115] J. Frischbier, A. Kraus, *Multiple stage turbofan bird ingestion analysis with ALE and SPH methods,* 17th International Symposium on Air Breathing Engines, Munich, Germany, September 4-9, 2005.

[116] M. Anghileri, L.M.L. Castelletti, V. Mazza, *Birdstrike: approaches to the analysis of impacts with penetration,* in: M. Alves, N. Jones (Eds.), Impact Loading of Lightweight Structures, WIT Press, Southampton, 2005, pp. 63-74.

[117] S.C. McCallum, C. Constantinou, *The influence of bird-shape in bird-strike analysis,* 5th European LS-DYNA Users Conference, Birmingham, UK, May 25-26, 2005.

[118] M. Anghileri, C. Bisagni, *Specific problems related to simulation of a bird impact against a turbofan inlet,* Proceedings of the International Crashworthiness Conference 2000, ICRASH 2000, London, UK, September 6-8, 2000, pp. 652-662.

[119] Z.H. Xie, W.J. Bian, H.S. Ang, X. Qiao, *The FEM analysis and simulation of bird impact radome with composite sandwich structure*, Explosion and Shock Waves 19, 3 (1999) 235-242.

[120] M. Hörmann, U. Stelzmann, M.A. McCarthy, J.R. Xiao, *Horizontal tailplane subjected to impact loading,* 8th International LS-DYNA Users Conference, Dearborn, MI, May 2-4, 2004.

[121] C. Shultz, J. Peters, *Bird strike simulation using ANSYS LS/DYNA,* 2002 ANSYS Users Conference, Pittsburgh, PA, April 22-24, 2002.

[122] R. Jain, K. Ramachandra, *Bird impact analysis of pre-stressed fan blades using explicit finite element code*, Proceedings of the International Gas Turbine Congress, Tokyo, Japan, November 2-7, 2003.

[123] J. Frischbier, *Bird strike - an impact load in the design of aero engines,* Vogel und Luftverkehr 27, 1 (2007) 4-21.

[124] Y. Pei, B. Song, Q. Han, *FEM analysis and simulation of bird striking aircraft structure,* in: Progress on Safety Science and Technology, Proceedings of the Asia Pacific Symposium on Safety, Shaoxing, China, November 2-4, 2005, pp. 379-384.

[125] A.G. Hanssen, Y. Girard, L. Olovsson, T. Berstad, M. Langseth, *A numerical model for bird strike of aluminium foam-based sandwich panels,* International Journal of Impact Engineering 32, 7 (2006) 1127-1144.

[126] J. Cheng, G.D. Roberts, W.K. Binienda, *Finite element simulation of soft projectiles impacting composite targets,* 45th AIAA/ASME/ASCE/AHS/ASC Structures, Structural Dynamics and Materials Conference, Palm Springs, CA, April 19-22, 2004.

[127] C.H. Tho, M.R. Smith, *Accurate bird strike simulation methodology for BA609 tiltrotor*, American Helicopter Society 64th Annual Forum, Montreal, Canada, April 29-May 1, 2008.

[128] X. Wang, Z. Feng, F. Wang, Z. Yue, *Dynamic response analysis of bird strike on aircraft windshield based on damage-modified nonlinear viscoelastic constitutive relation*, Chinese Journal of Aeronautics 20, 6 (2007) 511-517.

[129] H. Salehi, S. Ziaei-Rad, M.A. Vaziri-Zanjani, *Bird impact effects on different types of aircraft bubble windows using numerical and experimental methods*, International Journal of Crashworthiness 15, 1 (2010) 93-106.

[130] M. Nejad Ensan, D.G. Zimcik, M. Lahoubi, D. Andrieu, *Soft body impact simulation on composite structures*. Transactions of the CSME.32, 2 (2008) 283-296.
[131] I. Smojver, D. Ivancevic, *Coupled Euler Lagrangian approach using Abaqus/explicit in the bird strike aircraft damage analysis,* 2010 SIMULIA Customer Conference, Providence, RI, May 25-27, 2010.
[132] M. Anghileri, C. Bisagni, *New model of bird strike against aircraft turbofan inlet*, 3rd International KRASH Users Seminar, Tempe, AZ, January 8-10, 2001.
[133] M. Anghileri, L.M.L. Castelletti, L. Lanzi, F. Mentuccia, *Composite materials and bird-strike analysis using explicit finite element commercial codes*, in: N. Jones, C.A. Brebbia (Eds.), Structures under Shock and Impact VIII, WIT press, Southampton, 2004.
[134] R.A. Gingold, J.J. Monaghan, *Smoothed particle hydrodynamics: theory and application to non-spherical stars,* Monthly Notices of the Royal Astronomical Society 181 (1977) 375-389.
[135] J.J. Monaghan, *Smoothed particle hydrodynamics*, Annual Review of Astronomy and Astrophysics 30 (1992) 543-574.
[136] S. Hiermaier, K. Thoma, *Computational simulation of high velocity impact situations using smoothed particle hydrodynamics,* 9th DYMAT Technical Conference, Material and Structural Modeling in Collision Research, Munich, Germany, October 10-11, 1995.
[137] A. Kamoulakos, P. Groenenboom, *Moving from FE to SPH for space debris impact simulations - experience with PAM-SHOCK,* Proceedings European Conference on Spacecraft Structures, Materials and Mechanical Testing, Braunschweig, November 4-6, 1998.
[138] A.F. Johnson, M. Holzapfel, *Modeling soft body impact on composite structures,* Composite Structures 61, 1-2 (2003) 103-113.
[139] M.A. McCarthy, J.R. Xiao, C.T. McCarthy, A. Kamoulakos, J. Ramos, J.P. Gallard, V. Melito, *Modeling of bird strike on an aircraft wing leading edge made from fiber metal laminates - part 2: modeling of impact with SPH bird model*, Applied Composite Materials 11, 5 (2004) 317-340.
[140] W. Machunze, *Numerische Analyse des Restenergie- und Restfestigkeitskriteriums für eine CFK-Slat nach Vogelschlag-Impakt*, Diploma thesis, Technische Universität Dresden, 2005.
[141] V.K. Goyal, C.A. Huertas, T.R. Leutwiler, J.R. Borrero, *Robust bird-strike modeling based on SPH formulation using LS-DYNA,* 47th AIAA/ASME/ASCE/AHS/ASC Structures, Structural Dynamics, and Materials Conference, Newport, RI, May 1-4, 2006.
[142] C. Rössler, *Numerische Simulation des Vogelschlag-Impakts auf ein CFK-Flügelvorderkanten Konzept mit Hilfe der Finite-Elemente-Methode*, Diploma thesis, Technische Universität München, 2004.
[143] A. Letellier, H. Bung, P. Galon, M. Berthillier, *Bird impact on fan blade analysis using smooth particle hydrodynamics coupled with finite elements*, ASME Pressure Vessels and Piping Division, Symposium on Structures under Extreme Loading Conditions, Vol. 351, 1997, pp. 191-195.

[144] D. Chevrolet, S. Audic, J. Bonini, *Bird impact analysis on a bladed disk*, RTO AVT Symposium on Reduction of Military Vehicle Acquisition Time and Cost through Advanced Modeling and Virtual Simulation, Paris, France, April 22-25, 2002.

[145] R.L. Azevedo, M. Alves, *Numerical simulation of soft-body impact on GFRP laminate composites: mixed SPH-FE and pure SPH approaches,* in: H.S. da Costa Mattos, M. Alves (Eds.), Mechanics of Solids in Brazil 2009, ABCM Symposium Series, Rio de Janeiro, 2009, pp. 15-30.

[146] P. Starke, G. Lemmen, K. Drechsler, *Validierung von Verfahren für die numerische Simulation von Vogelschlag,* 4th German LS-DYNA Forum, Bamberg, Germany, October 20-21, 2005.

[147] S. Pilz, *Developing military vehicles and aircraft,* ANSYS Solutions 7, 3 (2006) 4-9.

[148] M.A. Lavoie, A. Gakwaya, M. Nejad Ensan, *Application of the SPH method for simulation of aerospace structures under impact loading*, 10th International LS-DYNA Users Conference, Dearborn, MI, June 8-10, 2008.

[149] W. Lammen, R. Van Houten, *Predictive simulation of impact phenomena for innovations in aircraft component design,* 6th EUROMECH Nonlinear Dynamics Conference, Saint-Petersburg, Russia, June 30-July 4, 2008.

[150] A.F. Johnson, M. Holzapfel, N. Petrinic, *Modeling soft body impact on composite structures,* European Conference on Computational Mechanics, ECCM-2001, Cracow, Poland, June 26-29, 2001.

[151] L.C. Ubels, A.F. Johnson, J.P. Gallard, M. Sunaric, *Design and testing of a composite bird strike resistant leading edge,* SAMPE Europe Conference & Exhibition, Paris, France, April 1-3, 2003.

[152] M. McCarthy, *Bird impact on leading edge structures*, CRAHVI Technical Workshop, Bristol, UK, March 17, 2004.

[153] T. Kermanidis, G. Labeas, M. Sunaric, L. Ubels, *Development and validation of a novel bird strike resistant composite leading edge structure*, Applied Composite Materials 12, 6 (2005) 327-353.

[154] R. Richter, *Validierung der Impakt-Resistenz einer CFK-Slat gegenüber Vogelschlag,* Diploma thesis, Universität der Bundeswehr München, 2005.

[155] M.A. McCarthy, J.R. Xiao, C.T. McCarthy, A. Kamoulakos, J. Ramos, J.P. Gallard, V. Melito, *Modeling bird impacts on an aircraft wing - part 2: modeling the impact with an SPH bird model*, International Journal of Crashworthiness 10, 1 (2005) 51-59.

[156] T. Kermanidis, G. Labeas, M. Sunaric, A.F. Johnson, M. Holzapfel, *Bird strike simulation on a novel composite leading edge design*, International Journal of Crashworthiness 11, 3 (2006) 189-201.

[157] G. Labeas, T. Kermanidis, *Impact behavior modeling of a composite leading edge structure,* Fracture of Nano and Engineering Materials and Structures, Proceedings of the 16th European Conference of Fracture, Alexandroupolis, Greece, July 3-7, 2006, pp. 1259-1260.

[158] A.F. Johnson, M. Holzapfel, *Numerical prediction of damage in composite structures from soft body impacts,* Journal of Materials Science 41, 20 (2006) 6622-6630.

[159] Y.C. Roth, *Composites in high lift applications for civil aircrafts: highlights in LuFo III,* IVW-Kolloquium 2006, Kaiserslautern, Germany, November 14-15, 2006, pp. 71-85.

[160] W. Machunze, P. Middendorf, R. Keck, Y.C. Roth, *Design, analysis and manufacturing of a thermoplastic slat*, 1st European Conference on Materials and Structures in Aerospace, EUCOMAS, Berlin, Germany, May 26-27, 2008.

[161] R. Keck, W. Machunze, W. Dudenhausen, P. Middendorf, *Design, analysis, and manufacturing of a carbon-fiber-reinforced polyetheretherketone slat*, Journal of Aerospace Engineering 223, 8 (2009) 1115-1123.

[162] M. Anghileri, L.M.L. Castelletti, D. Molinelli, F. Motta, *A strategy to design bird-proof spinners,* 7th European LS-DYNA Users Conference, Salzburg, Austria, May 14-15, 2009.

[163] J.L. Lacome, *Smoothed particle hydrodynamics method in LS-DYNA*, 3rd German LS-DYNA Forum, Bamberg, Germany, October 14-15, 2004.

[164] M. Selezneva, P. Stone, T. Moffat, K. Behdinan, C. Poon, *Modeling bird impact on a rotating fan: the influence of bird parameters,* 11th International LS-DYNA Users Conference, Dearborn, MI, June 6-8, 2010.

[165] M. Anghileri, L.M.L. Castelletti, F. Invernizzi, M. Mascheroni, *Birdstrike onto the composite intake of a turbofan engine,* 5th European LS-DYNA Users Conference, Birmingham, UK, May 25-26, 2005.

[166] J. Liu, Y.L. Li, F. Xu, *The numerical simulation of a bird-impact on an aircraft windshield by using the SPH method*, Advanced Materials Research 33-37 (2008) 851-856.

[167] L. Wu, Y.N. Guo, Y.L. Li, *Bird strike simulation on sandwich composite structure of aircraft radome,* Explosion and Shock Waves 29, 6 (2009) 642-647.

[168] F. Bianchi, *Bird strike simulations with RADIOSS on AW helicopter blade and rotor controls,* 3rd European Hyperworks Technology Conference, EHTC2009, Ludwigsburg, Germany, November 2-4, 2009.

[169] V.K. Goyal, C.A. Huertas, J.R. Borrero, T.R. Leutwiler, *Robust bird-strike modeling based on ALE formulation using LS-DYNA*, 47th AIAA/ASME/ASCE/AHS/ASC Structures, Structural Dynamics, and Materials Conference, Newport, RI, May 1-4, 2006.

[170] K.C. Chuan, *Finite element analysis of bird strikes on composite and glass panels,* BSc thesis, National University of Singapore, 2006.

[171] A.A. Ryabov, V.I. Romanov, S.S. Kukanov, Y.N. Shmotin, P.V. Chupin, *Fan blade bird strike analysis using Lagrangian, SPH and ALE approaches*, 6th European LS-DYNA Users Conference, Gothenburg, Sweden, May 29-30, 2007.

[172] S. Zhu, M. Tong, *Study on bird shape sensitivity to dynamic response of bird strike on aircraft windshield*, Journal of Nanjing University of Aeronautics & Astronautics 40, 4 (2008) 551-555.

[173] S. Zhu, M. Tong, Y. Wang, *Dynamic analysis of bird impact on aircraft windshield and bird shape sensitivity study,* Proceedings of the First International Conference on Modeling and Simulation, Nanjing, China, August 5-7, 2008, pp. 137-142.

[174] S. Heimbs, S. Heller, P. Middendorf, F. Hähnel, J. Weiße, *Low velocity impact on CFRP plates with compressive preload: test and modelling,* International Journal of Impact Engineering 36, 10-11 (2009) 1182-1193.

[175] Y. Li, Y. Zhang, P. Xue, *Study of similarity law for bird impact on structure*, Chinese Journal of Aeronautics 21, 6 (2008) 512-517.

In: Advances in Mechanical Engineering Research, Volume 3 ISBN: 978-61209-243-0
Editor: David E. Malach

Chapter 4

A Simple Model to Predict Ultra-Low Cycle Fatigue Fracture of Steel Bridges

Siriwardane Sudath Chaminda*[*1], *Mitao Ohga*[2], *Munindasa P. Ranaweera*[3], *Raveendra Herath*[3] and *Ranjith Dissanayake*[3]

[1]Department of Civil and Environmental Engineering, University of Ruhuna Hapugala, Galle 80000, Sri Lanka

[2]Department of Civil and Environmental Engineering, Ehime University, Bunkyo-cho 3, Matsuyama 790-8577, Japan

[3]Department of Civil Engineering, University of Peradeniya, Peradeniya 20400, Sri Lanka

Abstract

The concept of ultra-low cycle fatigue (ULCF) was recently originated with some of sudden failures of existing bridges, which were characterized by large scale cyclic yielding due to occasional loadings such as earthquakes, typhoons. Generally, experimental approaches are popular for ULCF fracture prediction. As for the authors view, only one theoretical study has been published in year 2007, and the observed failure mechanism is based on void growth process. The fracture is calculated to occur when cyclic void growth index (VGI_{cyclic}) exceeds its critical value. The VGI_{cyclic} demand is calculated based on complex integrations of a function, which depends on triaxiality and incremental plastic strain. However, it is required to modify commonly available FEM programs to cater this integration and finally it hindered the usage of general propose FEM packages as it is to estimate ULCF fracture. As a result, found applications of this fracture criterion are very less.

Therefore, this chapter presents a simplified model to assess the real ULCF fracture of steel bridges using available general-purpose FEM packages. However, this approach is limited to the situation where triaxiality remains relatively constant during its loading history. As highlighted in previous studies, in many realistic situations, this statement can

[*] siriwardane@cee.ruh.ac.lk

be applicable. The fracture mechanism of this model is also similar as previous model. But the fracture criterion is totally different from the previous approach such that the ULCF fracture is calculated to occur when significant plastic strain exceeds its critical value. As for alternative component of simplified ULCF fracture criterion, a simple non-linear hardening model was also proposed to employ in elasto-plastic analysis for the situations, where parameters of available mix hardening models are difficult to determine. The behavior of the proposed hardening model was verified with experimental behavior of few materials. The verification of the simplified fracture criterion was done by comparing the results with previous criterion-based estimations of some bridge components and hence importance of this model was clearly illustrated. Finally, chapter tends to conclude that the proposed model gives reasonably accurate prediction to ULCF fracture of steel structures where triaxiality remains relatively constant during the loading history.

1. INTRODUCTION

Most of the civil engineering structures, which are subjected to repeated cyclic or fluctuating loads, are usually affected from the high cycle fatigue. In contrast to high cycle fatigue assessment in service loading, modern structures are assessed for ductile fracture due to highly inelastic reverse cyclic loading, which is often the governing ultimate limit state in existing structures that were designed to resist large earthquake or typhoon by sustaining large inelastic deformation. Plastic hinging of beam-column connections (Stojadinovic et al., 2000) and inelastic cyclic buckling of steel braces (Fell et al., 2006) are two primary examples where low cycle ductile fracture is the ultimate limit state. The stress intensity-based classical fracture mechanics and strain-life curve based low cycle fatigue models are commonly accepted as providing accurate methods to assess ultimate limit state of structures (Kanvinde et al., 2007).

Ultra low cycle fatigue (ULCF) fracture initiation that occurs due to a few (generally, less than 20) cycles of large plastic strains (several times the yield strain) in ultimate limit state of structures cannot be accurately estimated using the commonly accepted fracture mechanics and low cycle fatigue models. First, the stress intensity based methods, typically, require the real or presumed presence of an initial sharp crack or flaw, which is absent in many structural details. This may reason to invalidate stress intensity based approaches during the large-scale yielding. Second, the induced loading histories are extremely random with very few cycles, making them difficult to adapt to conventional cycle counting technique such as rainflow analysis with strain-life curve. However, the void growth and coalescence-type mechanism based cyclic void growth model (CVGM) is recently proposed to estimate accurately the ULCF fracture initiation of structures (Kanvinde et al., 2007). The fracture is calculated to occur when cyclic void growth index (VGI_{cyclic}) exceeds its critical value. The VGI_{cyclic} demand is calculated based on implicit integrations of a function, which depends on triaxilaity (ratio between hydrostatic and effective stress) and incremental equivalent plastic strain. Therefore, it is required to modify commonly available finite element (FE) programs to cater to this integration in order to estimate ULCF fracture and it is difficult to use with the outputs of common FE programs. As a result, applications of this fracture criterion are found to be quite rare.

Therefore, the major objective of this chapter is to simplify the available cyclic void growth model (CVGM) to use with the outputs of general finite element method (FEM) employed programs. The major attention of this model is limited to estimate more realistic ultra low cycle fatigue (ULCF) fracture initiation life of structures under ultimate (extreme) loadings. The simplified model describes accurately the ULCF fracture of steel structures where triaxiality remains relatively constant during the loading history. As for alternative component of ULCF analysis, a simple non-linear hardening model is also proposed to employ in elasto-plastic analysis for situations where parameters of available mix hardening models are difficult to be determined.

Initially the details of simplified ULCF fracture criterion are comprehensively described. The fracture criterion associated ULCF failure prediction methodology is also clearly indicated. Then the hardening model is proposed and comparisons with experimental behaviors are made as verification. Then details, which are relevant to the verification of the simplified approach by comparing the results with previous approaches-based estimations of some bridge components, are comprehensively mentioned. Finally the validity and applicability of the simplified model are discussed.

2. Existing Ultra-Low Cycle Fatigue Models

The ultra-low cycle fatigue (ULCF) describes the mechanism which causes to crack initiation due to a few (generally, less than 20) cycles of large plastic strains (several times the yield strain) of a material. This section describes the details about existing ULCF models comprehensively. Initially, the section describes the commonly used ULCF model to estimate ultimate limit state of existing structures. Then the deficiencies are comprehensively discussed. Then, describes recent CVGM model, which are originated to overcome these shortcomings of previous model and their deficiencies related to the general application in comprehensive manner. Finally, this section describes the previous hardening models and their shortcomings in ULCF fracture prediction.

2.1 Commonly Used Models and their Deficiencies

Failure due to ductile fracture or low cycle fatigue in steel connections is an important limit state in earthquake-resistance design. The few approaches are commonly used for assessment purposes. The current standards and codes deal with this problem based on semi-empirical, experimental approaches. Conventional fracture indices (e.g. stress intensity factor-K_I, crack tip opening displacement (CTOD), and J integral) are limited to address the problem of crack initiation and growth based on far-field stresses. In situations of large scale yielding, these far field stresses have often lose correlation with near-tip stresses and strains, consequently invalidating the use of either linear-elastic or elastic-plastic fracture mechanics (Kanvinde, et al., 2005). Another key requirement for the application of conventional fracture theory is the presence of an initial sharp crack or flaw in the material. Many realistic situations in structural engineering (e.g. bolted connection details) have no such sharp cracks, forcing ad hoc and often unrealistic introduction of hypothetical flaws into analysis models.

Fundamental micromechanics-based models that employ detailed finite element analyses to capture interactions of crack tip stress and strain fields can provide more accurate means to evaluate ductile fracture initiation (Marini et al., 1985; Bandstra et al. 2004; Benzerga et al., 2004; Chi et al., 2006; Kanvinde et al., 2007). These micromechanical models are divided into two categories, one dealing with fracture due to microvoid growth and coalescence mechanism ("ductile" crack initiation), and the other category dealing with all other mechanisms, such as cleavage. Though, considerable research has been conducted to develop micromechanical continuum models, most of these developments have been highly theoretical and have not applied to investigate materials (mild steel) and structural details that are directly relevant to civil engineering construction. Since the structural engineering problems are usually subjected to multiaxial stresses and strains, the strain-based models which are used in structural engineering to predict fracture are relatively unsophisticated insofar.

Some of failure mechanism such as plastic hinging of beam-column connections (Stojadinovic et al., 2000) and inelastic cyclic buckling of steel braces (Fell et al., 2006) are two primary examples where low cycle ductile fracture is the ultimate limit state. The earthquake induced loading histories are extremely random with very few cycles, making them difficult to adapt to conventional cycle counting technique such as rainflow analysis (Kanvinde et al., 2007; Uriz et al., 2008). Further, it is observed that the Manson-Coffin law does not fit well in the range of very low life, i.e. about less than 100 cycles (Kuroda, 2002; Tateshi et al., 2007; Xue, 2008; Davaran et al., 2009). These reasons may cause to under predict the real ULCF failure of existing structures.

To bridge this gap, the relatively simple micromechanics based stress modified critical strain (SMCS) criterion was originated in year 2006 (Chi et al., 2006). As outlined in previous section, it is important to recognize the structures typically experience large inelastic cyclic deformation during earthquake. However, recently developed SMCS ductile failure criteria (Chi et al., 2006) is also unable to recognize such failure clearly since the ULCF failure mechanism is not exactly the ductile fracture but also it exhibits the fracture fatigue interaction behavior.

2.2 Cyclic Void Growth Model and its Deficiencies

As a result of that, the micromechanics based new ULCF criterion, which is named as the cyclic void growth model (CVGM), has been introduced to earthquake engineering practices (Kanvinde et al., 2007). The proposed model is an extension to previously published models that simulate ductile fracture caused by void growth and coalescence under monotonic loading. The underlying mechanisms of ULCF fracture involve cyclic void growth, collapse, and distortion, which are distinct from those associated with more conventional fatigue. The CVGM represents these underlying fracture mechanisms through plastic strain and stress triaxiality histories that can be modeled at material continuum level by FE analyses. The details regarding theoretical development of CVGM are comprehensively mentioned in the corresponding paper by following each step of development in logical manner (Kanvinde et al., 2007). Therefore, this section describes only the fracture criterion for brief understanding of the CVGM.

Two main aspects of the CVGM fracture criterion are level of void growth (demand), including the effects of void growth and shrinkage during reversed cyclic loading and critical level of the void growth (capacity), related to cyclic strain concentrations of the intervoid ligament material. In the previous paper (Kanvinde et al., 2007), the level of void growth is systematically described from cyclic void growth index (VGI_{cyclic}) as follows.

$$VGI_{cyclic} = \sum_{tensile\ cycles} \int_{\varepsilon_1}^{\varepsilon_2} \exp(|1.5T|)d\varepsilon_p - \sum_{compressive\ cycles} \int_{\varepsilon_1}^{\varepsilon_2} \exp(|1.5T|)d\varepsilon_p \quad (1)$$

where $T = \sigma_m / \sigma_e$ = stress triaxiality (ratio of mean stress σ_m to effective stress σ_e) and $d\varepsilon_p = \sqrt{(2/3)d\varepsilon_{ij}^p d\varepsilon_{ij}^p}$, is the incremental equivalent plastic strain. As mentioned in the preceding discussion, the ULCF fracture is calculated to occur when the VGI_{cyclic} exceeds its critical level of the void growth ($VGI_{cyclic}^{critical}$) over a characteristic length measure (l^*) in the region of high stresses and plastic strains, i.e.

$$VGI_{cyclic} > VGI_{cyclic}^{critical} \quad (2)$$

The critical level of the void growth parameter for cyclic loading, $VGI_{cyclic}^{critical}$, is reduced from its monotonic counterpart, $VGI_{cyclic}^{monotonic}$, due to void collapse and the cyclic straining of intervoid ligaments. To determine an appropriate mathematical relationship between these two parameters, various damage variables were evaluated against the observed test data. Based on the test data and the FE analyses, an exponential decay function was selected to relate these two critical parameters as shown below,

$$VGI_{cyclic}^{critical} = VGI_{monotonic}^{critical} \exp(-\lambda\varepsilon_p^{accumulated}) \quad (3)$$

The $\varepsilon_p^{accumulated}$ is defined as the equivalent plastic strain that has accumulated up to the beginning of each tensile excursion of loading. The λ is material dependent damageability parameter. The monotonic counterpart, $VGI_{monotonic}^{critical}$ is the critical value of the monotonic void growth index. The corresponding monotonic void growth index, which describes the level of void growth to predict the ductile fracture under monotonic loading, $VGI_{monotonic}$ is described as follows,

$$VGI_{monotonic} = \int_0^{\varepsilon^p} \exp(1.5T)d\varepsilon_p \quad (4)$$

The VGI_{cyclic} and $VGI_{monotonic}$ demands (Eq. (1) and (4)) are calculated based on implicit integrations of a function. Therefore, it is required to modify commonly available finite element (FE) programs to cater to this integration in order to estimate ULCF fracture and it is difficult to use with the outputs of common FE programs. As a result, applications of this fracture criterion are found to be quite rare.

2.3 Brief Description of Existing Elasto-Plastic Hardening Models

Today number of models is available to describe the hardening behaviour of materials. Major classes of these models include the multi surface models (Mroz, 1967, Krieg ,1975, McDowell, 1985, Ohno et al., 1986, Khoei et al., 2005) and the single surface models (Prager, 1956, Yanyao et al., 1996, Voyiadjis et al., 1998). However, the most of FEM programs consist of single surface model. Among these one of widely used hardening model is the Armstrong and Frederick model, which reasonably describes the material's non-linear kinematic hardening behaviours. This model is based on the assumption that the most recent part of the strain history of a material dictates the mechanical behaviour properly. The corresponding hardening rule is given as,

$$d\alpha_{ij} = \frac{2}{3}C_1 d\varepsilon_{ij}^p - C_2\alpha_{ij} d\bar{\varepsilon}^p \tag{5}$$

where α_{ij} is the back stress tensor, $d\varepsilon_{ij}^p$ is incremental plastic strain tensor and the effective plastic strain is $d\bar{\varepsilon}^p = \sqrt{2/3\, d\varepsilon_{ij}^p d\varepsilon_{ij}^p}$. The C_1 and C_2 are the constants, which can determine by material uniaxial test. But the major drawback is that this model produces somewhat inaccurate predictions to their behavior when it reaches to high values of total strains or effective plastic strains, (Navarro et al, 2005, Gozzi, 2004). However, some of mix hardening models solve this problem to some extent. Among these, Lemitre and Chaboche proposed non-linear mix (isotropic and kinematic) hardening model is very much popular among the most of FE programs (Lubliner, 1990). The evolution law of this model consists of two components, a nonlinear kinematic hardening component, which describes the translation of the yield surface in stress space through the back stress, α , and an isotropic hardening component, which describes the change of the equivalent stress defining the size of the yield surface, σ^0, as a function of plastic deformation. When temperature and field variable dependencies are omitted, the hardening law is,

$$d\alpha_{ij} = C\frac{1}{\sigma_0}(\sigma_{ij} - \alpha_{ij})d\bar{\varepsilon}^p - \gamma\alpha_{ij} d\bar{\varepsilon}^p \tag{6}$$

where C and γ are material parameters that must be calibrated from cyclic test data. The isotropic hardening behavior of the model defines the evolution of the yield surface size, R, as a function of the equivalent plastic strain $d\bar{\varepsilon}^p$. This evolution can be introduced by specifying *R*, directly as a function of $d\bar{\varepsilon}^p$ in tabular form as below.

$$R = \sigma_0 + Q_\infty (1 - e^{-b\bar{\varepsilon}^p}) \tag{7}$$

where σ_0 is the yield stress at zero plastic strain and Q_∞ and *b* are material parameters. However, this model is dependent on four hardening parameters and these are unknown for most of the materials. To fill this gap, the study has given effort to contribute new model to theory of plasticity by using single surface approach with making a little extension to Armstrong Fredrick hardening rule.

The existing damage models to predict ULCF failure of structures under ultimate loadings are clearly highlighted. The drawbacks, deficiencies, limitations and difficulties of existing models are comprehensively indicated. To overcome these key problems to some extent, this chapter proposes a simplified approach to assess the real ultra-low cycle fatigue (ULCF) failure of steel structures. The approach basically consists of i) different ULCF fracture criterion, which is obtained by simplifying the available cyclic void growth model (CVGM), and ii) different non-linear hardening model, of which parameters can be easily determined than the general mix-hardening models. However, this approach is limited to the situation where magnitude of triaxiality remains relatively constant during the history of loading. The fracture mechanism of this method is also similar to previous method. But, the fracture criterion is totally different from the previous approach such that the ULCF fracture is calculated to occur when accumulated equivalent plastic strain at cyclic loading exceeds its critical value.

3. Simplified Ultra-Low Cycle Fatigue Model

The CVGM based ULCF mechanism has been comprehensively described in the section 2. In this section, the CVGM based fracture criterion will be simplified to obtain a more simple form of fracture criterion, which can easily be employed together with commonly available finite element method (FEM) programs to assess ULCF fracture. However, usage of this fracture criterion is limited to situations where magnitude of the triaxiality (ratio between hydrostatic and effective stress) remains relatively (approximately) constant during the loading history. Initially section describes the mathematical relation of simplified ULCF fracture criterion. The evaluation method of material constants are mentioned at the end of the section.

3.1 Simplified Fracture Criterion

Two main aspects of the ULCF fracture model are level of void growth (demand), including the effects of void growth and shrinkage during reversed cyclic loading and critical level of the void growth (capacity), related to cyclic strain concentrations of the intervoid ligament material. In the previous CVGM model (refer section 2.2), the level of void growth is systematically described from cyclic void growth index (VGI_{cyclic}) as follows.

$$VGI_{cyclic} = \sum_{tensile\ cycles} \int_{\varepsilon_1}^{\varepsilon_2} \exp(|1.5T|) d\varepsilon_p - \sum_{compressive\ cycles} \int_{\varepsilon_1}^{\varepsilon_2} \exp(|1.5T|) d\varepsilon_p \tag{8}$$

where $T = \sigma_m / \sigma_e$ = stress triaxiality (ratio of mean stress σ_m to effective stress σ_e); and $d\varepsilon_p = \sqrt{(2/3) d\varepsilon_{ij}^p d\varepsilon_{ij}^p}$ = incremental equivalent plastic strain. Here the preceding cycles have been subdivided into tensile and compressive (based on the sign of triaxiality) at any point during the loading history. However this index cannot be a negative value, which would imply a negative void growth. When monotonic compression is applied to a material, the model would not predict a decrease (or negative increase of void volume) in index, but it would maintain the index as zero.

The situations where magnitude of triaxiality remains relatively constant during the loading history, the Eq. (8) can be simplified as,

$$VGI_{cyclic} = \exp(|1.5T|) \left[\sum_{tensile\ cycles} \varepsilon_p - \sum_{compressive\ cycles} \varepsilon_p \right]. \tag{9}$$

The $\sum_{tensile\ cycles} \varepsilon_p = \bar{\varepsilon}_p^t$ is defined as the equivalent plastic strain that has accumulated during every tensile excursion of loading. Similarly, $\sum_{compressive\ cycles} \varepsilon_p = \bar{\varepsilon}_p^c$ is also defined as accumulated equivalent plastic strain for every compressive excursion of cyclic loading. Subtraction of these two quantities describes equivalent plastic strain that has been accumulated during cyclic loading (both tensile and compressive cycles) and it is named as "accumulated equivalent plastic strain at cyclic loading ($\bar{\varepsilon}_p^{cyclic}$)".

$$\bar{\varepsilon}_p^{cyclic} = \left[\sum_{tensile\ cycles} \varepsilon_p - \sum_{compressive\ cycles} \varepsilon_p \right] = \left(\bar{\varepsilon}_p^t - \bar{\varepsilon}_p^c \right). \tag{10}$$

The value of $\bar{\varepsilon}_p^{cyclic}$ varies based on the loading history. However, it cannot be a negative value, which would imply a negative void growth. If the expression decreases to zero during the cyclic loading, it remains at zero until a subsequent tensile cycle increases its value above zero. When a monotonic compression is applied to a material, the model would not predict a decrease (a negative increase) in void volume, but rather it would maintain the void volume at zero. Therefore in such situations the quantity of $\bar{\varepsilon}_p^{cyclic}$ should also be zero.

Hence, the Eq. (9) is rewritten as,

$$VGI_{cyclic} = \exp(|1.5T|)\bar{\varepsilon}_p^{cyclic} \tag{11}$$

As mentioned in preceding discussion, ULCF fracture is calculated to occur when VGI_{cyclic} exceeds a critical value. Thus, it can be stated that the VGI_{cyclic} reaches to its critical value when the $\bar{\varepsilon}_p^{cyclic}$ reaches to its critical level, i.e.

$$VGI_{cyclic}^{critical} = \exp(|1.5T|)(\bar{\varepsilon}_p^{cyclic})_{critical} \tag{12}$$

where the $(\bar{\varepsilon}_p^{cyclic})_{critical}$ is critical value (capacity) of accumulated equivalent plastic strain at cyclic loading. Also this variable represents the critical level of void growth (critical void size) under cyclic loading as same as the $VGI_{cyclic}^{critical}$.

As mentioned in the corresponding paper (Kanvinde et al., 2007), the VGI_{cyclic} demand is calculated based on stress and strain outputs, which are obtained from elasto-plastic FEM analyses. The $VGI_{cyclic}^{critical}$ capacity is determined as a degraded fraction of critical monotonic index, $VGI_{monotonic}^{critical}$, i.e.

$$VGI_{cyclic}^{critical} = VGI_{monotonic}^{critical} \exp(-\lambda \varepsilon_p^{accumulated}) \tag{13}$$

The $\varepsilon_p^{accumulated}$ is defined as the equivalent plastic strain that has accumulated up to the beginning of each tensile excursion of loading. The λ is material dependent damageability parameter. The $VGI_{monotonic}^{critical}$ is the critical value of the monotonic void growth index. The corresponding monotonic void growth index, which describes the level of void growth to predict the ductile fracture under monotonic loading, $VGI_{monotonic}$ is described as follows (Kanvinde et al., 2007).

$$VGI_{monotonic} = \int_0^{\varepsilon^p} \exp(1.5T)d\varepsilon_p \tag{14}$$

Similarly in a cyclic loading situation, when the magnitude of triaxility remains relatively constant during the loading history, the Eq. (14) can be simplified as,

$$VGI_{monotonic} = \exp(1.5T)\bar{\varepsilon}_p^{monotonic} \tag{15}$$

where the $\bar{\varepsilon}_p^{monotonic}$ is named as "accumulated equivalent plastic strain at monotonic loading". Generally in plasticity theory this variable is called as cumulative equivalent plastic strain. Further it is an indicator of accumulated deformation at the considered location and it can be easily determined through continuum FEM analysis.

The ductile fracture under monotonic loading also occurs when $VGI_{monotonic}$ exceeds its critical value. Therefore, it is possible to state that the $VGI_{monotonic}$ reaches to its critical value when the $\bar{\varepsilon}_p^{monotonic}$ reaches to its critical level as similar to ULCF fracture situation, i.e.

$$VGI_{monotonic}^{critical} = \exp(1.5T)(\bar{\varepsilon}_p^{monotonic})_{critical} \tag{16}$$

where the $(\bar{\varepsilon}_p^{monotonic})_{critical}$ is critical value (capacity) of accumulated equivalent plastic strain at monotonic loading. Also this parameter represents the critical level of void growth (critical void size) under monotonic loading and it is determined by following expression (Kanvinde, et al. 2007).

$$(\bar{\varepsilon}_p^{monotonic})_{critical} = \alpha \ \exp(-1.5T) \tag{17}$$

The α is called as toughness index and it is a material constant that is determined by experimentally. Since it is only a function of triaxiality where higher triaxiality leads to lower $(\bar{\varepsilon}_p^{monotonic})_{critical}$ and vice versa, it can be easily determined for each location of interest.

Substituting Eqs. (12) & (16) to Eq. (13), the critical value (capacity) of accumulated equivalent plastic strain at cyclic loading $(\bar{\varepsilon}_p^{cyclic})_{critical}$ can be obtained as a degraded fraction of critical value (capacity) of accumulated equivalent plastic strain at monotonic loading $(\bar{\varepsilon}_p^{monotonic})_{critical}$ as below.

$$(\bar{\varepsilon}_p^{cyclic})_{critical} = (\bar{\varepsilon}_p^{monotonic})_{critical} \exp(-\lambda \varepsilon_p^{accumulated}) \tag{18}$$

As described by Eq. (18), the critical accumulated equivalent plastic strain at cyclic loading $(\bar{\varepsilon}_p^{cyclic})_{critical}$ reduces to critical monotonic limit for monotonic loading situation. Because the accumulated equivalent plastic strain $\varepsilon_p^{accumulated}$ is zero for monotonic loading situations since it is calculated at the beginning of each tensile excursion. In this way, the increment of cumulative equivalent plastic strain during current tensile cycle is not contributing to the damage that occurs within that loading increment. As a result the damage level is at a constant value within each tensile excursion. However, the accumulated equivalent plastic strain during current tensile cycles contributes only to void growth process, such that for each tensile cycle $\bar{\varepsilon}_p^{cyclic}$ is compared to a constant value of $(\bar{\varepsilon}_p^{cyclic})_{critical}$ which is calculated at the beginning of that cycle. This explanation confirms that ULCF fracture can only be initiated during tensile loading excursions. Hence the simplified ULCF fracture criterion is defined as,

$$\overline{\varepsilon}_p^{cyclic} > (\overline{\varepsilon}_p^{cyclic})_{critical} \quad (19)$$

The prediction of ULCF crack initiation is made when the $\overline{\varepsilon}_p^{cyclic}$ exceeds its critical value $(\overline{\varepsilon}_p^{cyclic})_{critical}$ over a characteristic length measure (l^*) in the region of high stresses and plastic strains. However, places where triaxiality varies significantly during loading histories, the mentioned simplified approach might produce a less accurate result than the CVGM approach.

3.2 Determination of Material Constants

The described ULCF failure criteria involve three parameters such as toughness index (α), damageability parameter (λ) and the characteristic length (l^*). The α and the λ are determined through testing and finite element analysis of circumferentially smooth-notched tensile specimen. The characteristic length (l^*) can be determined through the micro-structural measurements and observation of the fracture surface.

The toughness index (α) is obtained through testing and finite element analyses of circumferentially smooth-notched tensile specimens such as the one shown in Fig. 1. These tests referred to as SNTT , smooth-notched tensile tests have the same overall geometry as the smooth round bars with a circumferential notch machined into them to produce a triaxial stress condition and subjected to monotonic loading test. The triaxiality is varied by changing the notch size. Typically, the fracture condition is a function of the toughness index as well as the characteristic length.

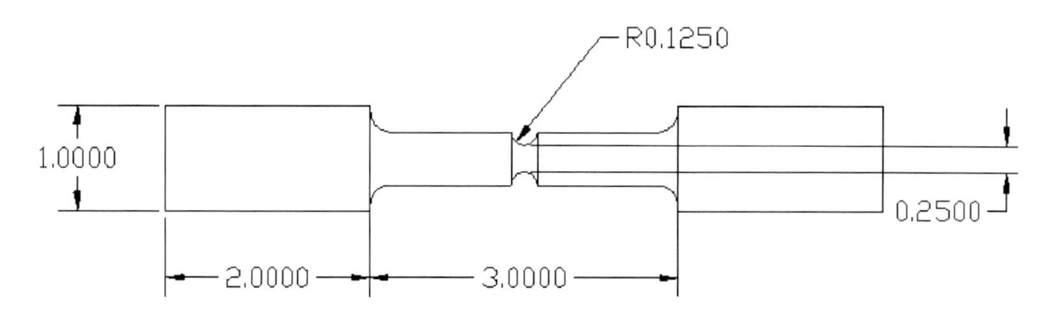

Figure 1. Notched round bar geometry with circumferentially machined notch.

However, ductile fracture initiation in these specimens appears to take place virtually simultaneously over most of the central portion of the bar cross-section area. Ductile fracture initiation in notched bars can be defined to coincide with sudden change in slope in the load versus deflection plot (see Fig. 2). One might question if the ductile crack initiation actually occurs prior to the load drop, but the flatness of the contours of the crack initiation parameters convinces us that a large crack would initiate all at once, causing the load drop. Using finite element simulations, the distributions of effective plastic strain, von Mises (or effective) stress and hydrostatic stress at ductile fracture initiation (i.e., load drop) can be examined to back-calculate the toughness index. Some of sample values for the toughness index (α) are as described in Table 1.

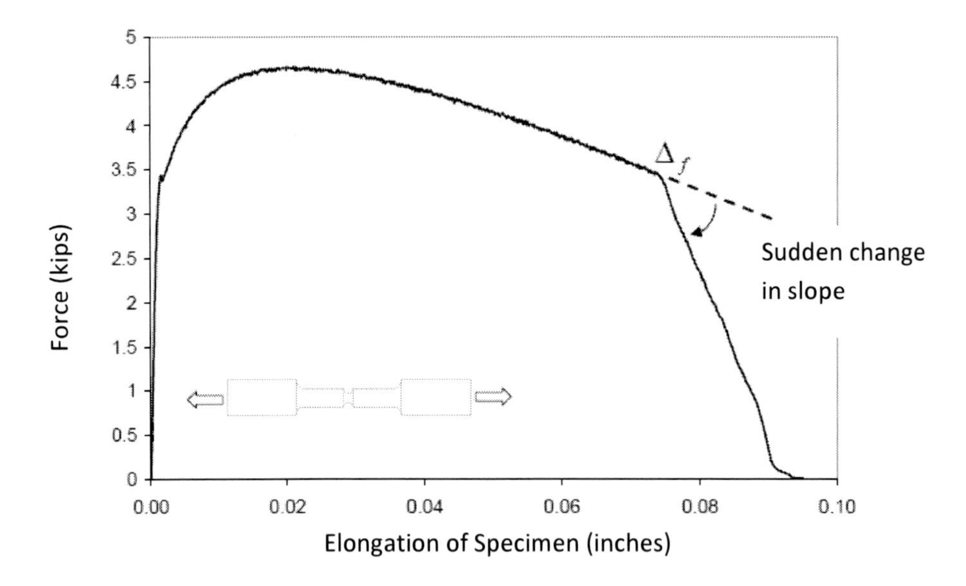

Figure 2. Force elongation curve for notched round bar specimens showing the point of ductile fracture.

The damageability parameter (λ) is calculated using a set of notched bar tests, which are similar to the notched bar tests discussed in previous sub section, except that here they are investigated subjected to cyclic loading. A mix of the different notch sizes as well as loading histories is used for the test program. The aim of varying the notch sizes is to study the effect of stress state or triaxiality on ULCF, whereas the aim of the different load histories is to calibrate and validate the model for general loading histories. Two types of loading histories are better to be used. The first involves cycling the applied displacements between two predetermined levels until failure occurs, typically during the last tensile cycle. The second involves cycling the specimen for a finite number of cycles at lower amplitudes and then applying a tensile excursion until failure. The second kind of loading history is thought to imitate earthquake loadings, where a few smaller cycles damage the component prior to a larger cycle which causes failure.

Table 1. Toughness index summary of few types of steels.

Designation of steel	Toughness index (α)	COV (%)
A572-Grade 50	1.18	15
A514-Grade 110	1.46	20
HPS 70W	2.90	7
JIS-SN490B Grade 50	2.89	9
JIS-SM490YBTMC-5L Grade 50	4.67	1

Generally, the load displacement curves of these specimens show a sudden loss in slope at a certain point during the loading history. This always occurs on a tensile cycle, and as in the case of monotonic loading, this is assumed to be the "point of ductile crack initiation" under cyclic loading. This assumption is reasonable because failure occurs almost simultaneously over the central part of cross section causing a sudden drop in load. The sharply falling load displacement curve beyond this point indicates tearing in the shear lips around the center where ductile crack initiation occurs first. The entire loading history leading up to point of ductile crack initiation is recorded for use in the finite element analyses and hence back calculates the damageability parameter. The sample values for the damageability parameter (λ) are as described in Table 2 for steel.

Table 2. Damageability parameter summary of few types of steels.

Designation of steel	Damageability parameter (λ)
A572-Grade 50	0.49
A514-Grade 110	0.48
HPS 70W	0.43
JIS-SN490B Grade 50	0.85
JIS-SM490YBTMC-5L Grade 50	0.25

To determine the macrocrack initiation stage, it is necessary to introduce a length scale l^* (characteristic length) to apply problems with steep stress and strain gradients. Recall that the notched bars are insensitive to the length scale because of the flat stress-strain gradients involved. This is why the notched bar tests are useful to calibrate the basic parameters for the failure criterion. However, for the same reason, the notched bar results do not provide the basis for determining the characteristic length. Though the notched bars do not provide mechanical response data to determine the characteristic length, they provide samples of fracture surfaces from which microstructural measurements can be made to calculate this length.

The procedure described under this failure criteria, is to have two bounds and a most likely value for the characteristic length. The lower bound would be twice the average dimple diameter. The reason is that it is defined the fracture as the coalescence of two adjacent voids in simple sense. Moreover, twice the dimple size has been shown to work reasonably by previous investigators. This lower bound value could be used as a conservative estimate for the characteristic length.

The upper bound on l^* would be the length of the largest plateau or trough observed in the angled fractograph. This would are the most liberal interpretation of the characteristic length, but would give an idea of the maximum likely toughness that a material would exhibit, and together, the lower bound dimple definition and upper bound plateau/trough should help explain some of the experimental scatter. The mean value of l^* would be arrived at by taking an average over roughly ten measurements of the lengths of the plateaus and troughs. This would be the most likely estimate of the l^* value. In Table 3 values of l^* for the upper bound, mean and the lower bound are tabulated for some types of structural steels.

Table 3. Characteristic length summary of few types of steels.

Designation of steel	Characteristic length (l^*) (inches)		
	Lower bound	Mean value	Upper bound
A572-Grade 50	0.0033	0.007	0.017
A514-Grade 110	0.0023	0.009	0.019
JIS-SN490B Grade 50	0.0028	0.009	0.014
JIS-SM490YBTMC-5L Grade 50	0.0022	0.005	0.009

4. Methodology to Predict Ultra-Low Cycle Fatigue Fracture

This section clearly describes the methodology to predict the ULCF fracture of a real structure using previously discussed simplified ULCF fracture criterion (Eq. (19)). This method consists of four major steps and these are described under separate sub sections given below.

4.1 Critical-Accumulated Equivalent Plastic Strain at Monotonic Failure

It is necessary to initially run a monotonic FE analysis to obtain the critical value (capacity) of accumulated equivalent plastic strain ($(\bar{\varepsilon}_p^{monotonic})_{critical}$) under monotonic loading prior to the cyclic analysis. The critical value is a function of triaxiality (Eq. (17)), which tends to deviate from point to point. Simultaneously, the effective plastic strain also deviates from point to point depending on geometry, loading as well as the boundary condition too. Therefore, it cannot be strictly decided that the highly stressed or highly strained location may be subjected to ductile failure at first. Due to this reason, it is important to consider a zone that has higher probability of fracture. In finding of this zone, one can select areas where effective stresses are present and such a high effective stress area is called a "critical zone". In most of the local stress concentration problems, the critical zone can be represented by a straight line. Having found this zone, the triaxiality versus accumulated equivalent plastic strain should be plotted for each gauss points in the critical zone. There it has to be confirmed that triaxiality should be relatively constant during plastic strain for each gauss points. Unless or otherwise it is not advisable to use this simplified fracture criterion for ULCF assessment. The situation where triaxilaity remains relatively constant, the $(\bar{\varepsilon}_p^{monotonic})_{critical}$ can be estimated for each Gauss points of this critical zone using Eq. (17).

4.2 Critical- Accumulated Equivalent Plastic Strain at Cyclic Failure

Once the critical value (capacity) of accumulated equivalent plastic strain ($(\bar{\varepsilon}_p^{monotonic})_{critical}$) for each sampling locations in critical zone are determined for monotonic loading, the cyclically degraded values of the critical accumulated equivalent plastic strain, $(\bar{\varepsilon}_p^{cyclic})_{critical}$ at that zone is calculated at the beginning of each tensile cycle for each sampling location (interest gauss points) using Eq. (18). For clearness of understanding the behavior, it is better to plot the $(\bar{\varepsilon}_p^{cyclic})_{critical}$ versus locations in particular critical zone for considered loading steps. Just for an example, when the critical zone becomes a nearly straight-line, plot can be easily done $(\bar{\varepsilon}_p^{cyclic})_{critical}$ versus distance along the line.

4.3 Accumulated Equivalent Plastic Strain at Cyclic Loading

The accumulated equivalent plastic strain under cyclic loading ($\bar{\varepsilon}_p^{cyclic}$) at the critical zone can be calculated during the FE analysis simultaneously with section 4.2 using following relations. The T is triaxiality and n is the loading step number,

when $T > 0$; $(\bar{\varepsilon}_p^t)_{(n+1)} = (\bar{\varepsilon}_p^t)_n + (d\varepsilon_p)_n$ (20)

when $T < 0$; $(\bar{\varepsilon}_p^c)_{(n+1)} = (\bar{\varepsilon}_p^c)_n + (d\varepsilon_p)_n$. (21)

Subtracting Eq. (20) from Eq. (21) the applied significant plastic strain at sampling location in critical zone is calculated as,

$$(\bar{\varepsilon}_p^{cyclic})_{n+1} = (\bar{\varepsilon}_p^t)_{n+1} - (\bar{\varepsilon}_p^c)_{n+1}. \quad (22)$$

4.4 The $\bar{\varepsilon}_p^{cyclic} - (\bar{\varepsilon}_p^{cyclic})_{critical}$ Plot and Determination of Crack Initiation

This $\bar{\varepsilon}_p^{cyclic} - (\bar{\varepsilon}_p^{cyclic})_{critical}$ is the difference between accumulated equivalent plastic strain at cyclic loading (demand) and the critical value (capacity) of it. Fracture at a considered location is assumed to have occurred when this quantity is greater than zero. A prediction of ULCF crack initiation is made when this quantity exceeds zero over the characteristic length (l^*) along the critical zone.

To utilize this methodology, it is needed to determine the previously mentioned plastic variables against the multiaxial cyclic loading. Here, it is compulsory to perform a proper elasto-plastic analysis using a proper cyclic hardening model, which is compatible to complex structures with a reasonable accuracy. Therefore, following section outlines the details of a new elasto-plastic hardening model to predict ULCF fracture of real structures.

5. Proposed Hardening Model for Elasto Plastic Analysis

The hardening model described in this section is proposed to predict the real behaviors of non-linear kinematic hardening materials using the fundamentals of plasticity theory such as yield criteria, flow rule, hardening rule etc. The proposed model consists of new hardening rule and it has been suggested by making a little extension of previous Armstrong – Fredrick rule (Jiang et al., 1996). The new term is a function of the stress increment and it is used to define the direction of yield surface. However, authors would like to recommend usage of this hardening model in situations where, i) the nonlinear mix hardening models (combination of isotropic and kinematic hardening models) are not available in used FEM program and ii) difficult to determine the parameters of mix hardening models. Initially section describes the mathematical relation of proposed hardening model. Then the details relevant to finite element method (FEM) formulation are motioned. Methodology to estimate the hardening parameters of particular materials is clearly described under a separate sub section. Finally the verification of the new hardening model is conducted by comparing the predicted behaviour with experimental behaviour of few materials.

5.1 Mathematical Formulation

In plasticity theory (Lubliner 1990), for a material exhibiting kinemetic hardening behavior as shown in Fig. 3, the von Misses yield criteria is defined as,

$$\phi = \left(\sigma'_{ij} - \alpha'_{ij}\right)\left(\sigma'_{ij} - \alpha'_{ij}\right) - \frac{2}{3}\sigma_0^2 = 0 \tag{23}$$

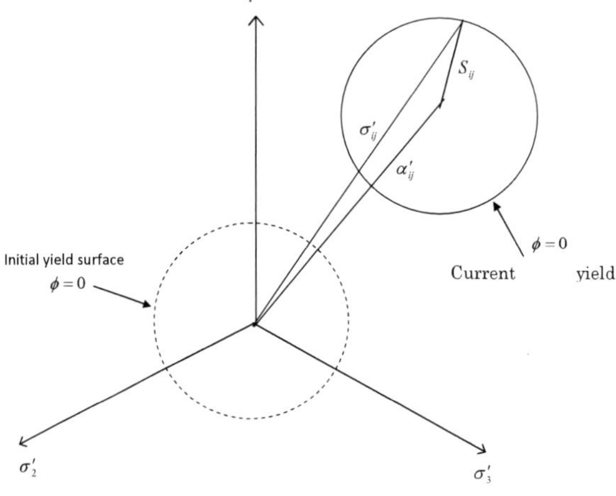

Figure 3. Initial and current loading surfaces in π - Plane.

where ϕ is the yield function, σ'_{ij} is deviatoric stress tensor and α'_{ij} is the deviatoric component of back stress (residual stress) tensor which controls the kinematics hardening behaviour, σ_0 is the initial yield stress in uniaxial tension. The above equation can be rewritten as

$$\phi = S_{ij}S_{ij} - \frac{2}{3}\sigma_0^2 = 0 \tag{24}$$

where $S_{ij} = \sigma'_{ij} - \alpha'_{ij}$.

When the plastic flow is assumed to be associated, the incremental plastic strain ($d\varepsilon_{ij}^p$) is found for an increase of load in the plastic region from the Drucker's normality condition, and it is mathematically described as,

$$d\varepsilon_{ij}^p = d\lambda\left(\frac{\partial\phi}{\partial\sigma'_{ij}}\right) \tag{25}$$

where $d\lambda$ is the plastic multiplier, which is zero in the elastic domain. This relation implies that the plastic flow vector $d\varepsilon_{ij}^p$, if plotted as a free vector in the stress space (Fig. 4), is directed along the normal to the yield surface for a stable plastic material.

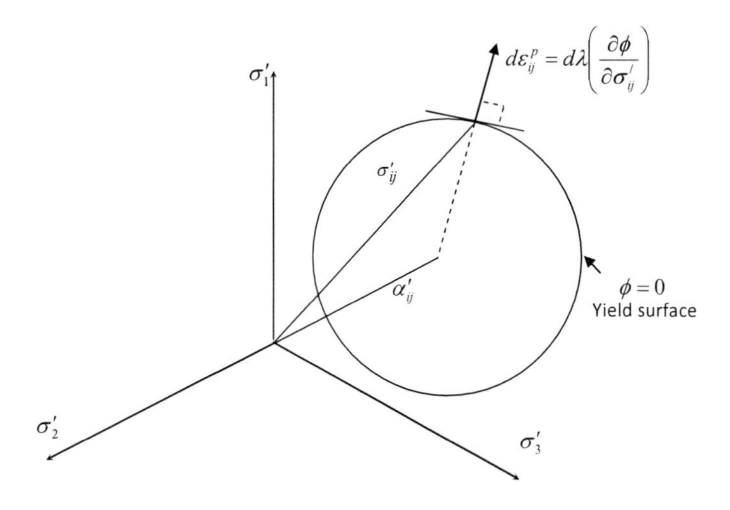

Figure 4. Graphical representation of direction of plastic strain increment vector.

Using Eq. (23), the above flow rule simplifies to,

$$d\varepsilon_{ij}^p = 2d\lambda\left(\sigma'_{ij} - \alpha'_{ij}\right). \tag{26}$$

When material is subjected to the multiaxial strain state, the equivalent uniaxial plastic strain, which is similar to the material uniaxial stress strain curve, is called as effective plastic strain or von Mises equivalent plastic strain. For particular load increment in the plastic region the incremental effective plastic strain is defined as,

$$d\bar{\varepsilon}^{p} = \sqrt{2/3\, d\varepsilon_{ij}^{p} d\varepsilon_{ij}^{p}} \quad (27)$$

Substituting Eq. (26) into (27) effective plastic strain is eliminated as,

$$d\bar{\varepsilon}^{p} = 4/3\, d\lambda \sigma_{0} \quad (28)$$

where the σ_0 is the initial yield stress in uniaxial tension.

After the elastic limit is reached, the state of stress lies on the yield surface. In this model, hardening follows the kinematic hardening behaviour when it is subjected to further continuation of loading. The new hardening rule is proposed here by making a further extension to previous Armstrong-Fredrick rule and it can be written as,

$$d\alpha_{ij} = \frac{2}{3} c d\varepsilon_{ij}^{p} - \gamma \alpha_{ij} d\bar{\varepsilon}^{p} + \beta d\sigma_{ij} \quad (29)$$

where c, γ, β are the plastic constants which can be determined by the uniaxial experimental results of particular material. The $d\alpha_{ij}$ is the incremental back stress tensor which perfectly describes the manner of yield surface translation in stress space.

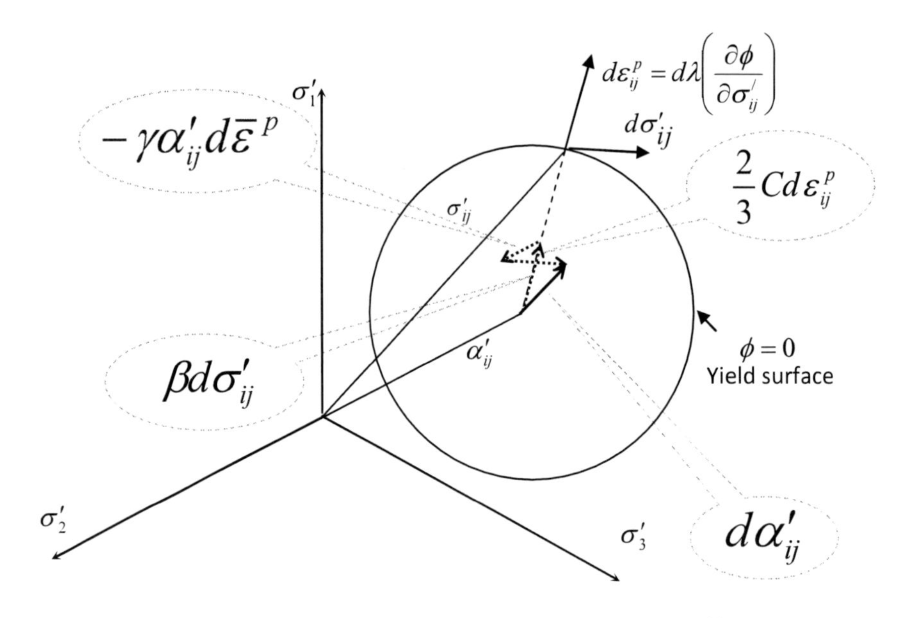

Figure 5. Graphical explanation to the three components of hardening rule in π -Plane.

The first term ${}^{2}\!/_{3}\, cd\varepsilon_{ij}^{p}$, describes the linear behaviour of yield surface translation with the change of plastic strain. The direction of this vector is perpendicular to the yield surface as shown in Fig. 5. The second term $-\gamma\alpha_{ij} d\bar{\varepsilon}^{p}$ called as material dependent dynamic recovery term, consists back stress vector. Since the back stress α_{ij} increases with the increment of plastic loading, the second term $\gamma\alpha_{ij} d\bar{\varepsilon}^{p}$ also changes in both magnitude and direction. As a consequence, the present yield surface translation manner exhibits non-linear behaviour. The direction is shown in Fig. 5. The new term $\beta d\sigma_{ij}$ is a function of the stress increment ($d\sigma_{ij}$) and its direction is same as the direction of stress increment (see Fig. 5). Therefore, the third term is also increasing with the plastic loading and finally it contributes to increase the effect of non-linear behaviour more especially in large strain range. Since this term physically increases the non-linearity with plastic loading as second term, the new term $\beta d\sigma_{ij}$ is also considered as material dependent dynamic recovery term. The graphical representation of the back stress increment vector in π -plane is shown in Fig. 5.

5.2 Elasto-Plastic Tangent Stiffness

Finite element formulation of the elasto-plastic stiffness matrix for incremental loading is described below. The function ϕ is such that the material is elastic for $\phi \leq 0$ and $(\partial\phi/\partial\sigma'_{ij})d\sigma'_{ij} < 0$. In this case material is described by,

$$d\sigma_{ij} = E_{ijkl} d\varepsilon_{kl}^{e} \tag{30}$$

where E_{ijkl} is material elastic stiffness matrix and $d\varepsilon_{kl}^{e}$ is the incremental elastic component of total strain.

If $\phi = 0$ and $(\partial\phi/\partial\sigma'_{ij})d\sigma'_{ij} \geq 0$ then the process is said to be plastic and the incremental stress is decomposed as,

$$d\sigma_{ij} = E_{ijkl}\left(d\varepsilon_{kl} - d\lambda\left(\frac{\partial\phi}{\partial\sigma'_{kl}}\right)\right) \tag{31}$$

For continued yielding beyond the initial yielding, the consistency condition in plasticity defines,

$$d\phi = \frac{\partial\phi}{\partial\sigma_{ij}} d\sigma_{ij} + \frac{\partial\phi}{\partial\alpha_{ij}} d\alpha_{ij} = 0 \tag{32}$$

Implementing the deviatoric components to Eq. (32),

$$d\phi = \frac{\partial \phi}{\partial \sigma'_{ij}} \frac{\partial \sigma'_{ij}}{\partial \sigma_{ij}} d\sigma_{ij} + \frac{\partial \phi}{\partial \alpha'_{ij}} \frac{\partial \alpha'_{ij}}{\partial \alpha_{ij}} d\alpha_{ij} = 0 \tag{33}$$

Then partially differentiating the yield function in Eq. (23)

$$\frac{\partial \phi}{\partial \sigma'_{ij}} = -\frac{\partial \phi}{\partial \alpha'_{ij}} \tag{34}$$

Considering the relation between deviatoric stresses to total stress, the following relation can be obtained,

$$\frac{\partial \sigma'_{ij}}{\partial \sigma_{ij}} = \frac{\partial \alpha'_{ij}}{\partial \alpha_{ij}} = \frac{2}{3} \tag{35}$$

Substituting Eq.(34) & (35) to Eq. (33), it can be decomposed as,

$$\left(\frac{\partial \phi}{\partial \sigma'_{ij}}\right)\left(d\sigma_{ij} - d\alpha_{ij}\right) = 0\,. \tag{36}$$

Substituting Eq. (28) to Eq. (30) the hardening rule can be simplified as,

$$d\alpha_{ij} = \frac{2}{3} c\, d\varepsilon^{p}_{ij} - \frac{4}{3}\gamma\sigma_0 d\lambda \alpha_{ij} + \beta d\sigma_{ij}\,. \tag{37}$$

Hence, by substituting Eq. (28) and (37) to Eq. (36) plastic multiplier has been derived as,

$$d\lambda = \frac{(1-\beta)\left(\frac{\partial \phi}{\partial \sigma'_{ij}}\right) E_{ijkl} d\varepsilon_{kl}}{\frac{2}{3}c\left(\frac{\partial \phi}{\partial \sigma'_{ij}}\right)\left(\frac{\partial \phi}{\partial \sigma'_{kl}}\right) - \frac{4}{3}\gamma\sigma_0\left(\frac{\partial \phi}{\partial \sigma'_{ij}}\right)\alpha_{kl} + (1-\beta)\left(\frac{\partial \phi}{\partial \sigma'_{ij}}\right) E_{ijkl}\left(\frac{\partial \phi}{\partial \sigma'_{kl}}\right)} \tag{38}$$

By substituting $d\lambda$ in Eq. (38) to the Eq. (32) and considering the corresponding plastic modulus based stress strain relation as shown in Eq. (39), the material elasto-plastic tangent stiffness is deduced as shown in Eq. (40).

$$d\sigma_{ij} = D_{ijkl} d\varepsilon_{kl} \tag{39}$$

where the D_{ijkl} is the material elasto-plastic tangent stiffness for particular loading stage.

$$D_{ijkl} = E_{ijkl} - \frac{(1-\beta)E_{ijrs}\left(\frac{\partial\phi}{\partial\sigma'_{rs}}\right)E_{klmn}\left(\frac{\partial\phi}{\partial\sigma'_{mn}}\right)}{\frac{2}{3}c\left(\frac{\partial\phi}{\partial\sigma'_{ij}}\right)\left(\frac{\partial\phi}{\partial\sigma'_{kl}}\right) - \frac{4}{3}\gamma\sigma_0\left(\frac{\partial\phi}{\partial\sigma'_{ij}}\right)\alpha_{kl} + (1-\beta)\left(\frac{\partial\phi}{\partial\sigma'_{ij}}\right)E_{ijkl}\left(\frac{\partial\phi}{\partial\sigma'_{kl}}\right)} \tag{40}$$

Finally the obtained D_{ijkl} is used to formulate global elasto-plastic stiffness matrix for whole structure in FEM program code.

5. 3 Determination of Hardening Constants

Determination of hardening constants associated with any proposed material model is one of the most challenging issues for researchers in order to obtain better representation of their hardening behavior. The identification procedure for the material constants involved in the described back stress evolution equation is based on available experimental results. If limited data is available, c, γ, and β can be based on the stress- strain data obtained from the half cycle of the uniaxial tension or compression experiments. The details of the procedure to determine the material constants are outlined below.

For continued yielding beyond the initial yielding, the consistency condition in plasticity defines as same as Eq. (32),

$$d\phi = \frac{\partial\phi}{\partial\sigma_{ij}}d\sigma_{ij} + \frac{\partial\phi}{\partial\alpha_{ij}}d\alpha_{ij} = 0 \tag{41}$$

Considering the relations between deviatoric stress to total stress, Eq. (41) can be simplified as,

$$\left(\frac{\partial\phi}{\partial\sigma'_{ij}}\right)\left(d\sigma_{ij} - d\alpha_{ij}\right) = 0 \tag{42}$$

From the uniaxial specialization of the Eqs. (41) and (27),

$$d\alpha = d\sigma \text{ and } d\bar{\varepsilon}^p = d\varepsilon^p \tag{43}$$

where the $\sigma, \alpha, \varepsilon^p$ are the corresponding uniaxial component of stress, back stress and plastic strain respectively.

Utilizing Eq. (43), Eq. (29) is rewritten as to obtain the view of uniaxial behavior,

$$d\alpha = \frac{2}{3} c d\varepsilon^{p} - \gamma \alpha d\varepsilon^{p} + \beta d\sigma . \tag{44}$$

Integrating the above expression over a half cycle of the stress strain data yields the following expression,

$$\alpha = \frac{2c}{3\gamma} + (\alpha_0 - \frac{2c}{3\gamma}) \exp\left[\frac{-\gamma}{1-\beta} (\varepsilon^{p} - \varepsilon_0^{p}) \right] \tag{45}$$

and the $(\varepsilon_0^p, \alpha_0)$ results from the previous flow.

Using the finite set of points in the uniaxial back stress plastic strain curve (Fig. 6), one can approximate the curve in the form of Eq. (45).

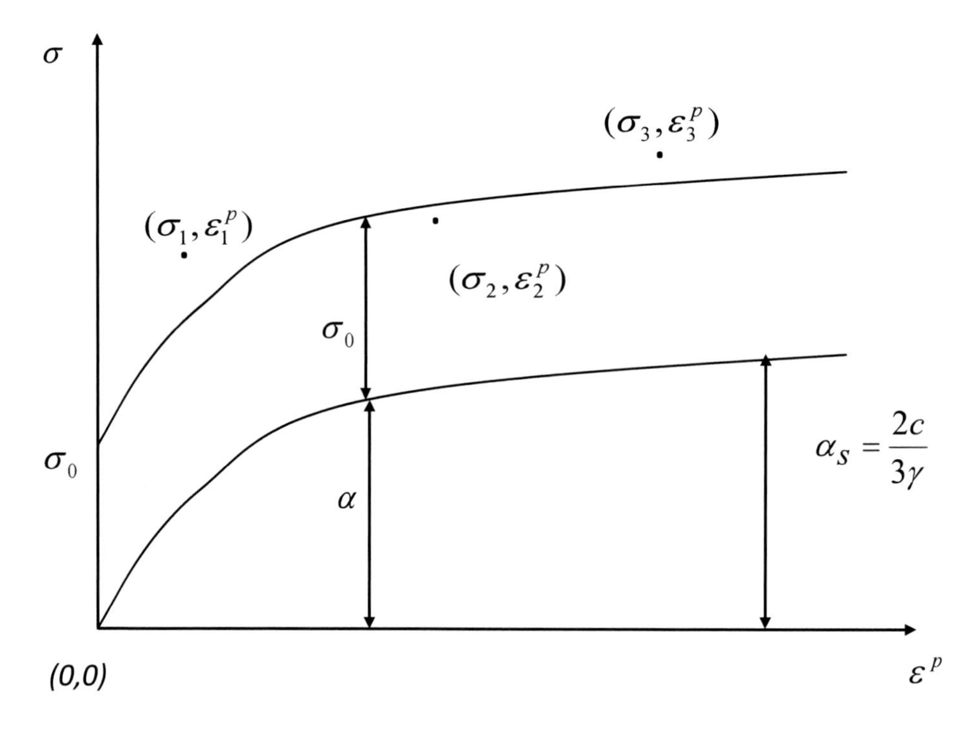

Figure 6. Half cycle of uniaxial stress strain data representing the non-linear hardening.

Close to the saturation point of hardening process, the hardening increments tends to zero and hence the saturation back stress,

$$\alpha_S = \frac{2c}{3\gamma} \tag{46}$$

hence, Eq. (45) is simplified as,

$$\ln\left[\frac{\alpha_s - \alpha_0}{\alpha_s - \alpha}\right] = \frac{\gamma}{1-\beta}(\varepsilon^p - \varepsilon_0^p). \tag{47}$$

The Eq. (47) is of the form of y=*ax* with,

$$y = \ln\left[\frac{\alpha_s - \alpha_o}{\alpha_s - \alpha}\right],\ a = \frac{\gamma}{1-\beta},\ \text{and}\ x = \varepsilon^p - \varepsilon_o^p \tag{48}$$

which is the equation of straight line. The application of least squares fit of the transformed variables in the forgoing form the c and γ can be obtained as,

$$c = \frac{3}{2}\alpha_s a(1-\beta)\ \text{and}\ \gamma = a(1-\beta). \tag{49}$$

Then the value of the β is determined by minimizing the squares of errors as Eq. (50).

$$e^2 = \sum_{i=1}^{n}(\overline{\alpha} - \alpha)^2 \tag{50}$$

where, e is the error, $\overline{\alpha}$ is back stress value from the actual data at the n data points and α is back stress value illustrates as a function of β. The value of the β is determined when the error e, is the smallest.

Finally the obtained constants are subjected to trial and error investigation with the experimental data until its provide best fit to the real behavior of the material and hence determine the final c, γ and β values for the corresponding material.

The behavior of the proposed hardening constants is analyzed against different types of situations. The c, γ and β are evaluated and discussed independently. Their individual importance and contribution to the model is highlighted and illustration is presented in the form of graphs.

5. 4 Verification of Proposed Hardening Model

In favor of verification, the models predicted behavior is compared with previously proposed models and the experimental behavior of two materials 316L stainless steel (Rashid 2004) and wrought iron (Ranaweera et al., 2002) under both uniaxial monotonic and cyclic loading cases. The modified in-house FEM code is used to simulate the uniaxial tensile loading and the mean stress zero cyclic loading behaviors as shown in Fig. 7 to 10 respectively.

Comparisons of uniaxial monotonic loading results reveal that the model prediction has better agreement with real behavior of the material than previous models and it exhibits more non-linearity than others. When the material is subjected hardening in large strain ranges (Fig. 8) the correlation between experiment and both proposed models are well confirmed.

Fig. 9 & 10 illustrate that the model correlate reasonably with the experimental behaviour of the material under reverse and cyclic loading. The deviation of the models from the experimental behaviour are mainly due to the small amount of isotropic hardening shown by the wrought iron material which is not captured in proposed models.

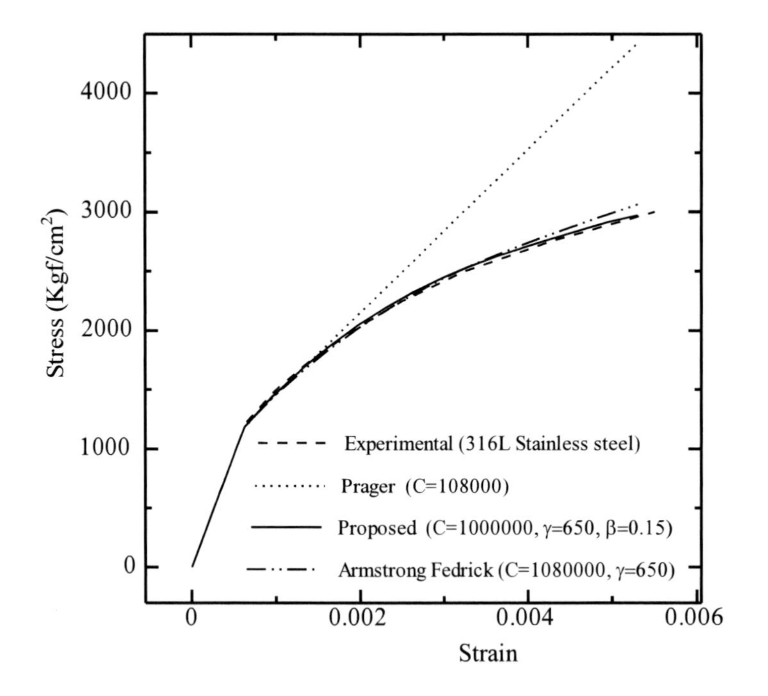

Figure 7. Comparison of uniaxial tensile loading behavior of stainless steel 316 L.

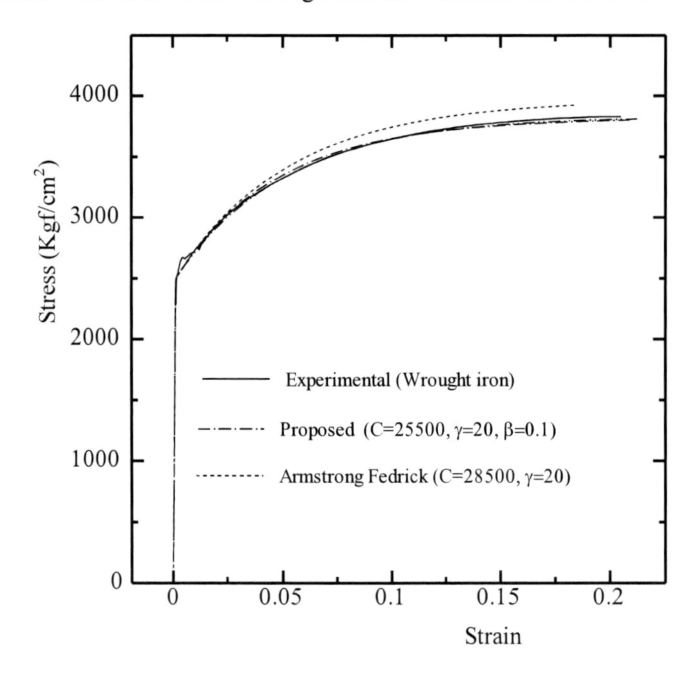

Figure 8. Comparison of uniaxial tensile loading behavior of wrought iron material.

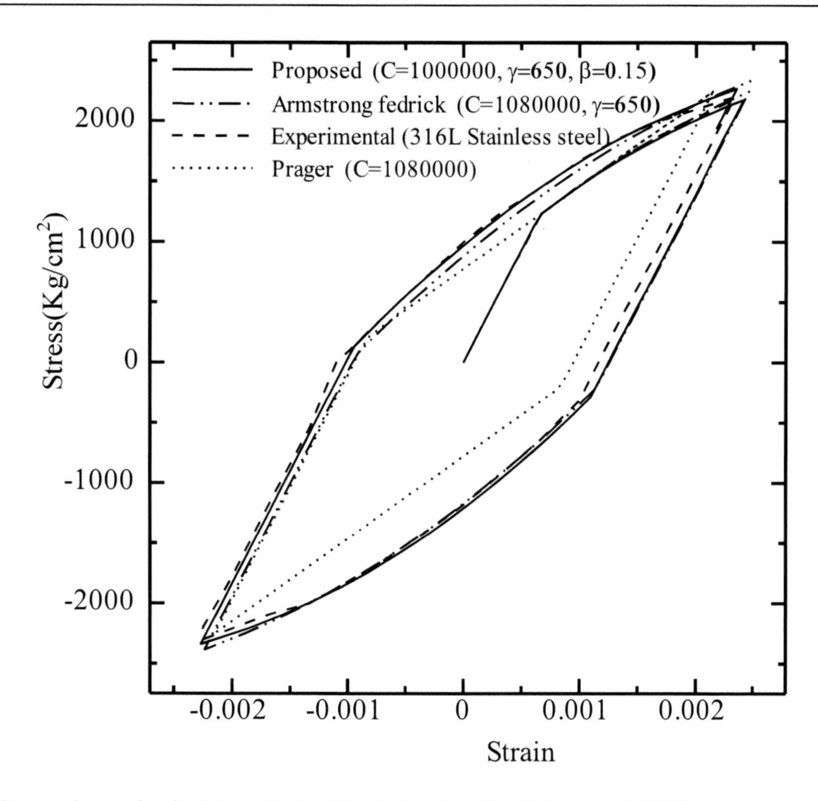

Figure 9. Comparison of uniaxial cyclic loading behavior of stainless steel 316L.

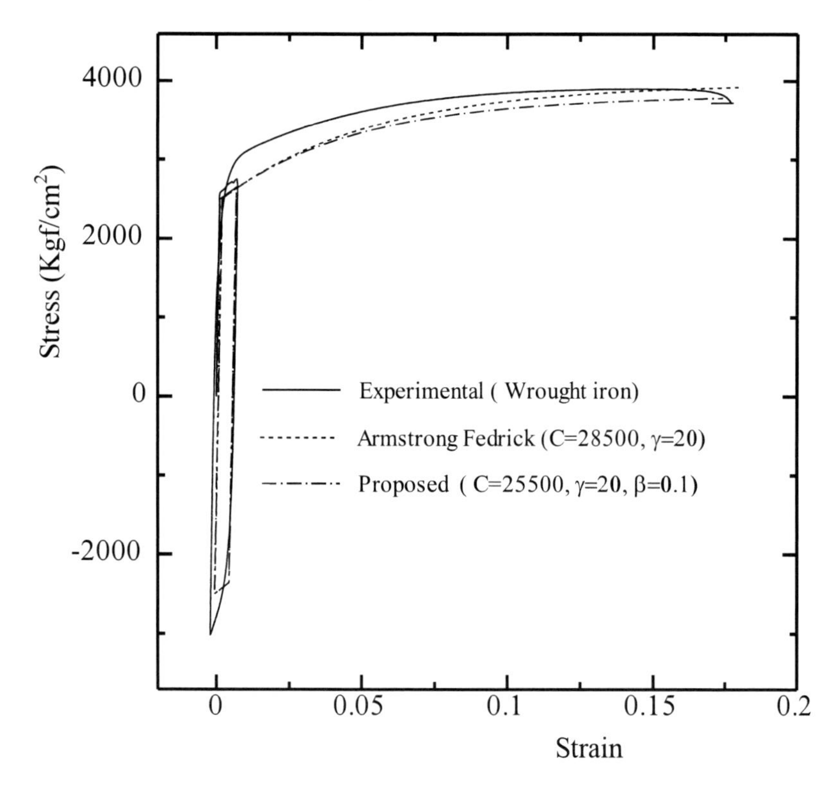

Figure 10. Comparison of uniaxial cyclic loading behavior of wrought iron material.

5. 5 Significance Analysis of Proposed Hardening Model

To exhibit the significance and the applicability of the proposed models in the usage of real elasto-plastic applications, two numerical examples are presented as the end of chapter.

Example 1. Two-Dimensional Plate Subjected to Ratcheting Load

The Fig. 11 illustrates the stress-strain curves of the stainless steel 316L plate under uniaxial stress cycling with non-zero mean stress (ratcheting load) with a cycle number 6. The geometry of the problem and applied displacement time history is clearly indicated in Fig. 12. This analysis is subjected to quasi-static sense of loading. The stress strain comparison (Fig. 11) does not produce considerable difference between proposed model and Armstrong Fedrick model. In advance, Fig. 13 exhibits the corresponding behaviour of effective stress versus effective plastic strain for same ratcheting operation and it reveals that there is a significant difference among the Armstrong Fedrick models and proposed model. The obtained effective plastic strains are tabulated in Table 4 and it shows 6% increments of effective plastic strain in proposed model compared to previous model. Since the effective plastic strain is the internal state variable, which governs the failure (ductile or ULCF), it can be concluded the Armstong-Ferdrick model based elasto-plastic analysis overestimate the failure life than proposed model. As a result of that significance of proposed hardening model could be clearly confirmed in the case of failure analysis.

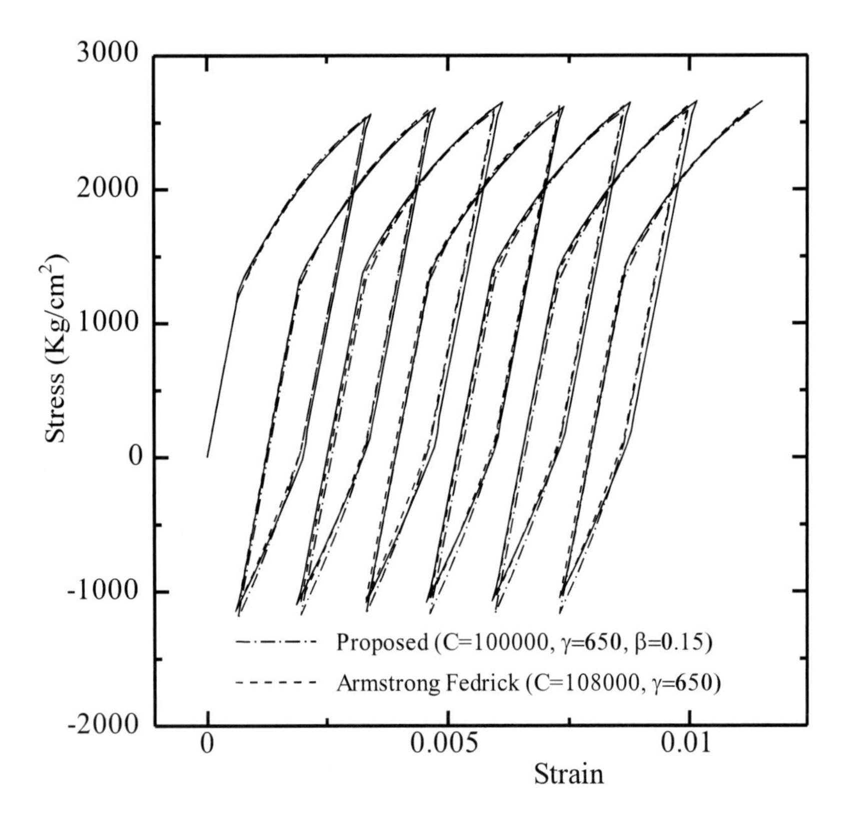

Figure 11. Comparison of uniaxial ratcheting behavior of stainless steel 316 L.

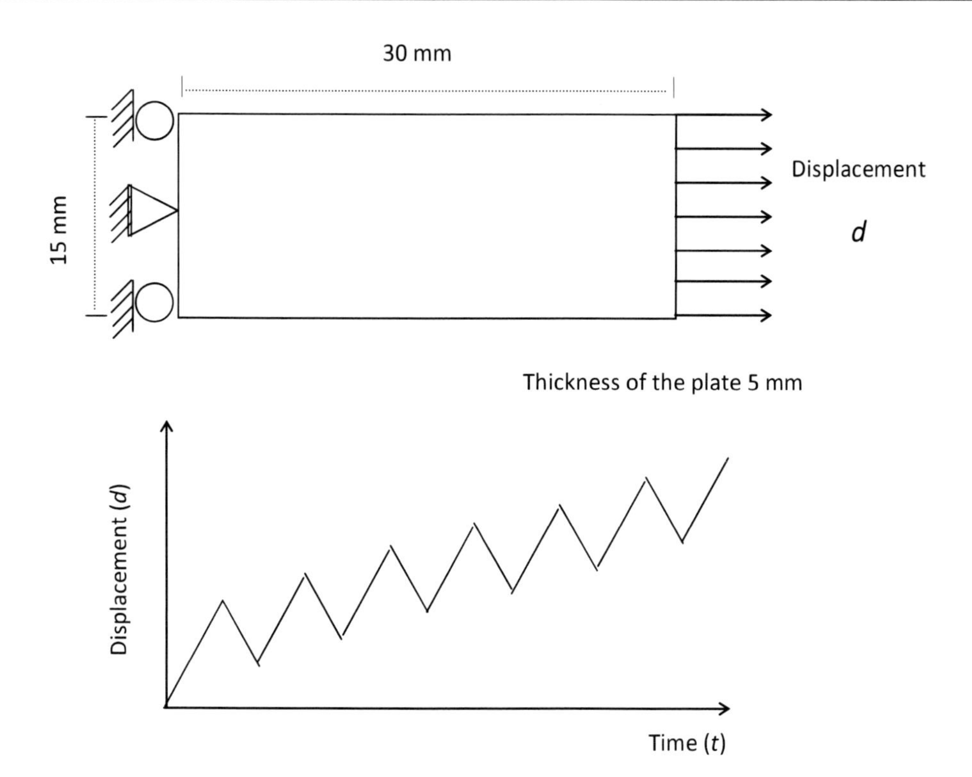

Figure 12. Geometry of the plate and the applied displacement time pattern for ratcheting load.

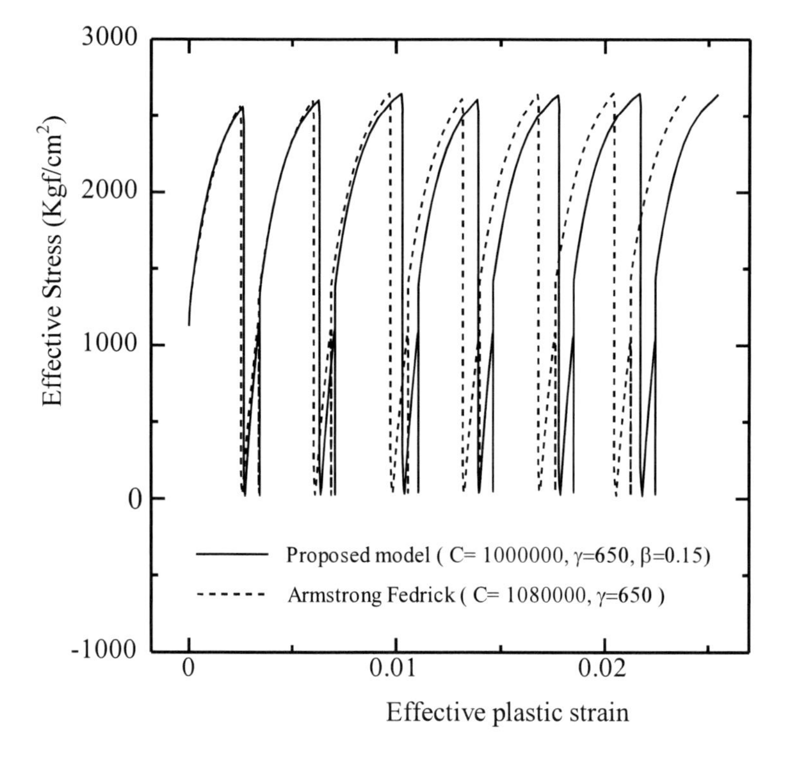

Figure 13. Comparison of effective stress vs effective strain behavior for ratcheting load of stainless steel 316.

Table 4. Comparison of accumulated effective plastic strain at the end.

Hardening model	Effective plastic strain $\bar{\varepsilon}^{p} * 10^{-2}$	
	Example 1	Example 2
Armstrong Fredrick Model	2.395	3.320
Proposed model	2.548	3.670

Since above all application limited to the uniaxial stress states with uniform elastic to plastic strain transmission, the comparison with multiaxial elasto-plastic application is presented as the next step.

Example 2. Riveted Plate Subjected to Uniform Cyclic Load

Stainless steel 316L riveted plate member which is subjected to constant amplitude uniform cyclic load in the range of +9 Tons (Fig. 14), applied to the far end from the rivets, is considered elasto-plastic analysis. The geometry consists of 120*10 mm plate cross section with two symmetrically located rivet holes apart from 60 mm with 20 mm diameter as shown in Fig. 14. By considering the symmetricity of the geometry, loading as well as boundary condition of the member, half of the member with one rivet hole was subjected to analysis using the modified in-house FEM code considering quasi-static condition of loading. The nine node isoperimetric shell element were used for the FE mesh as shown in Fig. 15. To simulate the unilateral contacts between rivet and plate, fully bonding restraint condition was employed.

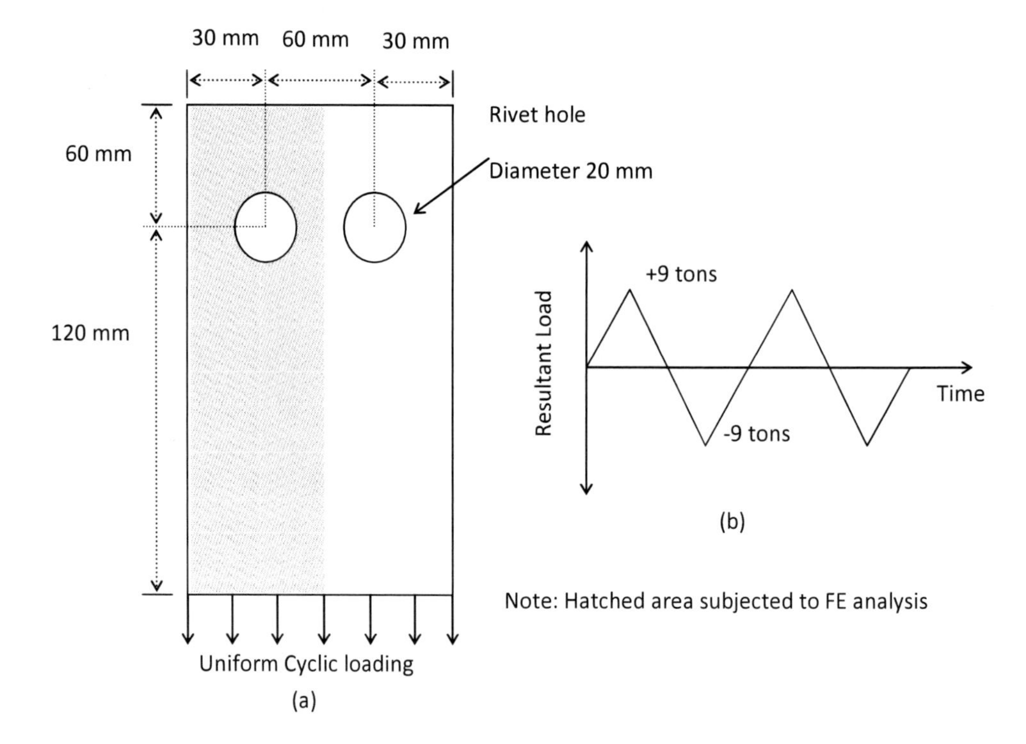

Figure 14. Stainless steel 316 L riveted plate member (a). Geometric details (b). Load time history.

The relevant von Misses stress contour which assists to identify the highly stressed location of the member for plastic loading is as illustrated in Fig. 15. The obtained effective stress versus effective plastic strain variation for single complete cycle at critically stressed location is plotted in Fig. 16. The corresponding results reveal that there is a major difference between these two models in the effective plastic strain and obtained results are mentioned in the Table 4. This shows that the 10% deviation of effective plastic strain and it is also extended to express that Armstrong Fredrick prediction overestimate the failure of real structures when compared to proposed model. Further it is emphasized that the significance of the proper hardening model in the vicinity of ULCF failure analysis.

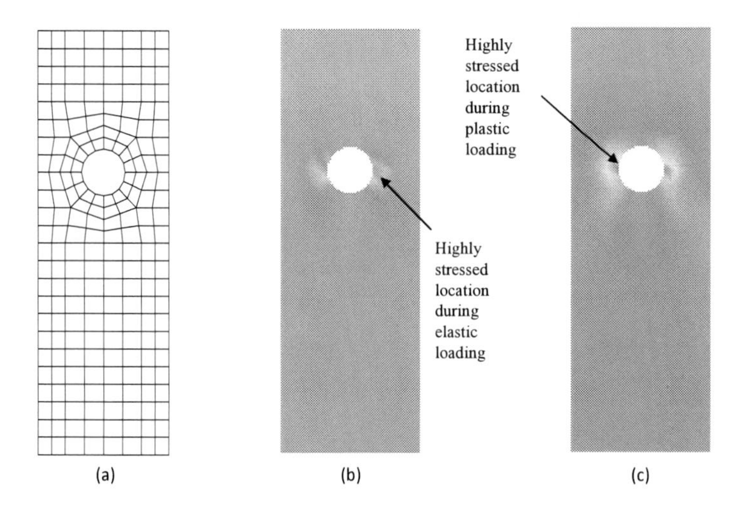

Figure 15. FE analysis results (a) FEM mesh (b) von Misses stress distribution at elastic loading (c) von Misses stress distribution at plastic loading.

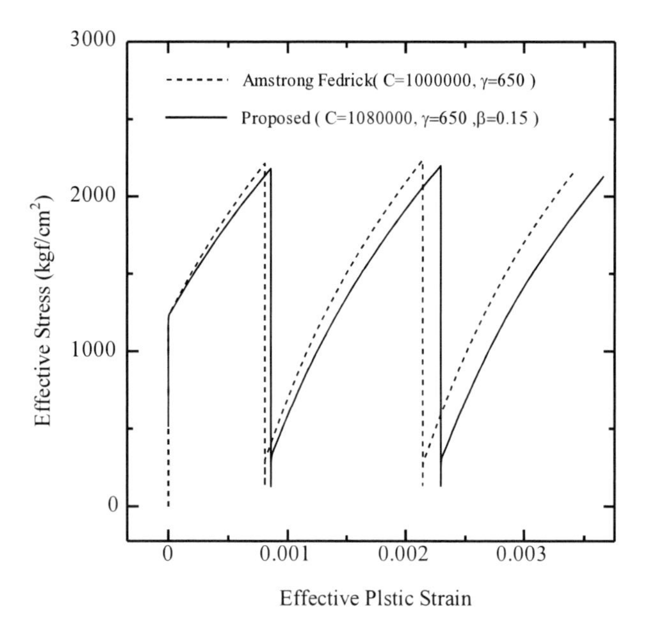

Figure 16. Comparison of effective stress versus effective plastic strain behavior with previous models at highly stressed location.

The mentioned two numerical examples reveal that contribution of proposed models provides more optimized prediction to the real problems of failures such as ductile, ULCF and etc. Further it confirms the significance of the proposed hardening model in cyclic elasto-plastic analysis especially for complex structures.

6. VERIFICATION OF SIMPLIFIED ULTRA-LOW CYCLE FATIGUE MODEL

Simplified ULCF fracture criterion is verified by comparing the simplified criterion predicted fracture displacements with CVGM (Kanvinde et al., 2007) predicted results of three different structural models. In this comparison, proposed hardening model is utilized with both fracture criteria (CVGM and Simplified criterion). Similarly Amstrong Fedrick hardening model is also utilized with both fracture criteria (CVGM and Simplified criterion). Addition to these predictions, the ULCF fracture displacements are again obtained for all these models based on mix hardening behavior (Lemaitre and Chaboche 1990) when simplified criterion governs the fracture. The A572-grade 50 steel is considered as the constructed material of all four models. The ULCF material constants, toughness index (α), damageability coefficient (λ) and characteristic length (l^*) are taken from the Tables 1 to 3 as 1.18, 0.49 and 0.18 mm respectively and considered material exhibits nearly non-linear kinematic hardening behavior.

6.1 Fracture Prediction of a Plate with a Opening (Model 1)

The geometry of the considered structural model is shown in Fig 17. Considering symmetry of the geometry, loading and boundary conditions, the one-fourth of the geometry was subjected to FE analysis. The nine-node isoperimetric shell element was used for FE mesh as shown in Fig 19. Following the procedure in section 4, initially considered geometry is subjected to the monotonic load analysis. By observing the stress distribution at ductile failure (stress contour is shown in Fig. 20) it is able to conclude that the critical zone lies along the transverse centerline of the specimen as shown in Fig. 20. From the monotonic load analysis, variation of triaxilality (*T*) versus effective plastic strain was plotted for sampling Gauss points at critical zone (Fig. 21). These variations reveal that the triaxiality *T* does not illustrate significant variation with increment of plastic loading and the average value was considered for future calculations. Hence, the critical values of accumulated equivalent plastic strain ($(\bar{\varepsilon}_p^{monotonic})_{critical}$) at monotonic loading are calculated for each sampling Gauss points along the transverse centerline of the model. Then FE elasto-plastic analysis was conducted for cyclic loading based on proposed hardening model. The applied displacement history is indicated in Fig. 18. Finally, cyclically degraded values of the critical accumulated equivalent plastic strain, $(\bar{\varepsilon}_p^{cyclic})_{critical}$ variations are plotted at different loading stages as shown in Fig. 22. Simultaneously, the demands of accumulated equivalent plastic strain are also calculated for sampling Gauss points. Hence $\bar{\varepsilon}_p^{cyclic} - (\bar{\varepsilon}_p^{cyclic})_{critical}$ variations along the transverse centerline were determined as shown in Fig. 23. Finally, ULCF macro crack initiation was made when $\bar{\varepsilon}_p^{cyclic} - (\bar{\varepsilon}_p^{cyclic})_{critical}$ exceeds zero over the characteristic length at loading stage seven. The applied displacement corresponding to this loading stage is recorded as the failure displacement at column 2 of Table 5. The model was subjected to ULCF prediction more four times based on CVGM criterion with proposed hardening model, simplified criterion with Armstrong Fedrick model, CVGM criterion with Armstrong Fedrick model and simplified criterion with mix hardening model separately. The obtained fracture displacements are also recorded in the columns 3, 4, 5 and 6 of Table 5.

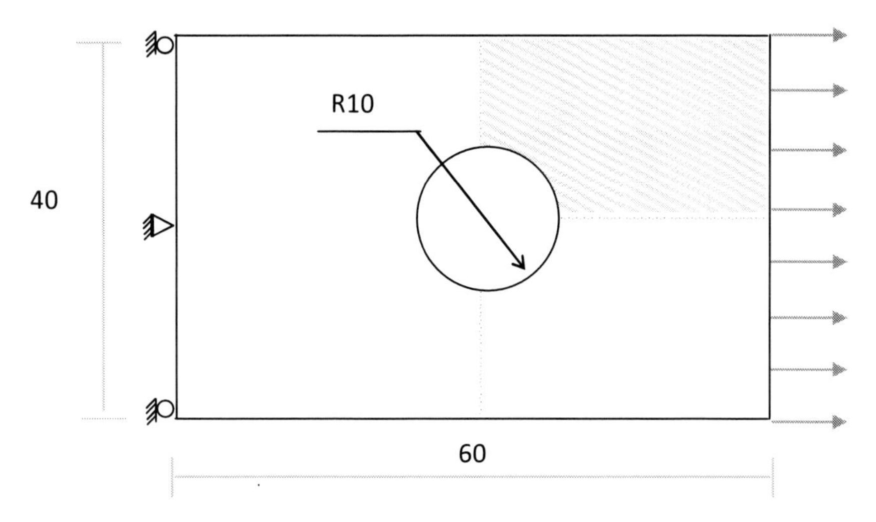

Figure 17. Geometrical details of Model 1.

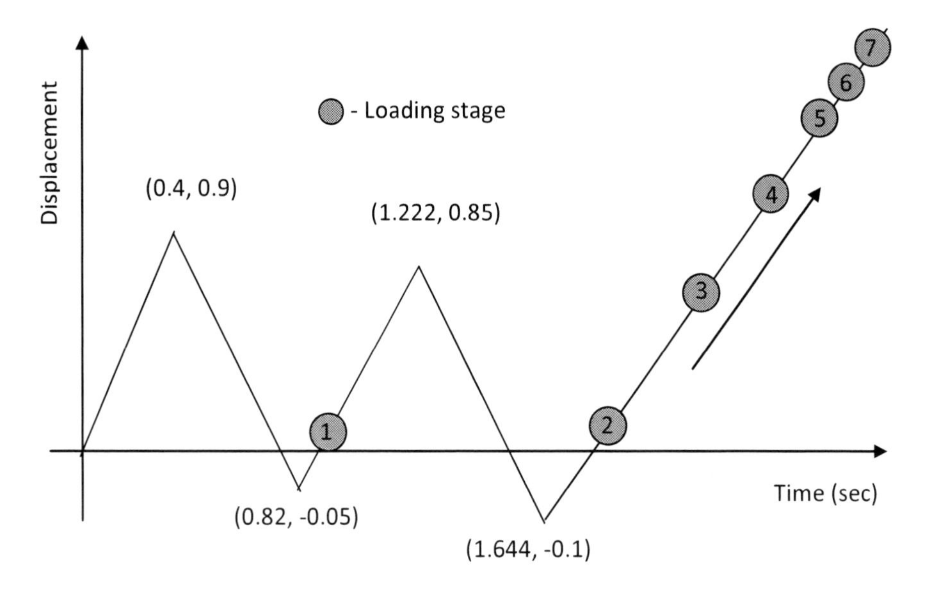

Figure 18. Loading history used for Model 1.

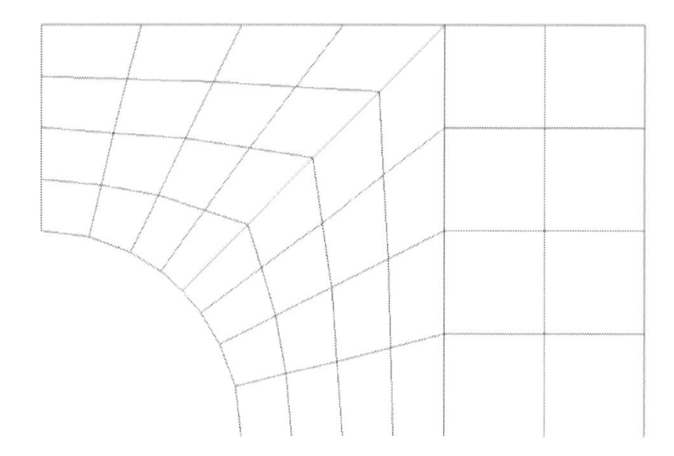

Figure 19. FEM mesh for Model1.

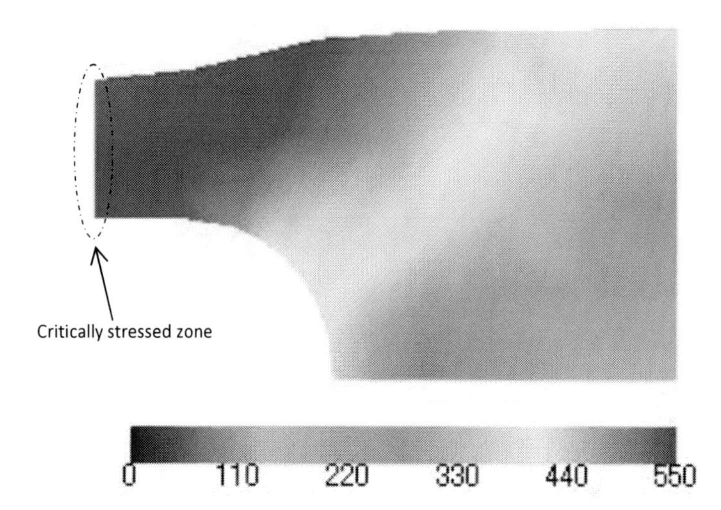

Figure 20. von Mises stress contour of Model 1 at ULCF fracture.

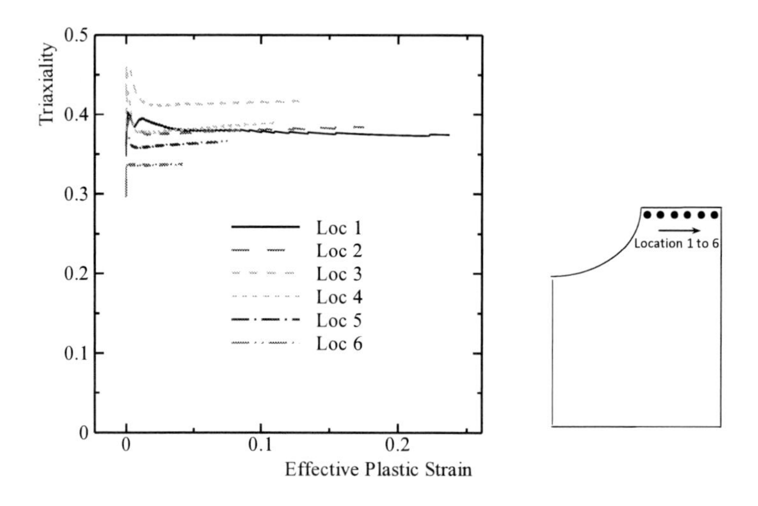

Figure 21. Triaxiality variation versus plastic strain at different location for Model 1.

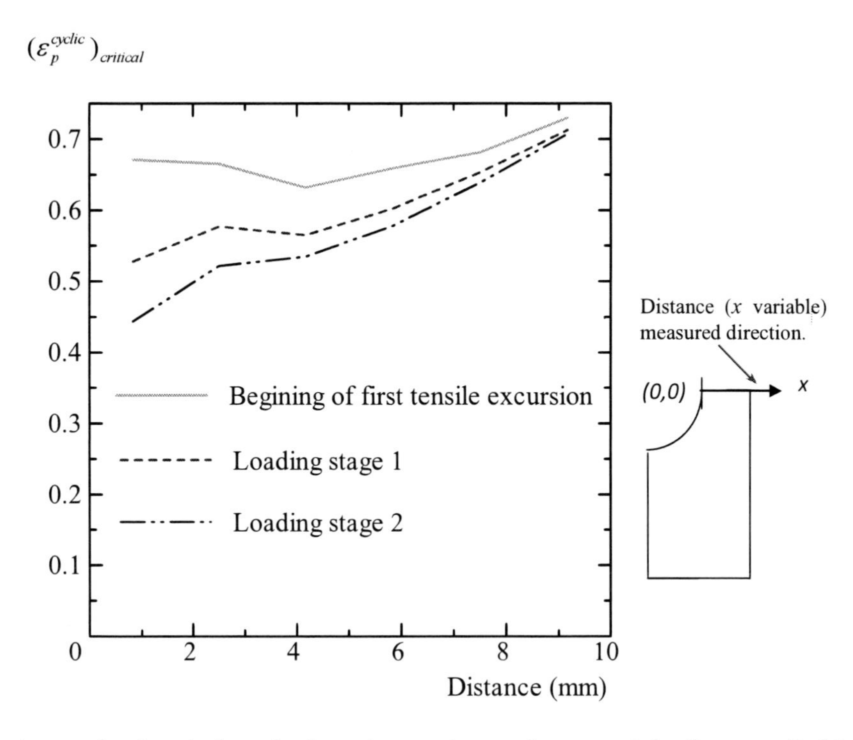

Figure 22. Accumulated equivalent plastic strain capacity envelope at each loading stage for Model 1.

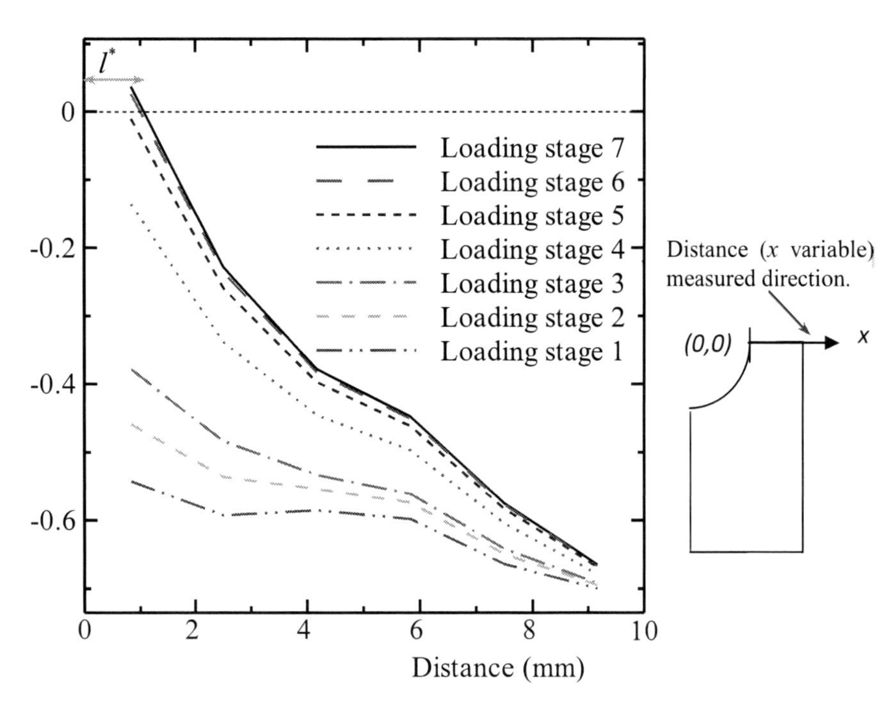

Figure 23. The $\bar{\varepsilon}_p^{cyclic} - (\bar{\varepsilon}_p^{cyclic})_{Critical}$ variation along the length of critical zone for Model 1.

Table 5. Comparison of ULCF fracture displacements of structural models.

Structural Model	ULCF fracture displacement (mm)				
	Proposed hardening model		Armstrong Fredrick model		Mix hardening model
	Simplified criterion	CVGM criterion	Simplified criterion	CVGM criterion	Simplified criterion
1	2.35	2.43 (3.4%)	2.89 (23.0%)	2.97(26.4%)	2.19 (6.8%)
2	213.30	190.20 (10.8%)	232.40 (8.9%)	267.30 (25.3%)	201.70 (5.4%)
3	62.05	65.62 (5.8%)	72.3 (16.5%)	75.3 (21.4%)	58.35 (6.0%)

6.2 Fracture Prediction of a Hollow Cylindrical Pier (Model 2)

The corresponding geometric details are shown in Fig. 24. The horizontally applied uni-directional ground displacement is considered for this model as shown in Fig. 25. The whole geometry was subjected to FE analysis using nine-node isoperimetric shell element (Fig. 26 & 27). The followed procedures for ULCF fracture prediction are similar to the case of Model 1. The corresponding illustrations are shown in Fig. 28 to 30. The displacements at fracture are recorded in Table 5.

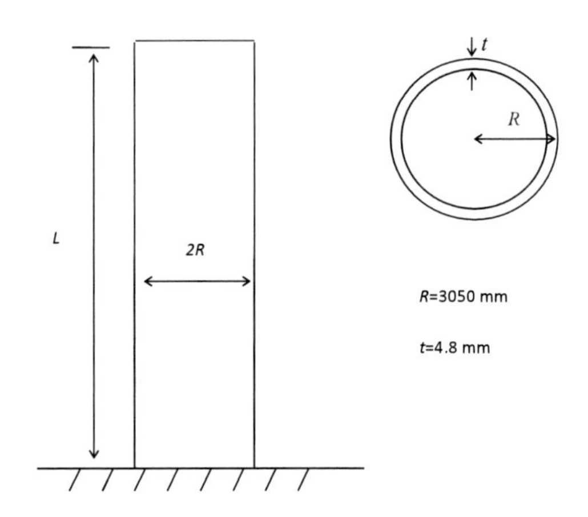

Figure 24. Geometrical details of Model 2.

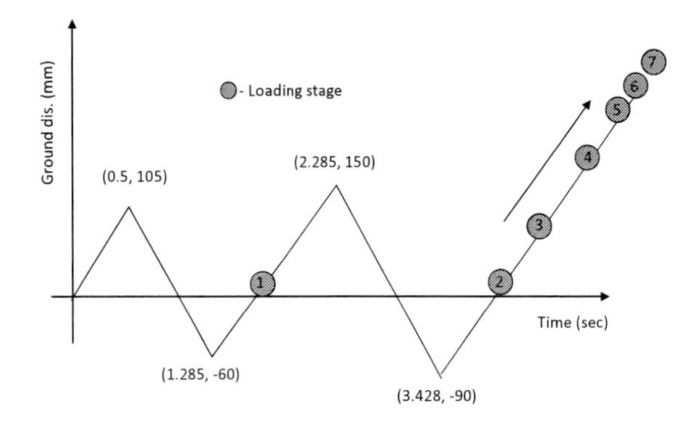

Figure 25. Loading history used for Model 2.

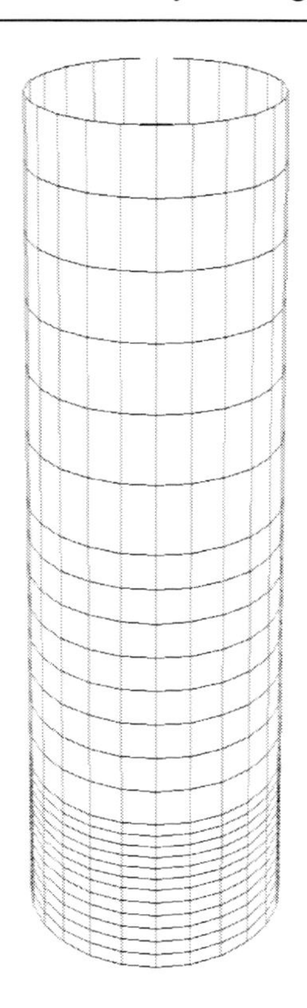

Figure 26. FEM mesh of Model 2.

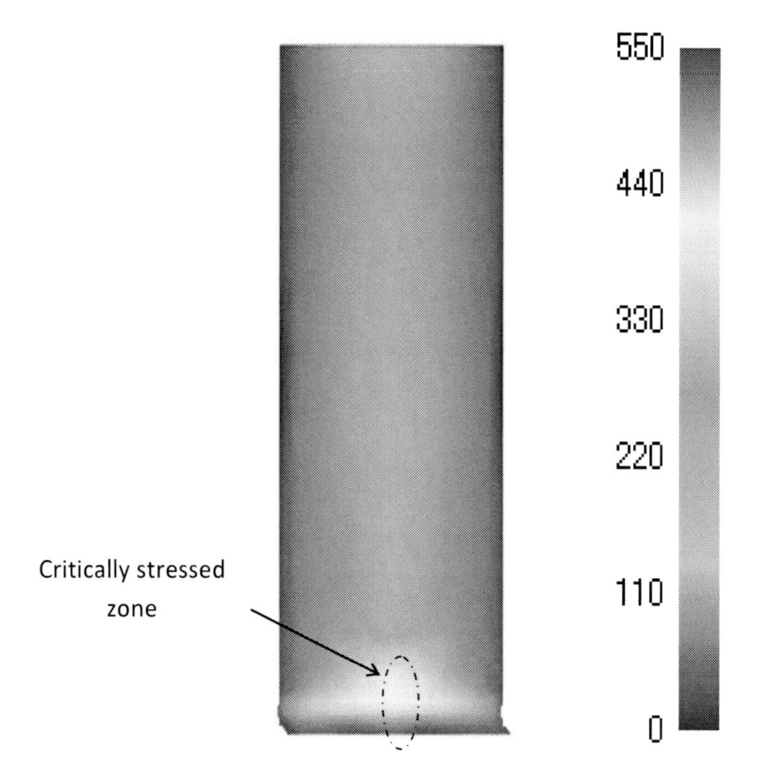

Figure 27. von Mises stress contour of Model 2 at ULCF fracture.

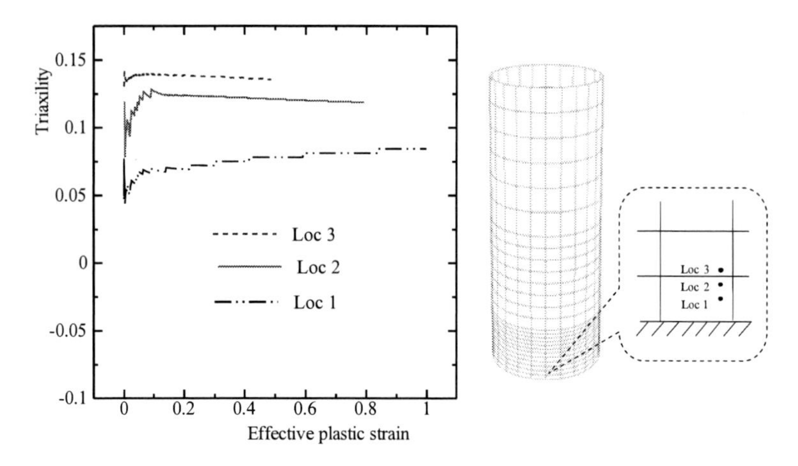

Figure 28. Triaxiality variation versus plastic strain at different location for Model 2.

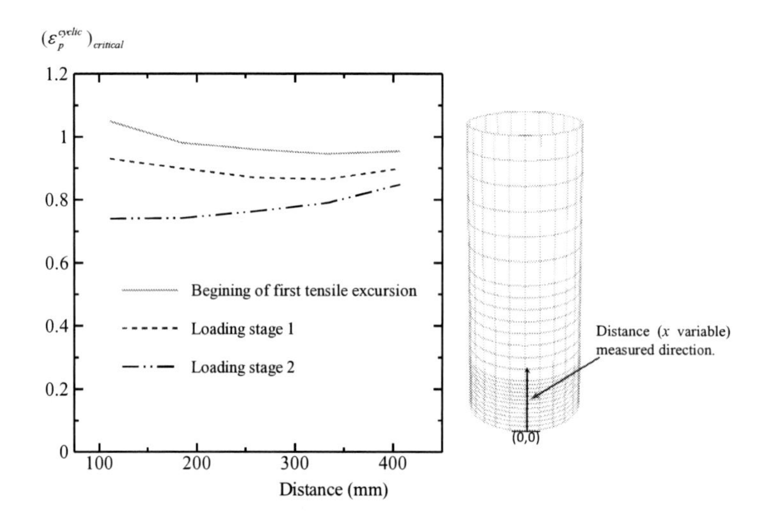

Figure 29. Accumulated equivalent plastic strain capacity envelope at each loading stage for Model 2.

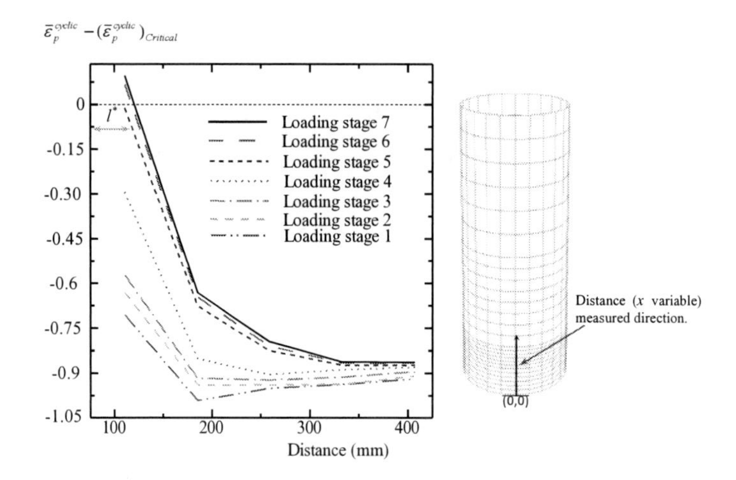

Figure 30. The $\bar{\varepsilon}_p^{cyclic} - (\bar{\varepsilon}_p^{cyclic})_{Critical}$ variation along the length of critical zone for Model 2.

6.3 Fracture Prediction of a Hollow Squared Pier (Model 3)

The geometric details are shown in Fig. 31. In this case also the horizontally applied uni-directional ground displacement is considered for this model as shown in Fig. 32. The FE analysis was conducted using nine-node isoperimetric shell element (Fig. 33 & 34). The followed procedures for ULCF prediction are as same as Model 1. The corresponding illustrations are shown in Fig. 35 to 37. Table 5 shows the applied displacements of ULCF fracture.

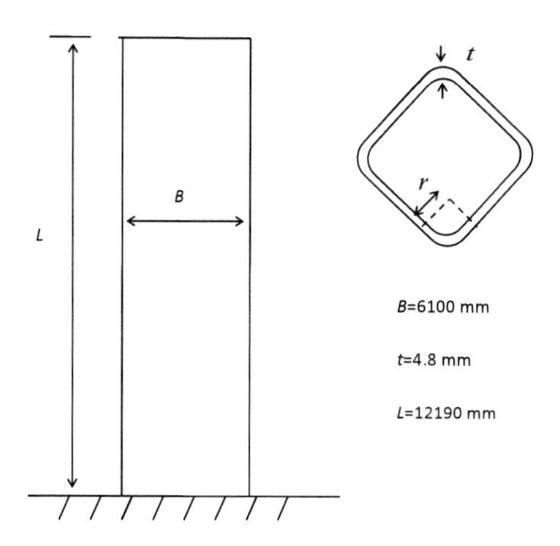

Figure 31. Geometrical details of Model 2.

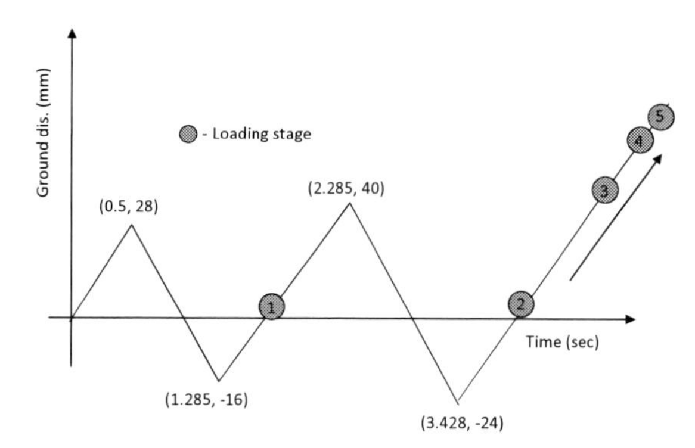

Figure 32. Loading history used for Model 2.

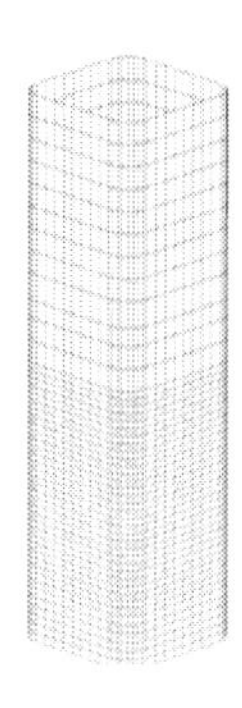

Figure 33. FEM mesh of Model 3.

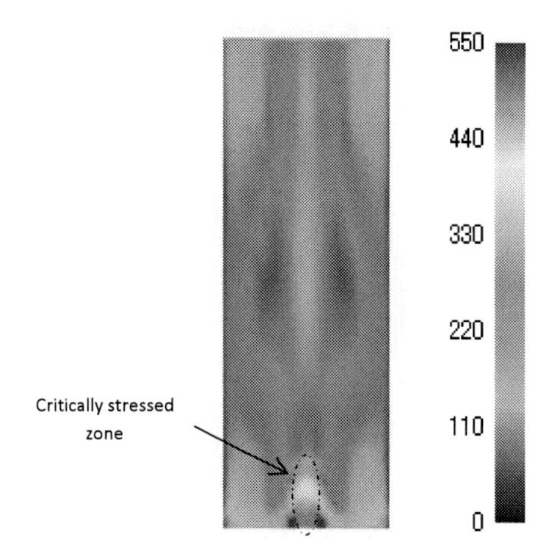

Figure 34. von Mises stress contour of Model 3 at ULCF fracture.

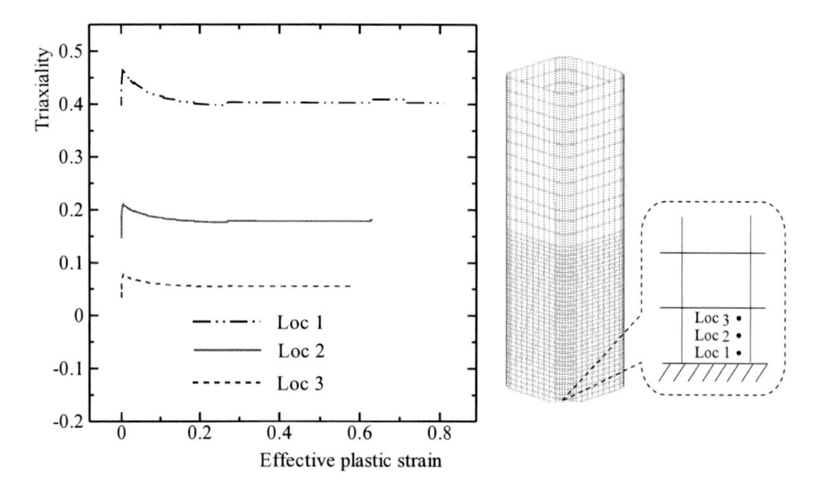

Figure 35. Triaxiality variation versus plastic strain at different location for Model 3.

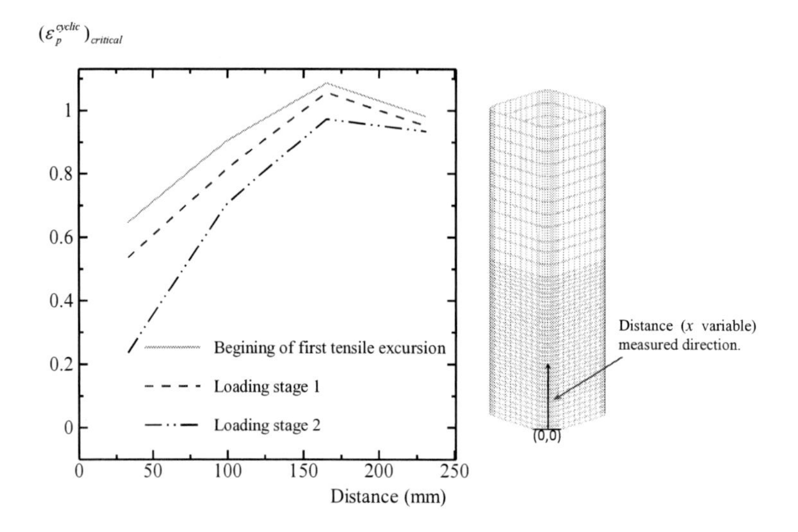

Figure 36. Accumulated equivalent plastic strain capacity envelope at each loading stage for Model 3.

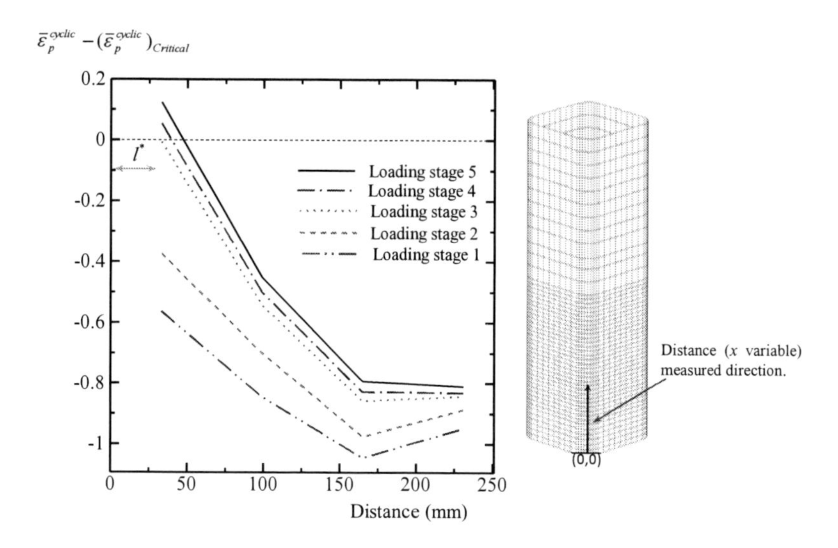

Figure 37. The $\bar{\varepsilon}_p^{cyclic} - (\bar{\varepsilon}_p^{cyclic})_{Critical}$ variation along the length of critical zone for Model 3.

Comparison of fracture displacements of both criterions (columns 2, 3 and 4, 5 separately) in Table 5 shows that there is a similarity between fracture criterions. However, the percentage wise difference of fracture displacement of Model 3 is slightly higher than other models. Assumed reason is that, even though the Triaxility (T) of the Model 3 is relatively constant, it shows slight rate of increment during plastic loading (Fig. 28). However, this comparison reveals that simplified ULCF fracture criterion produces reasonable accurate predictions to structures where the Triaxility is relatively constant.

The fracture displacements corresponding to the both hardening models (column 2 and 6) in Table 5 exhibit that there is no significant change of results depending on hardening model for considered material. But the results of Amstrong-Fedrick model deviates from the other two hardening models (proposed and mix hardening models). Therefore it can emphasize that proposed hardening model is applicable for ULCF failure estimation of steel, which exhibits closer relation to non-linear kinematic hardening behavior.

Conclusion

The general CVGM (cyclic void growth model) based ultra-low cycle fatigue (ULCF) fracture criterion was simplified for steel structures where magnitude of the triaxiality remains relatively constant during their loading history. A simple non-linear hardening model with less number of hardening parameters was proposed and corresponding finite element method (FEM) formulation was also clearly indicated. Hardening model predicted behaviors of some materials were compared with experimental results to check the validity of the proposed hardening model. Then the ULCF fractures of few bridge components were separately predicted by both simplified and CVGM fracture criterion and results were compared. Finally the applicability and validity of the proposed models are confirmed.

The comparisons of ULCF fracture initiation life of simplified model (section 3) with previous CVGM model of some structural components reveal that proposed model produces reasonable accurate prediction to steel structures where magnitude of the triaxiality remains relatively constant. The case study exhibits that the simplified ULCF fracture criteria work well with proposed hardening model to obtain the more realistic predictions to ULCF fracture. Further, it determines the location of fracture accurately. Considering all these reasons, it is advisable to use simplified ULCF model to describe ultimate limit state of steel structures in seismic design practice. The main advantage behind this simplified criterion is that it can be easily utilized with commonly available elasto-plastic finite element (FE) packages. Because most of elasto-plastic FE programs produce outputs of proposed model dependent plastic variables such as accumulated equivalent plastic strain, stress components and effective stress. In overall view, it is concluded that these models are widely applicable for most of civil engineering structures since these models depend on commonly available material properties or facilities. Finally, these reasons conclude that this study provide a new theoretical platform for structural design and maintenance communities by contributing convenient, precise and reliable ULCF models to describe real life of bridge structures.

If indeed it is feasible to apply these ULCF proposed models at a larger scale, e.g. in large-scale beam column connections under cyclic loading. Large scale modeling of specimens often requires transition from smaller to larger elements to capture local (microstructure-level) as well as global stress effects. As a result, large-scale modeling, especially with cyclic loading, can be computationally very demanding. Advances in computational technology suggest that it would be worthwhile to explore the feasibility of applying such models at the larger scale. Even though some options are available, sub-structuring method becomes as more convenient and such option based ULCF fracture predictions of large-scale structures are more appropriate for future studies.

REFERENCES

Bandstra, J.P., Koss, D.A., Gelmacher, A., Matic, P. and Evertt, R.K. (2004). *"Modeling void coalescence during ductile fracture of steel."* Material Science and Engineering A, 366(2), 269-281.

Benzerga, A.A., Besson, J. and Pineau, A. (2004). *"Anisotropic fracture Part II: Theory."* Acta Meterialia, 52(15), 4639-4650.

Chi, W.M., and Deierlein, G.G., (1999). *"Integrative analytical investigations on the fracture behavior of welded moment resisting connections."* Report SAC/BD-95/15, Stanford University, USA.

Chi, W.M., Kanvinde, A.M. and Deierlein, G.G. (2006). *"Prediction of ductile fracture in steel connection using SMCS criterion."* Journal of Structural Engineering, 132(2), 171-181.

Davaran, A. and Far, N.E. (2009). *"An inelastic model for low cycle fatigue prediction in steel braces"* Journal of Constructional Steel Research, 65(3), 523-530.

Fell, B.V., Myers, A.T., Deierlein, G.G. and Kanvinde, A.M. (2006). *"Testing and simulation of ultra-low cycle fracture and fatigue in steel braces."* 8th National Conference on Earthquake Engineering, San Francisco, USA.

Gozzi, J. (2004). *"Practical behavior of steel (experimental investigation and modeling)."* Licentiate Thesis, Lulea University of Technology, ISSN: 1402-1757.

Kanvinde, A.M. and Deierlein, G.G. (2006). *"Growth model and stress modified critical strain model to predict ductile fracture in structural steels."* Journal of Structural Engineering, 132(12), 1907-1918.

Kanvinde, A.M. and Deierlein, G.G. (2007). *"Cyclic void growth model to assess ductile fracture initiation in structural steels due to ultra-low cycle fatigue."* Journal of Engineering Mechanics, 133 (6), 701-712.

Khoei A.R. and Jamali N. (2005). *"On implementation of multi-surface kinematic hardening plasticity and its application."* International Journal of Plasticity, 21(2), 1741-1770.

Krieg, R.D. (1975). *"A practical two surface plasticity theory."* Journal of Applied Mechanics, 42, 641-646.

Kuroda, M. (2002). *"Extremely low cycle fatigue life prediction based on a new cumulative fatigue damage model."* International Journal of Fatigue, 24(6), 699-703.

Lubliner, J. (1990). *Plasticity theory.* New York, Macmillan Publishing Company.

Marini, B., Mudry, F. and Pineau, A. (1985). *"Experimental study of cavity growth in ductile rupture."* Engineering Fracture Mechanics, 22(6), 989-996.

McDowell, D.L. (1985). *"A two surface model for transient nonproportional cyclic plasticity."* Journal of Applied Mechanics, 52, 298-308.

Mroz, Z. (1967). *"On the description of anisotropic work hardening."* Journal of the Mechanics and Physics of Solids, 15, 163-175.

Navarro, A., Giraldez, J.M., and Vallellano, C. (2005). *"A constitutive model for elasto-plastic deformation under variable amplitude multiaxial cyclic loading."* International Journal of Fatigue, 27 (8), 838-846.

Ohno, N., and Kachi, Y. (1986). *"A constitutive model of cyclic plasticity for nonlinear hardening materials."* Journal of Applied Mechanics, 53, 395-403.

Prager, W. (1956). *"A new method of analyzing stress and strain in work hardening plastic solids."* Journal of Applied Mechanics, 23, 493-496.

Ranaweera, M.P.R., Aberuwan, H., Mauroof, A.L.M., Herath, K.R.B., Dissanayake, P.B.R., Siriwardane, S.A.S.C. and Adasooriya, A.M.N.D. (2002). *Structural appraisal of railway bridge at colombo over kelani river;* Engineering Design Center, University of Peradeniya, Sri Lanka.

Rashid, K.A.A. (2004). *Material length scales in gradient-dependent plasticity/damage and size effect, Theory and computation.* PhD thesis, Louisiana State University, USA.

Stojadinovic, B., Goel, S. and Lee, K.H. (2000). *"Development of post-Northridge steel moment connections."* Proceeding of 12th World Conference on Earthquake Engineering, New Zealand, Paper 1269.

Tateishi, K., Hanji, T. and Minami, K. (2007). *"A prediction model for extremely low cycle fatigue strength of structural steel."* International Journal of Fatigue, 29(5), 887-896.

Uriz, P, Filippou, F.C and Mahin, S.A. (2008). *"Model for cyclic inelastic buckling of steel braces"* Journal of Structural Engineering, 134(4), 619-628.

Voyiajis, G.Z. and Basuroychowdhury, I.N. (1998). *"A plasticity model for multiaxial cyclic loading and ratcheting."* Acta Mechanica, 126, 19-35, 1998.

Xue, L. (2008). *"A unified expression for low cycle fatigue and extremely low cycle fatigue and its implication for monotonic loading"* International Journal of Fatigue, 30(10-11), 1691-1698.

Yanyao, J., and Peter, K. (1996). *"Characteristics of the Armstrong-Fredrick type plasticity models."* International Journal of Plasticity, 12(3), 387-415.

In: Advances in Mechanical Engineering Research, Volume 3 ISBN: 978-61209-243-0
Editor: David E. Malach

Chapter 5

MARINE ENGINES EMISSIONS

Juan Moreno-Gutiérrez*[*1], *Zigor Uriondo*[†2], *Vanessa Duran*[‡1], *Manuel Clemente*[#2] and *Francisco José Jimenez-Espadafor*[¶3]

[1]Departamento de Máquinas y Motores Térmicos. Universidad de Cádiz, Spain
[2]Departamento de Máquinas y Motores Térmicos. Universidad del País Vasco, Spain
[3]Departamento de Máquinas y Motores Térmicos. Universidad de Sevilla, Spain

ABSTRACT

Emissions originating from maritime transport in European waters continue to increase at such a rate that the foreseeable evolution may exceed the levels previously forecast. MARPOL, in ANNEX VI, establishes certain conditions for marine engines constructed from the year 2000 with the goal of reducing emissions of NO_x[1]. Although the manufacturers of marine diesel engines make considerable efforts to design and manufacture low emission engines, it is very possible that a maladjustment of the injection system may cause an increase of emissions that has not been seen when the engine is subjected to bench trials. However, no reference at all is made to the limits that are likely to be exceeded when the engine is actually in operation with poor maintenance conditions.

This chapter describes the engine parameters which affect the in-cylinder formation of emissions. It also describes how maladjustment in some parameters like the injection system, changes in the injection pressure, the state of conservation of the nozzles and changes in the scavenging air pressure and temperature influence the emissions. With this objective, some studies have been described on particular engines with both normal or maladjustment conditions of all of parameters mentioned above.

Keywords: NO_x Emissions. Injection Marine Diesel Engines. Combustion Marine Diesel Engines

---

*juan.moreno@uca.es
† zuriondo@azti.es
‡ vanesa.duran@uca.es
# manuel.clemente@ehu.es
¶ fcojjea@esi.us.es

1. Diesel Engine Combustion Products

Emissions formed during burning of the heterogeneous diesel air/fuel mixture depend on the conditions during combustion, during the expansion stroke, and especially prior to the exhaust valve opening. NO_x emissions can be formed through a number of mechanisms during both premixed and diffusion burning. PM is generated in diesels primarily during the diffusion flame. In general terms, most of the NO_x is produced during medium to heavy load conditions; when combustion pressure and temperature are at their highest. However, a small amount of NO_x can also be produced during cruise, light loading and light throttle operations.

The working fluid released to the atmosphere (exhaust gases), originating from the combustion, still contains the energy that has not been transformed into mechanical work or heat; this gas consists of several different kinds of molecules. For example, part of the oxygen entering the engine at the air intake, by suction, is consumed for oxidation of the fuel during combustion, while the rest of this oxygen that enters with this air and is not burned [2] is present in the exhaust gases emitted.

Under conditions of high pressure and temperature in the interior of the engine cylinders, the atoms of oxygen and nitrogen react to form, mainly, NO (90–95%) and NO_2, normally referred to as NO_x. NO is primarily formed by two mechanisms, namely the thermal (Zeldovich) and the prompt (Fenimore) mechanisms. The thermal mechanism is activated above 1600ºC and is responsible for more than 90% of emissions. In addition, some marine fuels (residual fuel oils) contain some nitrogen that will also be oxidized during the combustion. Therefore, it is easy to deduce that the formation of NO_x will be less when the temperature at which the combustion takes place is lower, and when the combustion process takes place more rapidly. But both conditions have a negative influence on the power output of the engine. Therefore, in principle, any action proposed to reduce the emissions of NO_x should be based on reducing the maximum temperature of combustion and reducing the time needed for the combustion by applying measures that do not involve reduction of performance. However, most of these measures tend to reduce the maximum temperature of combustion, since the formation of NO_x is an exponential function of temperature [3]. In fact, the proportion of NO can be increased by a factor of 10 for every 100K increase in temperature [4].

In order to establish a link between the combustion processes in a diesel engine, the NO_x production and its relationship with what we have described so far, we need to take into account three phenomena: the ignition retard, the premixture combustion and the controlled combustion propagation.

The ignition retard (i.e. the time spent between the injection and the combustion) depends on the characteristics of the fuel ignition (cetane number) and other variables such as the combustion chamber temperature, the fuel atomization rate, air turbulence, etc. This makes the combustion gases stay at high temperatures during this period of time. The second combustion phase takes place at high temperatures and pressures. It has an effect on the peak combustion temperature. These two phases, combined with the amount of oxygen, have a direct impact on the formation of NO_x emissions and must be monitored during the engine's operation. We have to add that "poor fuel combustion" tends to have an impact on the increase of NO_x formation, as an excess of oxygen reacts with the nitrogen contained in the air at high temperature and pressure. This means that we also have to take into account a high level of air as a factor in the formation of thermal NO_x.

We must not forget the performance at partial loads. The injection advance angle and the injection length are parameters that cause some modifications in that working environment. Furthermore, they can also alter NO_x and particle production. Partial loads take place on any engine, taking into account torque conditions and speed. Also, engine power and crankshaft speed vary within extensive parameters [5].

The reduction in the load causes the temperature to drop in the combustion chamber and, consequently, there is a reduction in NO_x. This is due to control of thermal NO_x concentration through the molar concentration of nitrogen and oxygen, as well as the combustion temperature.

On the other hand, the poor process of the mixture formation which takes place during partial loads results in a deterioration of the combustion. It has a negative effect on the efficiency and fuel consumption. It also contributes to increasing the particle formation.

In summary, diesel engines (the type of thermal engine currently in common use for marine power plants), without some means of controlling emissions, issue substantial quantities of nitrogen oxides, particles and low concentrations of carbon monoxide and hydrocarbons, all of them being products resulting from the combination process parameters such as thermal output, high pressure, high temperature and heterogeneous combustion that take place in these engines.

Among all the variables that intervene in the process, there are three phenomena intrinsic to the working fluid once the mixture has been formed, and others related to the mechanical features of the engine that have a direct influence on the formation of these oxides. The first includes the time-lag or retardation of the ignition, the second includes combustion by inflammation, and the third includes controlled circulation. The mechanical characteristics with a major influence include the synchronization of valves of suction and exhaust, injectors, nozzles, fuel pumps and their actuation shaft.

It can be stated that among the most important and significant variables in the formation of NO_x are the amplitude of the period of ignition delay, which is a variable exclusive of the fuel that will depend on factors such as the ignition characteristics of the fuel (cetane number), the temperature in the combustion chamber, the degree of pulverization of the fuel, etc.; and has a direct incidence on the quantity of fuel consumed in the second phase, which takes place at high pressures and temperatures. Similarly, the mechanical feature with the most influence on the production of this contaminant is the angle of advance to the injection. These two variables (ignition retard and angle of advance) have a very similar effect because they influence the length of time that the mixture remains at high temperatures, which is known to be directly proportional to the flow of emissions of NO_x. On the other hand, the influence that the injection nozzles has on the formation of contaminants such as NO_x and CO is very important too, since the nozzle is the device that injects the fuel into the combustion chamber and that gives the discharge flow curve its characteristic shape. The nozzle thus affects both the pressure and temperature, as well as the residence time of the gases at high temperatures. The tendency of the injection pressure will be to decrease with the hours of use, although there are other changes in the system of injection that also influence the formation of contaminants, including a drop in the opening pressure of the injector nozzle and wear of the injection nozzle. Nozzle wear will lead to an increase of the diameter of the injection orifices, an increase of the clearance in the needle/nozzle assembly, causing increased losses and wear of the surface of the needle seat in the nozzle. All of these consequences will lead to a drop in the maximum injection pressure, with the consequent loss of both jet penetration and jet speed; such loss will directly affect the atomization of the fuel and the diameter of the fuel droplets that are formed. In addition, the defective sealing of the injector needle against its seating can produce leaking, as well as fuel leaks from the nozzle. According to the engine manufacturer, the estimated life of the nozzles is approximately 4,000 to 5,000 hours of operation.

Furthermore, in order for the stoichiometric value of the mixture to be reached in all parts of the combustion chamber, the rate of injection must be conjugated with the characteristics of the pulverization and the movement of the air, so as to achieve a good speed of mixing in the cylinder, which will also serve to obtain an optimum result in the speed of combustion, which is very significant in the formation of NO_x and particles. It should be remembered that the period of combustion (between 40º and 50º of the rotation of the crankshaft) is of the order of tens of milliseconds, since the reaction must be rapid and complete. As an example, an engine that operates at a speed of 1500 min^{-1} will have a time of approximately 11 ms available for combustion. This will be the necessary time to make the 50º that lasts the entire combustion process. As a consequence of this, some zones rich and others poor in fuel are formed during the process of injection in the engine's combustion chamber. Greater quantities of CO, HC, unburnt substances and particles of carbon are usually generated in the rich zones; whereas in the poor zones a greater quantity of NO_x is produced, although CO, HC and carbon dust formed in the rich zone also continue being burned in the poor zone before being eliminated through the exhaust valve. This means that oxides of nitrogen commonly termed "thermal NO_x" that are independent of those originating from the fuel, are formed in both zones, albeit with different intensity.

2. ENGINE DESIGN FOR LOW EMISSIONS

The state of conservation of the injection system is fundamental, due to its influence both on the temperature of combustion and on its duration. This has already been known since the time when the latest designs of marine engine began to be constructed; these are based on retarding the injection time to reduce the maximum temperature of the flame (among other technological developments), even though it is known that this causes a decrease of output and consequently an increase in consumption of fuel and in the emissions of PM. To recover part of the lost output, the pressure of injection is increased, which improves the pulverization of the fuel and the conditions of its mixture with the air; which results in a more complete combustion. Therefore it is evident that the injection nozzle is very important in determining what the performance of the system of injection during its operational life must be.

Manufacturers are working on the development of NO_x low emission nozzles. It has been proven that for any cylinder there is an optimum number of orifices in each nozzle which help decrease the NO_x[6] emissions. Some of them have developed a combustion chamber for medium power engines which decreases the NO_x and particle [7] formation.

Some other manufacturers have developed a system for low-speed engines which decreases the fuel admission in the combustion chamber when injection stops [8,9]. This system reduces NO and HC emissions. The nozzle is also optimized to avoid NO_x formation.

For example, this combined method has been tested on an engine (55 MW at 94 m^{-1}) with a 90% load. The outcome of the test was a reduction of 23 % of NO_x with a 1% increase in fuel consumption. The NO_x low emission injectors reduce the emissions from 18.5 to 15 g/kWh. However, there is a 2% [10] increase in fuel consumption.

Finally, the second generation of the "common rail system" is being successfully used. The common rail injection pressure is independently generated from the engine rating. In this way, it is possible to obtain the maximum injection pressure with any load on. The injection pressure reaches a value of 1.600 bar in some cases. At that high pressure there is a first-rate fuel atomization; the fuel-air mixture is very thin and effective, and clean fuel combustion is obtained. In other words, there is a more dynamic combustion with less fuel consumption and with fewer emissions. This multiple fuel injection type (previous, main and late injection) reduces the combustion noise, the emission of contaminants and the formation of black smoke during the working stroke [11].

The homogeneous charge compression ignition (HCCI) is a new combustion process [12] that has the potential to achieve low emissions. It is not in common usage in maritime transport. The main drawbacks are too many unburned hydrocarbons (HC) and carbon monoxide (CO) emissions, combustion control at high load and then limited operating range and power output.

The principle of HCCI combustion engines consists of preparing a highly diluted air/fuel mixture by burned gases, in achieving its simultaneous ignition in the whole space of the combustion chamber and in precisely controlling such combustion for the best performance in terms of efficiency and thermal NO_x formation; with soot production lower than with the typical conventional diesel combustion diffusion flame. There are three families of HCCI concepts [12]: with port injection, with in-cylinder injection and concepts using both port and in-cylinder injections.

1) Port Injection: These engines were developed by the University of Wisconsin, Madison, the Southwest Research Institute (SWRI), the Lund Institute of Technology (LIT), and the Tokyo Gas Corporation. There are two types of engine using port-injection, one working with liquid fuel and other working with gaseous fuel. They generally use an electronically driven gasoline injector in the intake pipe, and in some cases, two injectors with two different fuels in 2 intake pipes (LIT). The Tokyo Gas Corporation uses a mixing chamber before the intake pipe, to promote gas and air mixing.
2) In-Cylinder Injection: This concept uses one or many direct injectors in the main chamber. Two different types of injection timing are possible: early injection is the most common way and the new concept of late injection.

2.1. Early in-Cylinder Injection with one Injector

- HIMICS (Homogeneous Charge Intelligent Multiple Injection Combustion System).

The injection system used is common rail (with an injection pressure of 90 MPa), and one single center injector (6 holes with a diameter of 0.23 mm and a spray cone angle of 125° or

30 holes diameter 0.1 mm and three spray cone angles: 12 · 155°, 12 · 105°, 6 · 55°). The injection strategy is divided into four injections: early injection (called E), pilot injection (P), main injection (M) and late stage injection (A).

- UNIBUS: Uniform Buckly Combustion System and PCI: Premixed compressed ignited combustion.

These two concepts are very close to each other as they use one single center injector: a piezzo injector with a hollow cone spray (60°), a spray guide tip *(UNIBUS)* and an Impinged Spray Nozzle to limit penetration (PCI); and in both cases, a Common Rail system. The injection timing is very early: 120° BTDC.

2.2. Early in-Cylinder Injection with Two or More Injectors: Predic and Muldic

- Predic and Muldic mean premixed lean and multiple stage diesel combustion.

In this concept, developed by the *New Ace Institute* in 1996, the direct in-cylinder injection is realized thanks to three injectors, two on each side of the cylinder, and one in the center. The injection strategy is not simple: two injection pressures: 150 and 250 MPa, with each injected quantity controlled, and Predic uses two injection timings: an early injection: 80° BTDC (with side injectors) and a late injection: 40° BTDC (with centre injector), whereas Muldic adds a second stage injection with the center injector (2° BTDC to 30° ATDC).

2.3. Late in-Cylinder Injection: MK Concept

- MK Concept means modulated kinetics concept.

This concept uses a common rail injection system with one single injector (5 holes diameter 0.22 mm). The injection strategy is based on the concept: "low temperature, premixed combustion". So they use a high EGR rate (O2 15 %), a vigorous swirl (3 to 5) and a late injection (7° BTDC to 3° ATDC) to have a long ignition delay and no diffusion combustion; the goal is to inject all the fuel before the combustion starts.

3) Dual Fuel Introduction: HCDC Concept

- HCDC means Homogeneous Charge Diesel Combustion and was developed in 1997 by the Traffic Safety and Nuisance Research Institute.

The injection system consists of a gasoline injector in the intake pipe (injection pressure 5 MPa) and a DI Injector (4 holes diameter 0.21 or 0.14 mm with an injection pressure of 18 MPa) in the combustion chamber. The strategy is to use premixed injection and direct injection (15° BTDC) to ignite the mixture and to vary the premixed fuel ratio.

3. Effect of Biodiesel Fuels on Diesel Engine Emissions

The term biodiesel commonly refers to fatty acid methyl or ethyl esters made from vegetable oils or animal fats, whose properties are good enough to be used in diesel engines. Biodiesel fuels have higher lubricity than conventional fuels, but they can contribute to the formation of deposits, the degradation of materials or the plugging of filters, depending mainly on their degradability, their glycerol (and other impurities) content, their cold flow properties, and on other quality specifications [13] . Biodiesel is 100% renewable only when the alcohol used in the transesterification process is also renewable, but this proportion is reduced to around 90% - 95% when fossil alcohol is used. This makes biodiesel a powerful tool to reduce CO2 emissions [14].

In short, biodiesel does not cause any loss of power unless the maximum power is demanded. A surplus in fuel consumption would, in any case, compensate for the lower heating value of biodiesel as compared with diesel fuel. It is possible to have some decrease in rated power [15]. This decrease is lower than the reduction in heating value in volume basis as compared to diesel. The lower fuel leakage in the injection pumping system, the advance of the combustion process and the higher lubricity of biodiesel have been pointed out as contributing to the mentioned power recovery.

3.1. Effect on Nitric Oxides

Although most of the published literature shows a slight increase in NO_x emissions in certain operating conditions, others did not find differences between diesel and biodiesel fuels, and still others found decreases in NO_x emissions when using biodiesel. Experimental work carried out in a 7.3l Navistar engine running the 13-mode US Heavy-Duty test cycle using different soybean-oil biodiesel blends is described in the report [16].The increases in NO_x emissions obtained were in proportion to the concentration in biodiesel. An 8% increase was reached in the case of pure biodiesel.

A 200kW 6-cylinder was tested at 1200 and 2100 rpm and 50% and 100% load with 10%, 20%, 30% using 40% soybean-oil biodiesel blends, the NO_x emissions increased up to 15% in the case of the 40% blend. Other cases tested a Cummins L10E engine under transient conditions with diesel fuel and 20% and 30% biodiesel blends. They observed an increase in NO_x emissions of 3.7% with the 20% blend, while producing only 1.2% with the 30% blend. They also tested the engine with pure biodiesel under steady conditions (work collected in) reaching a 16% increase with respect to diesel fuel NO_x emissions. Other experiments measuring increases in NO_x emissions were also collected by Graboski and McCormick [17]. For example, Police et al. measured increases around 20%, while Rantanen et al. found 10%

increases, in both cases operating heavy-duty engines under the ECE R49 test cycle with pure rapeseed-oil biodiesel.

3.2. Effect on Particulate Matter and Smoke Opacity

Although some authors have occasionally reported some increases in PM emissions when substituting diesel fuel by biodiesel [14], a noticeable decrease in PM emissions with the biodiesel content can be considered an almost unanimous trend. PM emissions data collected from a number of laboratory studies were used by the EPA to adjust an equation providing a maximum reduction of PM emissions of close to 50% for pure Biodiesel. However, they noticed that PM emission reductions were lower (or there were no significant reductions) in heavy-duty engines than in light-duty engines (20–40% reductions), and maximum reductions (around 40%) were reached in the case of indirect injection engines.

3.3. Effect on THC Emissions

The EPA review (18) shows a 70% mean reduction with pure biodiesel with respect to conventional diesel. However, a few studies in the literature show no significant differences or increases in THC emissions when fuelling diesel engines with biodiesel instead of diesel [14]. Even decreases with high biodiesel percentages and increases at low percentages (synergic effects) have been reported. These surprising trends may be due either to the small content of biodiesel in the fuel or to the very low THC emissions, close to the lower detection limit of the detectors, as is typical in diesel engines. Rather similar reductions to the one supported by the EPA (70%) were reported by other researchers. Nwafor tested a research single-cylinder indirect injection engine with diesel and several blends of rapeseed-derived biodiesel.

3.4. Effect on CO Emissions

The EPA review [18], proposed an equation for the general trend, leading to mean CO reductions of almost 50% of biodiesel with respect to conventional diesel fuel. Other authors also found similar CO reduction values. Krahl et al., after testing biodiesel from rapeseed oil, obtained an approximate 50% decrease with respect to both low and ultra-low sulfur diesel fuels. Peterson and Reece fuelled a turbocharged engine with diesel and several biodiesel fuels, pure and differently blended. They concluded that the decrease in CO emissions with biodiesel was almost 50%. It is possible however to find lower reductions in other sources. The report of a research project [19] showed 28–37% reductions in CO emissions when using pure biodiesel.

Table 1 presents some of the procedures that are currently employed to reduce the formation of NO_x in marine diesel engines.

Table 1.

TYPE OF COMBUSTION	CHARACTERISTICS	EFFECTS
ADVANCED DIRECT INJECTION WITH ONE INJECTOR	Utilizes a common rail system Injection pressure = 90 MPa Central injector with two possibilities Compression of poor mixtures with multiple injections Part is injected advanced, and the rest in the PMS	
ADVANCED DIRECT INJECTION WITH TWO OR MORE INJECTORS	Two laterals (same quantity of fuel, same pressures, and same point of injection)	One tenth of the emissions of NOx in comparison with conventional diesel. Increase of 15% in fuel consumption, HC and CO.
ADVANCED DIRECT INJECTION WITH ONE CENTRAL INJECTOR	One central (nozzle with 30 orifices of 0.08mm diameter, injection pressure between 100 and 200 MPa). Range of injection between 80° and 20° before the PMS.	
MULTI-STAGE INJECTION	Two combustion stages (two lateral injectors). Premixed poor combustion mixture Combustion by diffusion. High temperature and low concentration of oxygen (one central injector). Maximum pressure of central injector: 250 MPa, of lateral injectors: 120 MPa.	Simultaneous reduction of NOx and smoke, even at full loading. Higher fuel consumption. Higher thermal efficiency.
RETARDED DIRECT INJECTION	Utilizes a common rail injection system. Low temperature, premixed combustion process. Low temperature of combustion. Increase in retard time. For higher loadings, the retard time must be longer than the injection time. High injection pressure, low compression ratio, with cold EGR.	Reduces NOx and soot simultaneously. Reduces NOx by 90%. Reduction in the emission of soot. Limited to partial loadings. Retard time greater than high injection pressure
DUAL INJECTION	Conventional form of operation at high loadings. Homogeneous or premixed combustion at partial loadings. Works with homogeneous combustion when idling or at partial loadings. Like conventional combustion for start-up and operation at full loading.	

4. The Importance of the Maintenance

4.1. The Influence of Fuel Injector Valves and Nozzles

The velocity of the air within the combustion chamber is relatively slow because the air is only displaced as a result of the inertia of its mass; that is, the air tends to maintain the velocity at which it enters the cylinder due to the whirl or vortex effect. This effect is assisted by the movement of the piston; the degree of vortex is greater the closer the position of the piston is to the upper dead point. In these engines the nozzle is a determining factor in the efficiency with which the mixture is formed and the combustion takes place due to the low speed of the air. Consequently the nozzle is a determining factor in the formation of various contaminants such as NO_x and CO, since the nozzle is the device that injects the fuel into the combustion chamber and that gives the discharge flow curve its characteristic shape [20].

Thus, the nozzle affects the pressure and temperature as well as the residence time of the gases at the high temperatures. All these are factors that influence the formation of these contaminants. The tendency of the injection pressure will be to decrease with the hours of use, although there are other changes in the system of injection that also influence the formation of contaminants, including:

- A drop in the opening pressure of the injector nozzle;
- Wear of the injection nozzle. Nozzle wear will lead to:

    1) an increase of the diameter of the injection orifices;
    2) an increase of the clearance in the needle/nozzle assembly, causing increased losses; and
    3) wear of the surface of the needle seat in the nozzle.

All of these consequences will lead to a drop in the maximum injection pressure, with the consequent loss of both jet penetration and jet speed; such loss will directly affect the atomization of the fuel and the diameter of the fuel droplets that are formed. In addition, the defective sealing of the injector needle against its seating can produce leaking, and fuel leaks from the nozzle. According to the engine manufacturer, the estimated life of the nozzle is approximately 4,000 to 5,000 hours of operation.

Therefore it is evident that the element that most determines the performance of the system of injection during its operational life must be the injection nozzle. The nozzle needs to be replaced 7 or 8 times for every change of piston/sleeve assembly of the injection pump, which amounts to around 20 replacements during the life of the pump. Therefore, incidents arising from the injection nozzles are much more likely than from other components. Some trials carried out cover most of the possible ways that the injection can affect the emissions of NO_x and CO.

Since the atomization of the jet of fuel has considerable influence on the production of thermal NO, we will record how the characteristics of the nozzle influence this atomization. In general terms, the atomization of the jet of fuel can be considered as the rupture of the jet as it leaves the orifice of the nozzle; the size of the droplets that are formed in the atomization are less than the diameter of the nozzle, and the degree of atomization increases as the jet moves down-stream. The rupture of the jet, and consequently the atomization, continue taking place for as long as the forces of inertia are capable of overcoming the opposition of the surface tension of the fuel; this tends to keep the droplets in their original condition [21] (with a Weber number higher than threshold value).

To evaluate the degree of atomization the parameter known as the SMD (Sauter Mean Diameter) is utilized; that is, the diameter of a droplet that has the same surface-to-volume ratio as the total of the jet of fuel. The lower the SMD, the better the atomization, and vice-versa. Better atomization means that the cloud of vaporized fuel will increase in surface area, and will give rise to the formation of a better fuel/air mixture at the periphery of the jet. It will be at this periphery where a diffusion flame is formed. In this diffusion flame the temperatures will be high (the combustion will be almost stoichiometric) and there will be a sufficient source of air. These conditions are almost ideal for the production of thermal NO; therefore significant quantities of NO will be generated at the periphery of the jet that is less rich in fuel [22]. Thus we will find that these quantities of NO will be greater the better the atomization, the better the mixture, and the bigger the diffusion flame at the periphery.

The nozzle is the component of the injection system that undergoes the most wear due to the cavitations that are produced at the instant of injection; due to the high injection pressure, the fuel reaches high velocities and zones of low pressure are generated in the various curves of the nozzle. These low pressure zones tend to appear at the edges where the fuel enters the nozzle orifice; then, as the injection pressure increases over the course of the cycle of injection the zones are displaced towards the interior of the nozzle orifices. With the passage of time, more and more imperfections are generated on the surface of the orifices of the nozzle, and the formation of cavitations tends to increase. This process generates progressive wear and an increase of the diameter of the orifices of the nozzle. Other very sensitive zones are those of the seat of the nozzle's needle and the inlet edge of the orifice. During the phases of opening and closing of the injector, the smallest section of flow is in the gap between the needle and its seat. It is here that the cavitation is generated, which later collapses in the interior of the nozzle orifices. In consequence, even nozzles with smooth orifices and well-rounded inlet edges rapidly accumulate small imperfections. The sensitivity of the flow to minute imperfections of the inlet edge, and the flows of very high velocity present in diesel injectors, lead to the conclusion that, even in nozzles with very well-rounded edges of orifice, cavitation will inevitably occur after a certain number of hours of operation [23].

In figure 1 we can see the operation results with both nozzles, new and old. From the "Injection curves with new nozzles (in red colour), and old nozzles and nominal opening pressure (in green colour)", it can be seen that the injection curves are similar at the start of the injection. The slope of the increase of injection pressure is similar in both cases, since they were measured in the same cylinder and with the same injection pump. The opening occurs at the same place, with the consequent drop of pressure, and the injection pressure then continues increasing; but although they follow parallel development paths, the curve of the old nozzle remains slightly below that of the new due to the larger orifice size of the old nozzle and greater discharge coefficient. The time of injection is very similar, and so the rack reading on the injection pump was similar for a human observer. The flow rate of fuel supplied in this cylinder was very similar; however, the increased consumption of the old nozzles compared with the new was very small, practically inappreciable in real operating conditions where the repeated variations of load complicate the evaluation of fuel consumption. The emission of smoke was visually appreciable, although only at partial loads and when the loading was varied; for this reason smoke emission could not be employed as a determining criterion for establishing the condition of the injection.

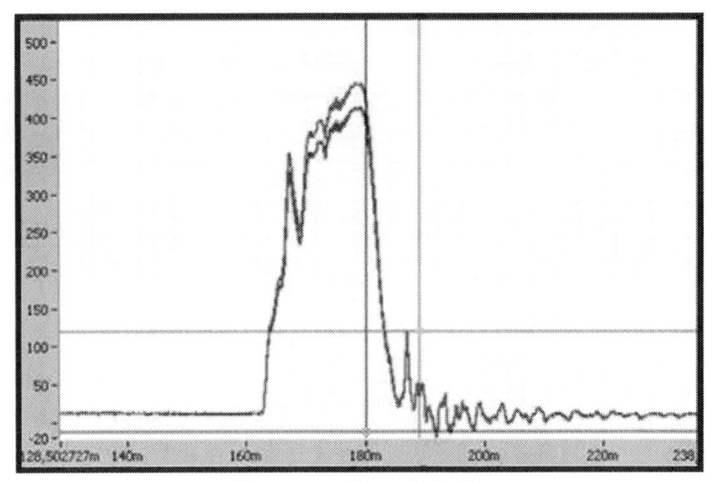

Figure 1. Pressure (bar) versus Degrees of crankshaft, 75% load. Injection curves with new nozzles (in red colour) and old nozzles and nominal opening pressure (in green colour).

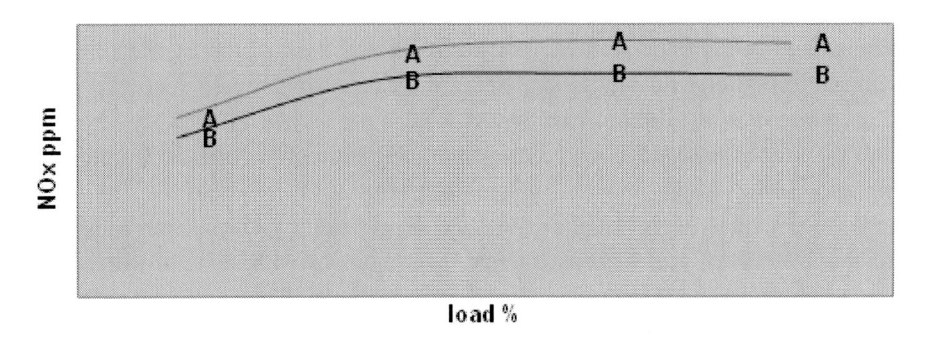

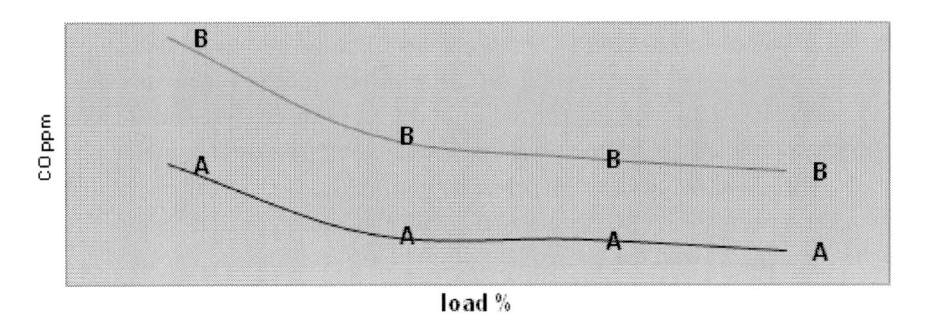

Figure 2. Differences between both situations (A normal conditions and B after 9000 hours with nominal opening pressure).

There were no appreciable differences in the temperatures of the exhaust gases; and the maximum combustion pressures were slightly higher for the trial with the used nozzles, however, this could also be attributed in part to the ambient temperature being slightly higher.

We can conclude that the parameters of the engine do not provide us with reliable information to determine the condition of the nozzles. However, measurements of the composition of the exhaust gases, in particular the measurements of NO_x and CO, do serve this purpose. In Fig. 2 we can see that the amounts of NO_x measured diminished. Emissions of CO differed in that they underwent a much more substantial increase.

Working with used nozzles and reduced opening pressure, the results and conclusions are similar to those with used nozzles at nominal opening pressure.

The conclusions that can be drawn are similar to those of the case of used nozzles with nominal opening pressure; the principal difference is that the nozzle opens sooner due to a lower opening pressure; but that does not imply notable differences in the behavior with respect to the opening at nominal pressure. The reductions of NO_x emissions are of the same order as well as the increases of CO emissions.

4.2. Influence of a Maladjustment in the Injection System

The burnt gas arising from the part of combustion which occurs before peak pressure is compressed due to the rising pressure in the combustion chamber. This means it remains at high temperatures for a long time compared with the burnt gas from the later stages of combustion. This allows more time for NO_x to form. Delayed injection leads to lower pressure and temperature throughout most of the combustion. Delayed injection increases fuel consumption due to the later burning, as less of the combustion energy release is subject to the full expansion process and gas temperatures remain high later into the expansion stroke, resulting in more heat losses to the walls. Smoke also increases due to reduced combustion temperatures and thus there is less oxidation of the soot produced earlier in the combustion

The most common engine tuning measure is increased compression ratio combined with retarded injection timing. A good option may be the combination of increased compression ratio and delayed injection timing for a slow speed engine. In this case, the peak pressure is the same as for the standard engine and occurs at about the same crank angle, even though combustion begins later than for the standard engine. This means that there is less after-compression of the earlier burnt gas, so it does not reach as high a temperature as in the standard case, and it resides at high temperature for less time. Increased compression ratio also tends to offset the increases in fuel consumption resulting from retarded injection timing [6].

The compression ratio can be increased by increasing the geometric compression ratio or advancing exhaust valve closing. Advancing exhaust valve timing would increase the charge mass. This would increase the amount of mass available to absorb the combustion energy, but would also increase the amount of oxygen available for NO_x production. If the geometric compression ratio is increased by reducing the clearance volume, the combustion space will be flatter, which could result in more cooling of the flame by the surfaces and thus increased soot with an additional decrease in NO_x due to the cooling. Combustion chamber shape and fuel spray geometry may need to be adjusted to compensate for reduced combustion chamber height. For 4 stroke medium speed engines, a very high compression ratio may require reduced valve overlap to avoid contact between the valves and the pistons [24]. This reduces scavenging efficiency and cooling of the exhaust valve. Reduced scavenging efficiency can lead to reduced NO_x.

Due to the mechanical adjusting mechanism of the engine it was very easy to make a mistake when adjusting it after an overhawl or in the event of the locking nuts getting loose, the first injection timing advance that the system would have. So with the injection timing modification we were provoking a realistic malfunction situation. We have to bare in mind that the measures adopted to date by the manufacturers of marine diesel engines for controlling the levels of NO_x emissions have been restricted to setting particular design parameters for engines (injection advance, compression ratio, mechanical injection, etc.); but the measures that are proposed for the near future could be considered more dynamic, totally changing in function of the operating conditions of the engine. A good example to clarify this might be the comparison between mechanical injection and electronic injection. Mechanical injection is adjusted when the engine is assembled, and is kept constant by means of the classic mechanism of distribution gears and cams. In contrast, it is easy to manipulate a system of electronic injection using appropriate means, which allows various adjustments to be made that are difficult to identify without those appropriate means, and it is even possible to make the adjustments with the engine in operation. In addition they have a complicated cartography for the operating conditions of the engine. So in the future it will be very complicated to determine the injection timing in the event of a small malfunction of electronic injection timing. For example, for a low speed two stroke diesel engine, running at 75% of its maximum loading, the variation of the injection profile from "ECONOMY MODE" to "SPECIAL EMISSION MODE" brings about a variation in the concentration of NO_x in the exhaust gases from 1150 ppm to 500 ppm [25]. Thus, once the engine has been installed in the vessel, there is some doubt as to what the injection profile of the engine is, and whether, from the economic point of view, it is really more advantageous for the final user to employ the "ECONOMY MODE".

As stated previously, the maximum temperature that is reached during combustion in an internal combustion engine depends, among other factors, on the air/fuel ratio, and if this is controlled, the amount of NO_x generated can be reduced. On the other hand, the synchronization of ignition/injection has an influence on the residence time of the fuel and air; hence adjustments in the synchronization of the valves can reduce the time of residence at the maximum temperature and, therefore, can also control the formation of NO_x.

It can therefore be deduced from the foregoing that inadequate maintenance, such as failing to detect and correct maladjustment in the injection, can provoke an increase in the flow of emissions of NO_x. The advancing of the injection produces an input of fuel when the pressure and temperature in the combustion chamber are lower; this retards the start of the combustion and increases the quantity of fuel that enters during the time of retardation; this fuel is then burned very rapidly in the premixing combustion phase. This will produce an increase of pressure and temperature, resulting in the formation of more NO_x. The injection starts at a low pressure, which initially produces fuel droplets of larger diameter; these need more time to evaporate and begin the process of autoignition. In figure 3 we can see the situation and the differences, A normal situation and B the same engine when the injection is advanced.

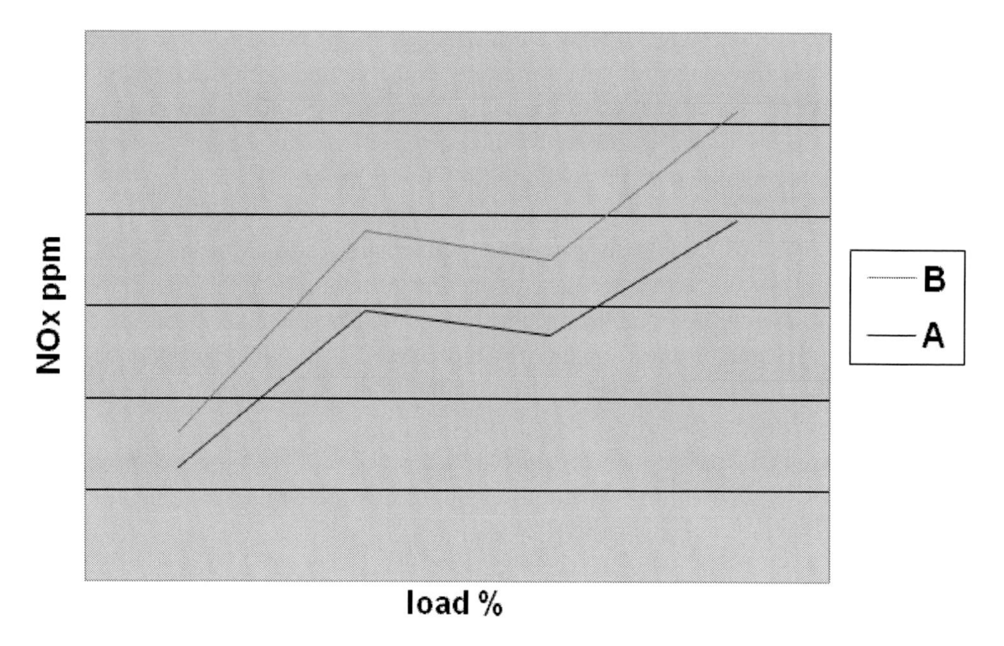

Figure 3. Differences between both situations. A, before maladjusted B, after maladjusted.

4.3. Injection Pressure

The usual operating conditions of an injection system with the hours of use are replicated to see the affects of emissions of NO_x. The tendency is for injection pressure to decrease with hours of use. The incidents that may take place in the system of injection include:

- Reduced aperture pressure of the injector nozzle.
- Wear of the injection nozzle.
- Wear of the piston, sleeve and delivery valves of the injection pump.
- Wear of the delivery valve, and retention of the injection pump.
- Maladjustment of the advance of the injection, i.e. advance or delay with respect to the optimum point.
- Wear of the injection cam, which will alter the profile of the cam and cause a drop in the injection pressure.

Maladjustment in the injection pressure can thus have serious effects on operating performance. Vanesa Durán et al. [26], tested for emissions of NO_x, detecting them by periodically analysing emissions with portable equipment.

Figure 4 shows the results of a comparative test between two situations. The subject was to compare the effect on the emissions of NO_x of the drop in aperture pressure that occurs over the course of the engine's operating hours. It was focused on the effect of the reduction of pressure on the emissions of NO_x, and the possibility of detecting this effect by analysing the exhaust gases. The nominal aperture pressure in the injectors of the engines tested is 35 MPa. From field experience of these engines, about 2,000 hours of use normally elapses before the injectors need to be repaired and reconditioned, these are found to have an aperture pressure reduced to around 30 to 31 MPa. During the life of the nozzles, some 4,000 to 5,000 hours, it is normal to find no significant variations in the geometry of the nozzle (i.e. in the shape and diameter of the orifices). As a result of this, no significant increase of the flow occurs due to the ageing of the nozzle, and the structure of the jet is hardly affected by this reduction of aperture pressure over this period of use of the nozzle .

The advancing of the injection produces an input of fuel when the pressure and temperature in the combustion chamber are lower; this retards the start of the combustion and increases the quantity of fuel that enters during the time of retardation; this fuel is then burned very rapidly in the premixing combustion phase. This will produce an increase of pressure and temperature, resulting in the formation of more NO_x. The injection starts with a low pressure, which initially produces fuel droplets of larger diameter; these need more time to evaporate and begin the process of auto-ignition.

The test performed corresponds to the type E2, according to the Technical Code of Annex VI of MARPOL 73/78, in which the revolutions of the engine are held constant and the loading applied to the engine is varied.

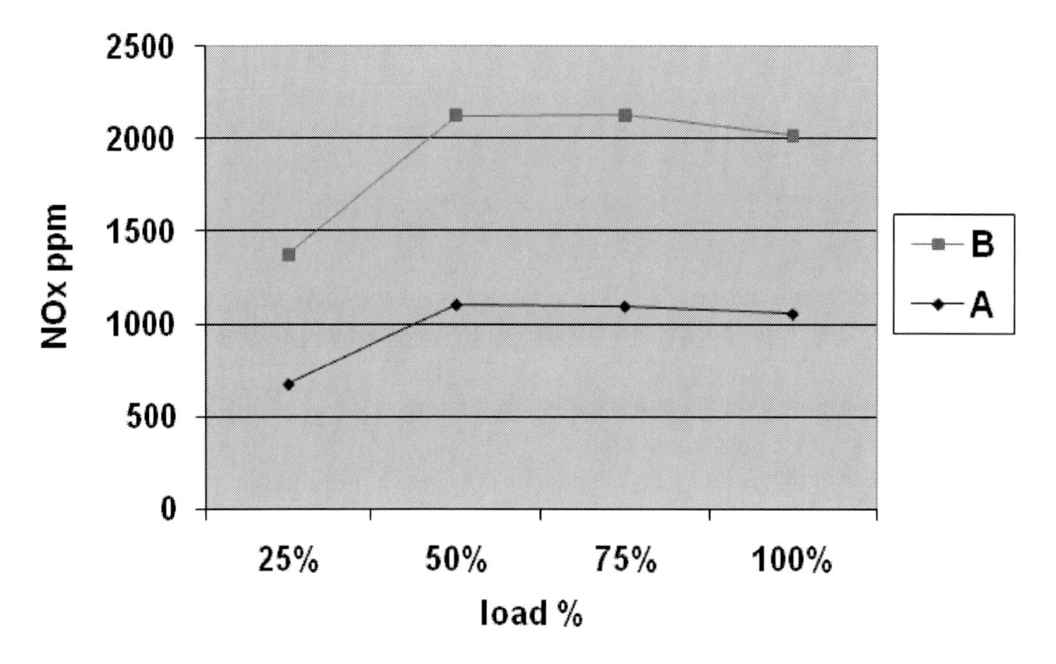

Figure 4. Emissions of NO_x against loading, with nominal pressure(A) and low injection pressure(B).

5. Cylinder Charge Air Pressure Effect in Emissions

The turbocharger performance is a key element for the combustion characteristics of a turbocharged diesel engine, but not only for the combustion, but also for the reliability and cleanliness of the combustion chamber thanks to its scavenging and cooling effect. The operation maintenance of a turbocharger is a key maintenance operation to keep the turbocharged diesel engine in good running order.

The charge in air pressure effect in NO_x emissions is not easy to predict. This is due to the NO_x formation conditions inside the cylinder. So it is not clear if the effect of the air supply conditions to the cylinder will affect the whole cylinder in the same way. It is considered that an improvement of the combustion will bring an increase in combustion flame temperature and formation of NO_x, so it is necessary to know the emissions performance of an engine when the air supply and the pressure of the air supply to the cylinder change [28]. The NO_x will be mainly formed in the flame zone, but also in the burnt zone. The burnt zone will mix with the unused air, cooling down the temperature of the mixture. A reduction of the air available for cooling down the burnt products will increase the time of the burnt products at high temperature and increase the formation of NO_x [29].

For the turbocharged diesel engines, once the engine is in operation the usual tendency is that the turbocharger looses its ability to supply the same amount of air, reducing the air supply pressure (charge pressure). In some cases, in operational conditions, burning heavy fuel oil is not uncommon since the nozzle ring gets dirty with combustion residues, reducing the nozzle area and increasing the turbine speed. This will cause an increase in charge pressure; but a reduction of air mass flow [30]. The reduction of flow is caused because there is a displacement of the turbocharger working point from its design point. The air is supplied to the cylinders in a higher temperature due to a higher compression ratio; but in a smaller quantity. So, the possible effects could be an increase of emission concentrations due to a higher cylinder charge air temperature, as previously described; but less contaminant mass flow due to a reduction of air and exhaust gas mass flow. This phenomenon is not usual when burning distillate oils because the combustion residues are not as abundant, and is not the most common problem with turbochargers. The most common phenomenon in the turbocharger is a loss of efficiency that translates to a reduction of charge air pressure and flow due to dirtiness and wear in the different turbocharger components: turbine blades, compressor blades, air filter, turbine housing and dirt in the cylinder charge air cooler that increases the pressure loss through the cooler reducing the charge pressure and air supply. In marine engine rooms, especially for auxiliary engines, it is common to have charge air pressure of lower values than the ones obtained in the test bench because the installation conditions are not as good as in the test bench due to the space constrains found in the vessels. The values are usually within 5% of the test bench result, so 10% of the test bench result is not an abnormal value in certain operating conditions after a running period and before maintenance operations.

If the charge air pressure has been reduced, the pressure loss in the turbocharger air filter increases. So, the exhaust gas temperature increases when the charge air pressure is reduced. This is caused by the reduction of air supply. From the other side the injection pump racks increase, what means there is a longer injection period and more fuel is sent into the cylinder. There is a loss of efficiency in combustion and the specific fuel oil consumption increases. The poor combustion inside the cylinder means that the fuel is not totally burned and the exhaust gas has more unburned fuel than the case with standard charge air pressure. It is also possible to appreciate an increase in visible black smoke.

The NO_x formation increases when the charge air pressure and the air supply to the cylinders are reduced. This could be seen in the NO_x concentration increase in the exhaust gases. Increasing the pressure loss reduces the charge air pressure, and the air supply to the cylinder is also reduced which leaves less air available for combustion inside the cylinder. A reduction of charge air pressure also causes a density reduction in the cylinder charge air, and the cylinder filling is done with a smaller air mass which will reduce the effective compression ratio. The air compression inside the cylinder practically follows an adiabatic compression. This means that the reduction of combustion chamber pressure in the time of injection has a strong dependence on the reduction of air density in the beginning of the compression phase. The diffusion controlled combustion process will be slowed due to reduced cylinder gas density and lower oxygen availability [31]. When the fuel injection process begins, the higher the density of the air inside the combustion chamber the better the atomization of the fuel spray will be, and as a consequence of the better atomization the mixture of air and fuel reduce the physical part of the ignition delay. The ignition delay reduction causes a reduction of the premixed combustion phase, which is the phase where the bigger formation of NO_x is due to its effect in the peak pressure and temperature. So a reduction of initial density will cause an increase of ignition delay, more fuel injected in the delay period and a longer premixed combustion phase, with a higher combustion peak pressure and temperature and more formation of NO_x. From another part, a better atomization and mixture of air with fuel will improve combustion efficiency, increasing the flame temperature and formation of NO_x. The flame temperature in the combustion chamber is distributed non-uniformly. The flame temperature, after reaching a peak value at a certain time in the combustion cycle, suffers a reduction of temperature resulting from expansion in the combustion chamber volume during the expansion stroke and heat exchange with cooler unused air in the cylinder [32]. If the air mass inside the cylinder is smaller another effect will be the reduction of available air for cooling the combustion products, so the combustion products will remain at a high temperature for a longer period of time, also increasing the NO_x formation. The combination of increased ignition delay and reduced air for cooling could cause the increase of NO_x formation when the air supply and pressure are reduced; but a loss of combustion efficiency due to poor air/fuel mixture could cause a reduction of NO_x formation. The two opposed phenomenon make the prediction of the emission performance difficult when the charge air pressure is reduced.

When the air supply to the cylinders is reduced the exhaust gas mass flow is also reduced. To obtain the final emission value it is necessary to combine the total exhaust gas flow together with the exhaust gas contaminant concentration. This means that although the NO_x concentration in exhaust gas increases with charge air pressure, reduction the total NO_x mass flow is not really increased. But when the load increases the exhaust gas mass flow suffers a very important reduction, and the same happens with the NO_x mass flow. The combustion time is also longer at high loads than at low loads, so the ignition delay is more important at low loads than at high loads, where the cylinder pressure and temperature are already very high at the time of injection. At high loads the premixed phase of combustion loses importance compared with the diffusion phase combustion.

In short, with temperature values within manufacturers' specifications the NO_x concentration increases in exhaust gases from standard values. This increment is not so big in the NO_x mass flow rate when the exhaust gas mass flow rate is considered. The cylinder charge air density is reduced with the temperature increase reducing the air supply to the cylinders and also the total exhaust mass flow.

If the charge air pressure is reduced for installation or operational reasons the NO_x concentration within the cylinder will increase; but the NO_x emissions to the atmosphere will be reduced due to a reduction of cylinder charge air flow rate and hence due to a reduction of exhaust mass flow rate. The NO_x emission's mass flow rate will be reduced. The other pollutant emissions (CO, THC and particulate matter) and the CO_2 emissions will be increased due to a loss of combustion efficiency.

6. Water Injection and Humidification

Water emulsification of the fuel is a technique which also lowers maximum combustion temperature without an increase in fuel consumption. Water has a high heat capacity, which allows it to absorb enough of the energy in the cylinder to reduce peak combustion temperatures. There are at least two ways to accomplish the emulsification during combustion, in the combustion chamber or in the fuel tank. Testing on a diesel engine has shown a 40 percent reduction in NO_x with a water-fuel ratio of 0.5, with only a slight increase in smoke [33]. Water dilution does have significant challenges. Combining water and fuel for the first time in the chamber requires significant changes to the cylinder head to add an injector. Using a single injector with stratified water and fuel adds complexity to the injection system. Combining water with the fuel in the tank may introduce combustion problems due to unstable emulsion. Also, this technique requires a significantly redesigned fuel handling system to overcome the potential risk of corrosion and to maintain power output. However, these problems may be overcome in the future as the strategy is refined. In any event, extra liquid storage availability is necessary to retain a similar range.

In practical applications, the NO_x reducing potential of the water–in-fuel emulsification method is only about 20-30%. This is because the maximum water/fuel ratio which is stable, and can therefore be used, is rather low.

Concepts for introducing water into the combustion process are [34]:

1) Direct water injection:

   a) 50-60%NOx reduction
   b) Optimization of injection timing and duration possible
   c) Good engine performance in "Non-Water operation mode"

2) Water in-fuel-emulsion

   a) 10-30%NOx reduction

Limitations:

   a) Emulsion stability
   b) Poor engine performance in "Non-Water operation mode"
   c) Cavitation risk in the injection system

3) Humidification of combustion air.

   a. 10-30% NO_x reduction

Limitations:

a) Water condensation in the air receiver
b) Erosion/corrosion risk in inlet ports and valves

References

[1] MARPOL 73/78 *.Anexe VI* . NO_x Technical code
[2] Hashizume, T., Miyamoto, T., Akagawa, H., and Tsujimura, K. *"Combustion and emission characteristics of multiple stage Diesel combustion". SAE Paper 980505.1998*
[3] *John B. Heywood, McGraw Hill* "Internal Combustion Engines"
[4] European Commission Directorate General Environment. *"Service Contract on Ship Emissions: Assignment, Abatement and Market based Instruments"*.2005
[5] Weisser, G., F. X. Tanner, K. Boulouchos, J. Kramer and R. Holtbecker. *"Integrating CRFD Simulations into the Development Process of Large Diesel Engine: A Status Report"*. CIMAC 98 Paper No. 05.09. 1998
[6] Laurie Goldsworthy. *"Design of ship engines for reduced emissions of oxides of nitrogen"*. Faculty of Maritime Transport and Engineering Australian Maritime College 2002
[7] Paro, D. *"Development of the Sustainable Engine"*. 23rd CIMAC Congress. 2001
[8] MAN B&W, *"Trends in the Volume and Nature of Propulsion Machinery Demand – the Low Speed Sector"*, 1999(368-99.12).
[9] Egeberg, C. and A. Ostergaard. *"The MC Engine and its Future Development"*. 23rd CIMAC Congress. 2001
[10] Sowman, C., *"Mitsubishi Engineers for the Environment"*, in Motor Ship. p. 45. Fankhauser, S. and K. Heim, "The Sulzer RT-flex: Launching the Era of Common Rail on Low Speed. Engines", 23rd CIMAC Congress, 2001.
[11] MAN Nutzfahreuge *AG. Feria Boot* . D̈usseldorf. 2004
[12] B. Walter and B.Gatellier. *"Near Zero NOx Emissions and High Fuel Efficiency Diesel: the NADI ™ Concept Using Dual Combustion"*. IFP (2003 Vol.58)
[13] Waynick JA. *"Characterization of biodiesel oxidation and oxidation products"*. USA SouthWest Research Institute 2005 [Project 08-10721].
[14] GHG. *"Highlights from greenhouse gas (GHG) emissions data for 1990–2004 for Annex I Parties"*. United Nations Framework Convention for Climate Change. 2006
[15] Lapuerta M, et al. *"Effect of biodiesel fuels on diesel engine emissions"*. Progress Energy Sci ,doi:10.1016/j.pecs.2007.07.01
[16] FEV. *"Engine Technology. Emissions and performance characteristics of the Navistar T444E DI engine fuelled with blends of biodiesel and low sulphur diesel"*. Final report to National Biodiesel Board 1994.
[17] Graboski MS, McCormick RL. *"Combustion of fat and vegetable oil derived fuels in diesel engines"*. Progr Energy Combust Sci 1998;24:125–64.
[18] Assessment and Standards Division (Office of Transportation and Air Quality of the US Environmental Protection Agency). *"A comprehensive analysis of biodiesel impacts on exhaust emissions"*. EPA420-P-02-001. 2002

[19] Handbook of biodiesel: *"Emissions reductions with biodiesel"* 1999.
[20] BOSCH Diesel Engine Management. *Systems & Components*. 3rd Edition
[21] *"EPA Proposal for More Stringent emissions Standards for Locomotives and Marine Compression-Ignition engines"* EPA420-F-07-015, March 2007
[22] John E. Dec. *"A Conceptual Model of DI Diesel Combustion Based on Laser-Sheet Imaging"* SAE Paper No. 970873, John E. Dec. 1997
[23] H. Chaves, M. Knapp, A. Kubitzek, F. Obermeyer and T. Schneider *"Experimental Study of Cavitation in the Nozzle Hole of Diesel Injectors Using Transparent Nozzles"* SAE Paper No. 950290. 1995
[24] Schlemmer-Kelling, U. and M. Rautenstrauch, *"The New Low Emissions Heavy Fuel Engines of Caterpillar Motoren (MaK)"*, 23rd CIMAC Congress, 2001
[25] Presentation *"ME Engines – the New Generation of Diesel Engines"*.2007
[26] J.Moreno- Gutiérrez, Ismael Rodríguez Maestre, Tarik Shafik, Cristina V. Durá n Grados, and Paloma Rocío Cubillas. *"The Influence of Injection Timing over Nitrogen Oxides Formation in Marine Diesel Engines"*. Journal Marine Environmental Engineering. Vol. 16 pag.1-10. 2006.
[27] Vanessa Duran, Zigor Uriondo; Manuel Clemente, and Juan Moreno. *"Correcting injection pressure maladjustments to reduce NOx emissions by marine diesel engines"*.Transpòrtation Research Part D. Volume 14,Issue 1, Pages 61-66. 2009
[28] Watson, N., Janota, M.S. *"Turbocharging the Internal Combustion Engine"*
[29] Yoshikawa, S., Ogawa, M., Inaba, H., Fujita, Y., Imamori, T., Yasuma, G., *"The Development of Low NOx Emission Diesel Engine"* CIMAC Congress 1995, Interlaken
[30] Hountalas, D.T., *"Prediction of marine diesel engine performance under fault conditions"* Applied Thermal Engineering 20 (2000) 1753 – 1783.
[31] Benajes, J., Molina, S., Martín, J., Novella, R., *"Effect of advancing the closing angle of the intake valves on diffusion-controlled combustion in a HD diesel engine"* Applied Thermal Engineering 29 (2009) 1947 – 1954
[32] Han, Y., Kim, K., Lee, K., *"The investigation of soot and temperature distributions in a visualized direct injection diesel engine using laser diagnostics"* Measurement Science and Technology 19 (2008) 115402
[33] Konno, M., Chikahisa, T., Murayama, T., *"Reduction of Smoke and NOx by Strong Turbulence Generated During the Combustion Process in D.I. Diesel Engines,"* SAE Paper 920467. 1992
[34] Milieu Press Release Wärtsilä. *"Direct Water Injection to be installed on Ro-Ro vessels"*. 2000

In: Advances in Mechanical Engineering Research, Volume 3 ISBN: 978-61209-243-0
Editor: David E. Malach

Chapter 6

COMPOSITE PROCESSED BY RTM: FATIGUE BEHAVIOR

***Maria Odila Hilário Cioffi*[1], *Herman Jacobus Cornelis Voorwald*[1], *Marcos Yutaka Shiino*[1] *and Mirabel Cerqueira Rezende*[2]**
[1]Fatigue and Aeronautic Materials Research Group - DMT/FEG/UNESP – Univ Estadual Paulista, Av. Ariberto Pereira da Cunha, 333, CEP 12516-410 Guaratinguetá/SP, Brazil
[1]Divisão de Materiais /IAE/CTA, Praça Marechal Eduardo Gomes, 50 Campus do CTA, Vila das Acácias, CEP 12228-904 São José dos Campos/SP, Brazil

ABSTRACT

Structural polymer composites have been widely applied in the aeronautical field. However, composite processing, which uses unlocked molds, should be avoided in view of the tight requirements in service and also due to the possibility of environmental contamination. To produce composite aeronautical structural components with low cost, the aircraft industry has shown interest in resin transfer molding process (RTM) as an adequate option to substitute for conventional process with the advantages of faster gel and cure times, besides the low percentage of voids and high fiber volume percentage, which are the essential parameters to design aircraft structures. Since the low viscosity resin is injected into the closed mold, in this process, the edge effect can promote incomplete wetting of the fiber reinforcement, dry spot formation and other defects in the final composite. Knowledge of material behavior is essential to design structures as aircraft landing gear, for example. Compared with isotropic materials, polymeric fibrous composites submitted to cyclic loading present a degradation phenomenon of mechanical properties as a consequence of residual stress redistribution into the structure. It was established that fatigue mechanisms associated to fibrous composites occur in four stages: nucleation of local damage, stable propagation of crack due to the cyclic load, local crack propagation which is dependent of fibers orientation, the matrix ductility and the interfacial adhesion and propagation of last loading cycle, which is analogous to the tensile test fail. Because of the many mechanisms involved during the degradation of the composites, scatter in fatigue is higher and special care need to be taken when approach the S-N curve. In this chapter, the review of data presented in the literature with focus in the fatigue behavior of polymeric composites reinforced with carbon fiber processed by

RTM are compared with some experimental data obtained during three years of study. As a specific aim this chapter proposes a new methodology for fatigue behavior on composites, as this field lack of reliable predictive methods, and the main drawbacks of composites applied in aircraft structures.

INTRODUCTION

Metals are the most widely material used for structural applications; however, the present interest in composites demonstrated by aircraft industries is increasing due to the requirement of high quality materials, essential to comply with product specifications and to reduce operational costs [1-2]. It is important to know that fiber-reinforced composites present an excellent weight saving and load-bearing capacity, especially in the spacecraft area, which has typical weight-sensitive structures [1, 3.]

The more extensively the use of aircraft composite material, carbon fiber/epoxy resin, proves a weight/strength gain that in most of the time is higher than 20% [4]. An outstanding example is the substitution of aluminum alloy by the NCF/RTM6 composite for landing gear component presented in Figure 1a, with a minimum of design modification. In Figure 1b, it was observed a reduction of 59% in the weight of the sub-component [5].

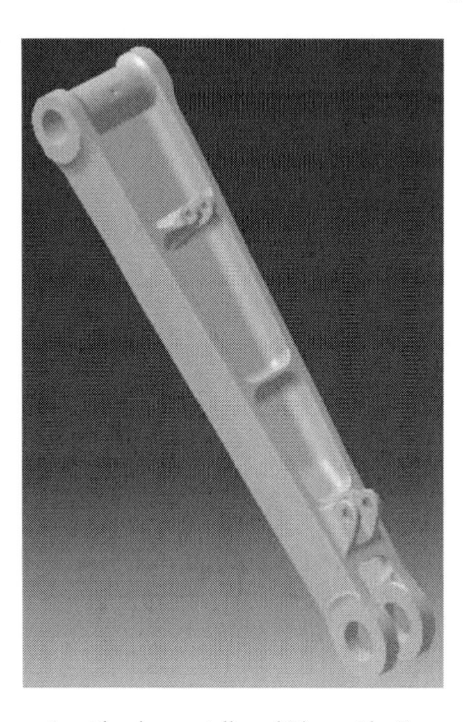

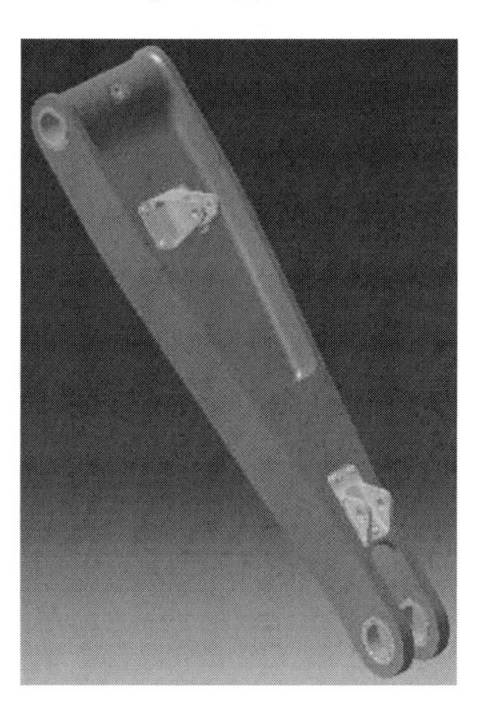

Figure 1a. Aluminum stalk and Figure 1b. Composite stalk.

Weight savings in aircraft components promote a lower fuel burning up, reducing gas emissions which is important not only because of the fuel economy but as well as the environmental issues. Another advantage for aircraft application is the reduction in the use of rivets, not needed for plastic components and in the case of RTM process, which uses lock molds, machining is not required and most of the components are integrated.

An important consideration to take apart is that material qualification for aerospace applications is expensive; time consuming that frequently contributes to delays in the application of new materials. It is thus imperative to develop new evaluation methods to speed up durability testing if composites are considered candidates for further applications in structural aerospace parts [6].

Considering that any new technology brings problems, in case of composites the important ones are regarding to the production automatic process, production systems developments and the establishment of a data base for design [7]. Components and architecture of composite manufacturing is not a simple action, the introduction of reinforcement into the matrix promotes properties changes then it is fundamental to consider microstructures variations, which develop residual stress due to the difference in coefficient of thermal expansion (CTE) during the manufacturing [8].

The RTM process is one of the most efficient technique to produce advanced fiber reinforced polymeric matrix composites [9] associated to an effective low cost technology to produce great scale composites components. It was developed as an economic method of high quality composites to produce more complex components than those obtained from the traditional methods [10], present excellent control of mechanical properties and shorter process cycle [11].

Since low viscosity resin is injected into the mold that is closed, in this process, the edge effect can promote incomplete wetting of the fiber reinforcement, dry spot formation and other defects in the final composite.

During the cure process, thermal, physical, rheological and mechanical properties of the resin varies making the analysis difficult. The modeling has to be improved for a better representation of properties variations and advances in experimental characterizations, especially for the new resin generation, have to be the aim of future studies [2].

A precise preform free from defects and distortions presents a permeability assumed to be uniform, which could avoid sections of unsaturated regions of the mold. In the opposite side there are defects and preform distortions that produce residual stresses and stress concentration to the components submitted to loadings [12].

These anomalies in the preform affect composites mechanical properties; consequently significant changes in stiffness, strength and fatigue life occur. As a result, components submitted to fatigue loads show a redistribution of residual stress into the composite structures [13].

Fatigue mechanisms associated to fibrous composites occur in four stages: nucleation of local damage, stable propagation of crack due to the cycling load, local crack propagation which is dependent of fibers orientation, the matrix ductility and the interfacial adhesion [14].

This caption describes the basic characteristics of resin transfer molding manufacturing composites with emphasis in a new methodology for composites fatigue behavior and the influence of polymeric resin relaxations.

RTM PROCESS

The resin transfer molding process (RTM) was developed in eighties for general applications, followed by a small aeronautical use. In 1946 the American Marine produced boats in glass fiber/polyester resin composite, molded by injection using vacuum. As this method was unavailable for aircraft components production, from 1952 on some patents have presented the development of methods using resin injection by pressure, guarantying the full impregnation [15]

Nowadays, the RTM process is considered as the more efficient and attractive techniques to advanced composites production using polymeric matrices reinforced by fibers [9]. It deals with an effective low cost technology to produce composites components in high scale. This development is justified by the economic manufacturing, which enables the complex components production in high quality compared to traditional methods [10], with an outstanding mechanical properties control in a short time cycle [11]. Besides it is a safety process due to the volatiles that do not spread in the environmental work. The components produced present chemical resistance and good electrical insulation [16]

The RTM process has been considered as an optimum method to develop structural composites for aircraft industry either for the quality degree or for capacity of producing complex parts. Due to the low percentage of voids in the final product, this method of laying up guarantees an outstanding performance regarding fatigue process.

The difference in respect to other ones is associated to the fact that resin flows for a long extension through the dry fiber reinforcement pores. During the process the resin in liquid state is injected into the closed metal mold, as indicated in Figure 2, under moderated pressure until the saturation of preform is reached, frequently with the vacuum pump as an auxiliary accessory connected to the outlet port in order to increase the flow; afterwards the cure is started [17].

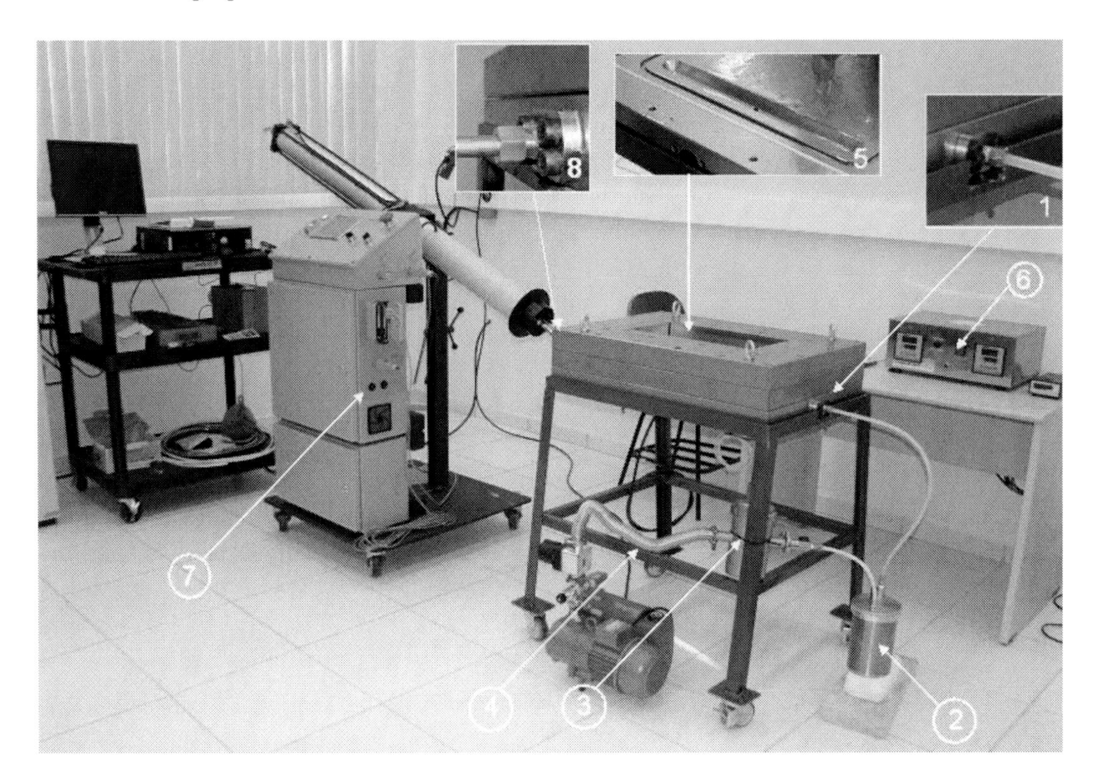

Figure 2. RTM system.

As indicated in Figure 2, the RTM system is composed by

- Injector outlet gate using Teflon tube, Figure 2 detail 1;
- Trap for pump protection against resin, Figure 2 detail 2;
- Trap for pump protection against volatile, Figure 2 detail 3;
- Vacuum system to guide resin flow and reduce entrapped gases, Figure 2 detail 4;
- Metallic mold with one injection chamber where resin flows into the outlet port direction, in order to reduce the risk of air bag formation and increase the resin distribution into the fabric, Figure 2 detail 5;
- Aluminum Heating system with temperature control, Figure 2 detail 6;
- Resin injection system with pressure, resin temperature and volume control, Figure 2 detail 7.
- Injector inlet gate using cooper tube, Figure 2 detail 8;

The quality of impregnation depends on the permeability of the reinforcement by the resin [10], which represents the resistance of the fiber to the resin flow expressed by Darcy's law [11].

The impregnation of dry reinforcement occurs through mold injection inlet port and the orientation requirement depends on the final component 18]. When the homogeneous resin pass through the injection outlet port, the injection port is turned off and the cure cycle begins [19]. With the increase of temperature, resin viscosity increases until the polymerization onset [20].

A Minimum variation of the reinforcement position or permeability causes resin flow deviation [18-21]. The velocity of flow through the porous medium is directly proportional to the pressure and to the gap size, which generates a proportionality coefficient called permeability [22-23]

The cure cycle is the more expensive stage and critical for the structural composites production. Many complications can occur in a viewpoint of a non-controlled process, which should consider the geometry of a component or materials constituents. In general this stages is based on a manufacturing recommendation, which takes as reference, not forced resin [24].

Unknown deviation in the cure cycle, as laminate thickness variation, temperature gradients and, as a consequence, variation in the cure degree can introduce residual stress into the component [8].

The process is carefully monitored with respect to the volume, Figure 3, pressure, Figure 4 and cure temperature Figure 5.

Through the RTM process an economy in materials cost is obtained as a result of the dry reinforcements, cheaper materials when compared to *prepregs*, besides friendly work environment in comparison to the traditional composites manufacturing. Furthermore promoting labor reduction, real time process control and high production velocity [25-26]

The time cycle reduction for RTM process is associated to high initial temperature, which removes the long time of heating mold step; the process time is lower, around 1.5 h. Data related to the laboratory environments, in which the process is suitable, due to the easy handling, small space required and the ability to product several kind of geometry, depends just of the mold.

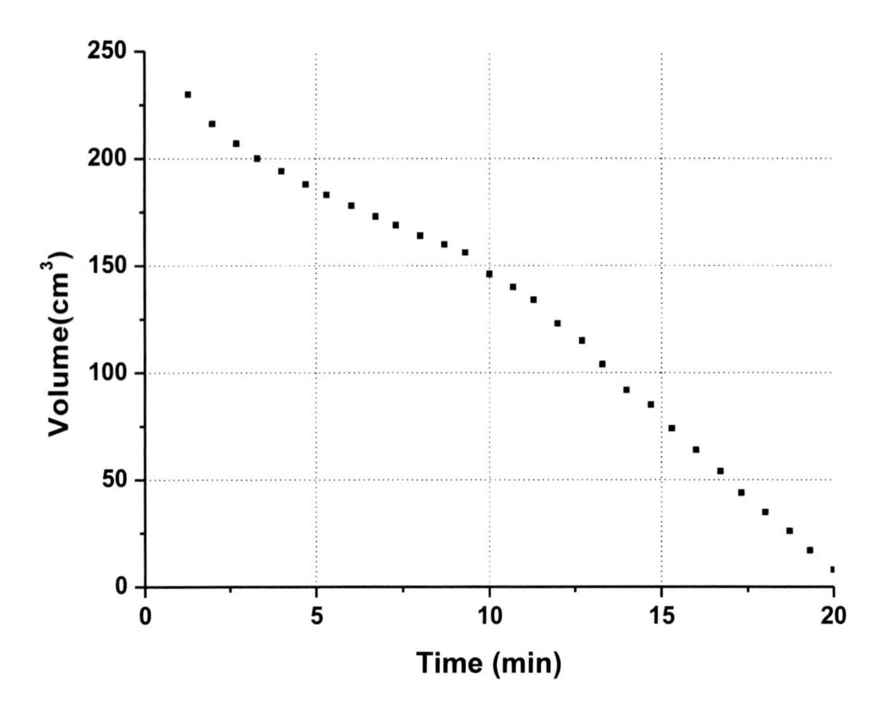

Figure 3. Acquisition of volume progress in real time.

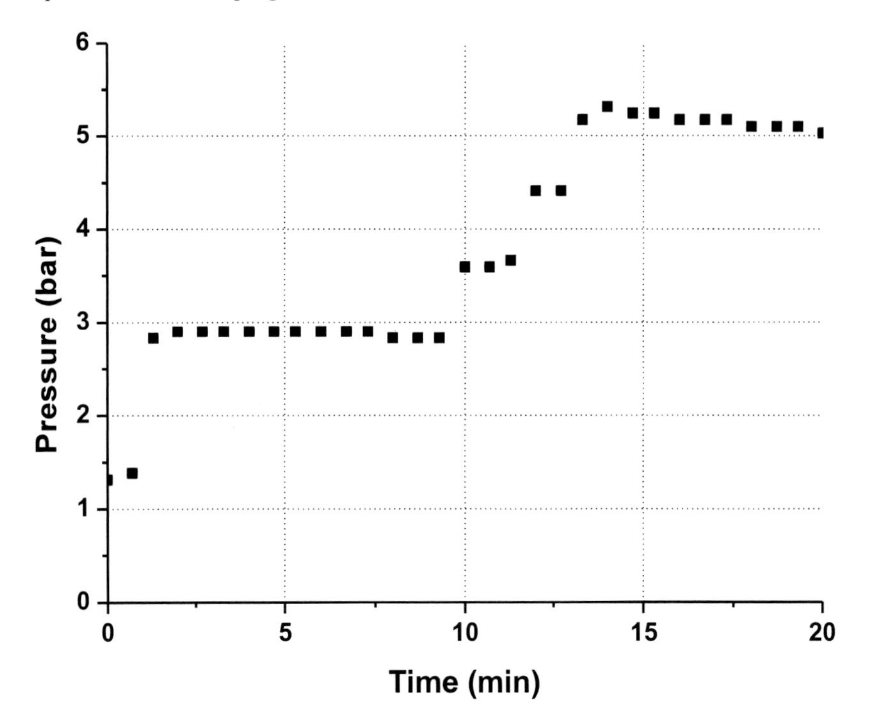

Figure 4. Acquisition of pressure progress in real time.

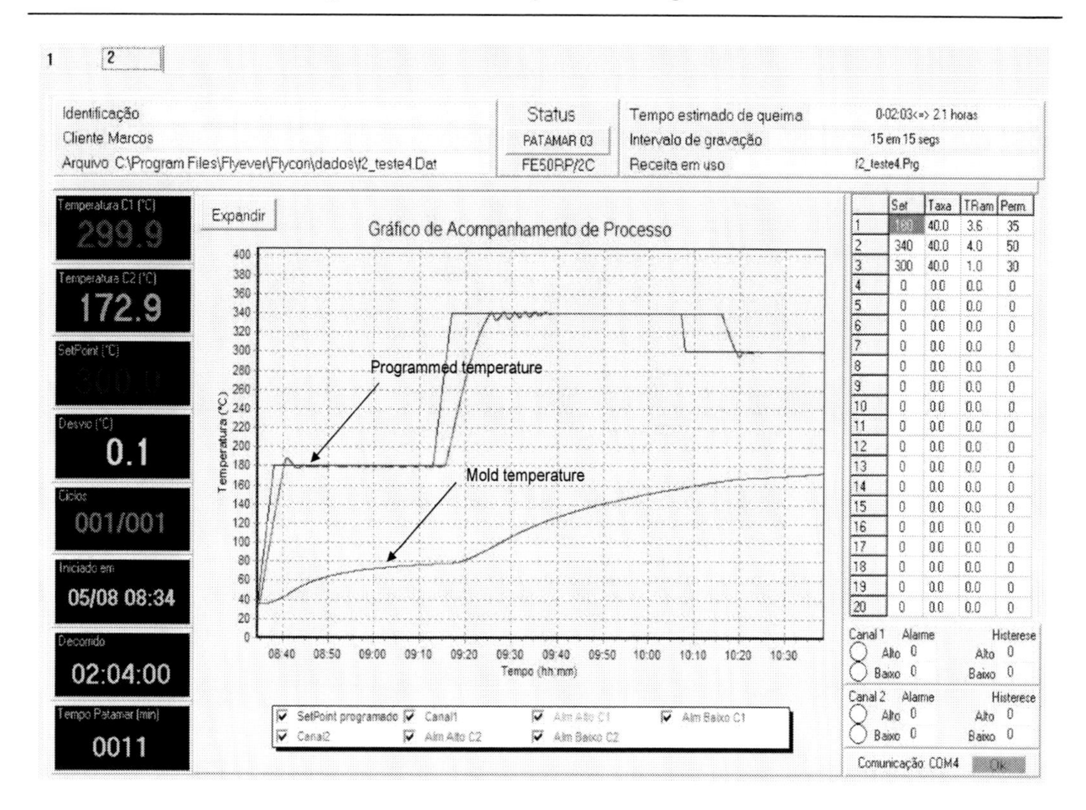

Figure 5. Cure temperature control.

Fatigue Behavior

The fatigue mechanism in metals is characterized by crack nucleation and propagation stages conducting the material to fracture. For composites, the fatigue process is more complex, concerning changes in the microstructure and different failure modes, which are dependent on the applied load [27]. It is a progressive mechanism initiated with the local damage nucleation; from stress concentrators as inclusion, microcracks or voids generated during the process. Nucleation is followed by stable crack propagation, influenced by fiber orientation, matrix ductility and fiber/matrix adhesion [28-29]. The propagation of last loading cycle is analogous to the tensile test fail [13].

To analyze fatigue behavior on composites, experimental tests conducted should demonstrate a minimum of residual stress after the load cyclic, thus generating a severe transition in the S-N curve between the initial cycles and the fatigue strength. For example, the fatigue resistance at constant amplitude under axial loading is high, presenting a smoothly transition, tending to a flat linear curve, in the S-N curve. This is translated as load cyclic insensitivity by composite material, however, damages presence is used to having great influence on materials behavior.

A suitable program to test the material is demonstrated through the characterization of carbon fiber non-crimp quadri-axial orientated +45/0/-45/90 with a mirror 90/-45/0/+45, stitched by a polyester yarn, as conducted by Cioffi et all [5]. This reinforcement is a combination of Hexcel intermediate modulus IM7-12k with areal weight 772g/m^2.

Fig. 6a shows the Carbon Fiber/RTM6 composite laminate produced by RTM technique and provided by Hexcel Composites and Fig. 6b represents the impregnation map from C-scan analysis, as the first step was conducted in order to drawing about defects, which will have further influences on the composites fatigue behavior.

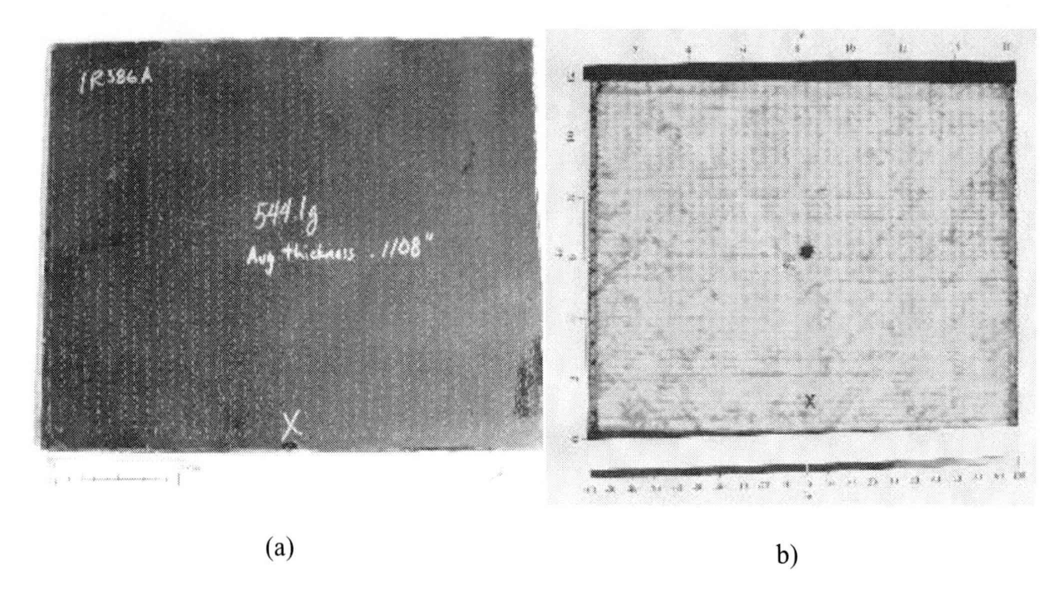

Figure 6. –NC2/RTM6 Composite a) Laminated; b) C-scan map.

After C-scan analysis, axial tensile tests provide the tensile strength, value used as a reference for axial fatigue test. Figure 7a represents axial specimens for tensile and fatigue tests produced based on ASTM D 3039 test method [30]. Tabs could be considered unnecessary based on the valid failure mode obtained; in this case, it was observed that the failure location occurred in the middle part of the specimen with delamination and in an explosive form, as observed in Fig. 7b.

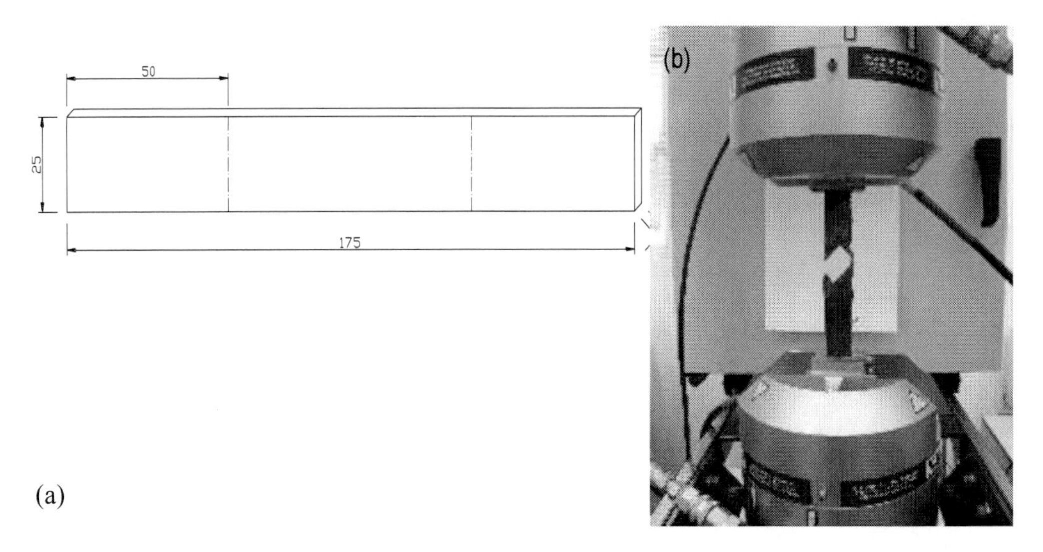

Figure 7. Composite tensile test.

For axial fatigue tests the ASTM D 3479 is used [31]. Experimental tests were performed with a sinusoidal load of 10 Hz frequency and load ratio R = 0.1 in a universal testing machine INSTRON 8801. Herein, R denotes the ratio of the minimum (S_{min}) and the maximum (S_{max}) nominal axial stress components in one load cycle [29].

In order to arrange the fatigue data into a S-N curve, the test, as an example for metals, with intention to establish the fatigue interval for composite, starts with a 75% of axial tensile strength, considering that carbon fiber/epoxy matrix has a brittle character. Depends upon the cyclic life desired, the tests range from higher to lower applied stress.

Figure 8 shows a comparison between the composite with an aeronautical aluminum alloy through normalized S-N curve in which the number of fatigue cycles is represented as a function of the ratio between maximum applied stress and the ultimate tensile strength of the tested materials. It is absolutely clear the highest fatigue strength for the Carbon Fiber/RTM6 composites in comparison to the Al 7050-T7451 alloy. For the composite the applicable stress ratio is in the interval 60-75%; on the other hand for the Al 7050-T7451 alloy the same analysis indicates results between 35-55%.

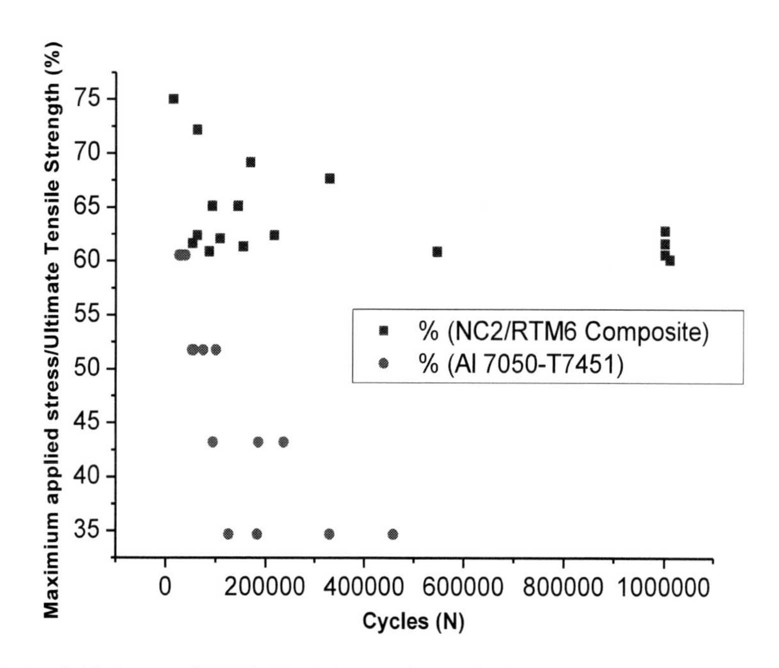

Figure 8. Relative S-N Curve of NC2/RTM6 Composites and Al 7050-T7451[5].

In designs, where lifetime ranges from 10^6 to 10^8 cycles are required, it is known that aluminum fatigue strength is substantially lower regarding the ratio indicated in Figure 8, in contrast, fibrous composite shows a wide load ability without decreasing the number of cycles, maintaining the cycles required for a safety life application [32].

It is important to present that in this experimental procedure, the residual stress and stress redistribution were considered and will discussed during the fatigue degradation issue. However, the delamination process was observed.

In another work, authors observed that fiber-reinforced composites show a significant degradation of the stiffness and strength during cyclic loading and categorized existing fatigue damage theories into [32]:

- General damage accumulation theories.
- Stiffness and strength degradation theories.
- Progressive damage models with damage initiation and evolution.

General damage accumulation theories such as the Miner rule have been investigated extensively for lifetimes accumulation of fibrous composite [33-34] with unsuccessful results.

The complex fatigue damage is based on four types of failure: matrix cracking, interfacial debonding, delamination and fiber breakage. Based on the wide experimental work, models were defined in view of strength degradation, stiffness degradation and energy dissipation. Still, the models are valid just for special composites, which means that fatigue damage mechanism of composites is a subject to be discussed [35].

Regarding energy dissipation during the applied cyclic load, knowledge about viscoelastic parameters is so important as the relaxation phenomena presented by the material, for example, when the polymeric composite is submitted in frequencies of 1, 5, 10, 100 Hz a correlation between viscoelastic parameters and the frequency variation is observed. As indicated in the Figure 9, where is presented the DMA curve, the performance of the laminate in high frequency increases, which is associated to non-time to macromolecules relaxation [36].

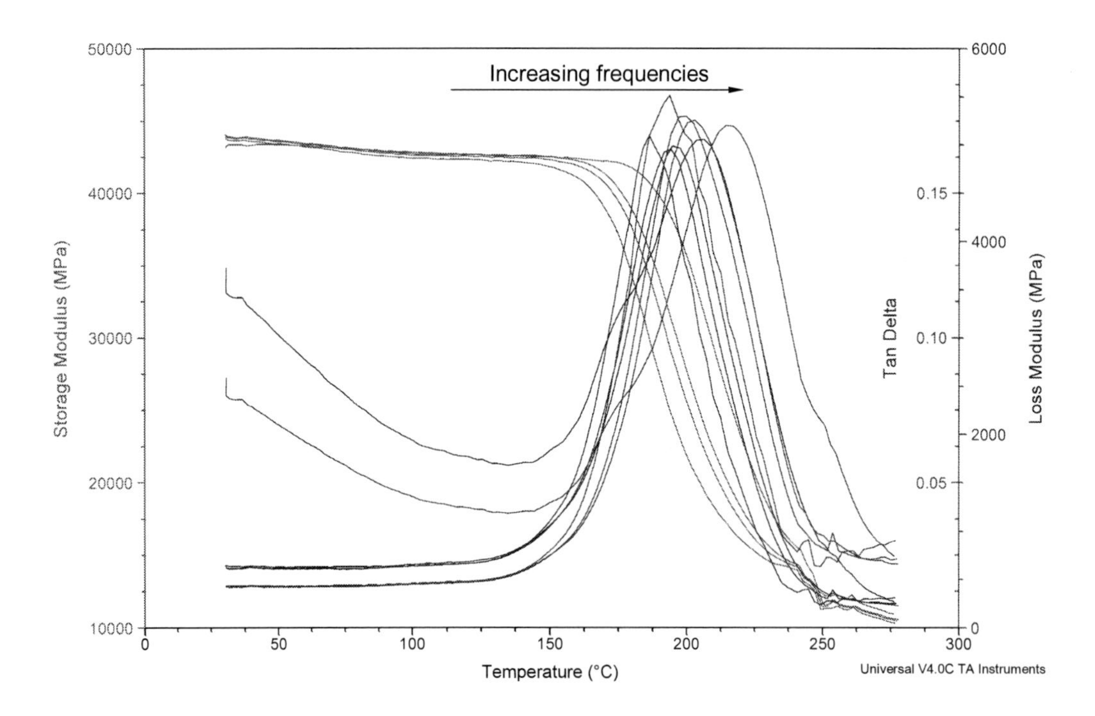

Figure 9. Composite DMA curves in frequencies range of 1, 5, 10, 100 Hz.

The characteristic of relaxation makes the polymeric composites a good candidate for application where energy dissipation is required and, in this case, it occurs by damage development. In fact in contrast to metals, composites present high resistance to the damage growing in repeated strain levels, even for a limit strain of 80%, with consequent small reduction of residual strengths [37].

Moreover, fatigue behavior of composites that are inhomogeneous and anisotropic is defined by different types of damages that can occur in this material: fiber fracture, matrix cracking, matrix crazing, fiber buckling, fiber matrix interface failure, delamination, interactions and different growth rates, which means that microestructural mechanism of damage accumulation or even if interactions occur [38].

Applied cyclic load degrades the mechanical strength and stiffness of the composite. Due to heterogeneity and anisotropy, under applied low cyclic load failures may occur in fibers or matrix in independent form or throughout defects interaction [38, 39]. In this condition the composite can bear the applied load since the stress is redistributed into the component structure [8, 13, 14, 38].

To explain stiffness and strength degradation theories, lets take as example a carbon non crimp fabric/RTM6 composite that presents multiaxial laminates processed by RTM and, as other manufacturing process, more defects than an unidirecional laminate is introduced, which make possible the degradation of the stiffness in fatigue process [40]. These results come from the multiple points of crack nucleation that deteriorate the capacity of the matrix to redistribute the load between the laminae, overcharging a preferential laminae and premature failure is expected.

By taking into account the exposed, some authors have tried to link the degradation of the material (loss of stiffness and strength) to crack density and loading cycles. For a multiaxial lay up the relationship between the parameters changes from ply to ply and the curve becomes non linear, thus the prediction of stiffness degradation usually does not match to experimental results [35,40].

Here it can be reported that, despite the well-described to stiffness degradation [41], the multidirectional laminates have rarely followed the approach of residual stress hold by the laminae of great elastic modulus, when this laminate has a significant void content (~ 5 – 10%). The resulting degradation process can follow a random process that could be difficult to predict the final failure or, in this case, should call premature failure, Figure 10 [5].

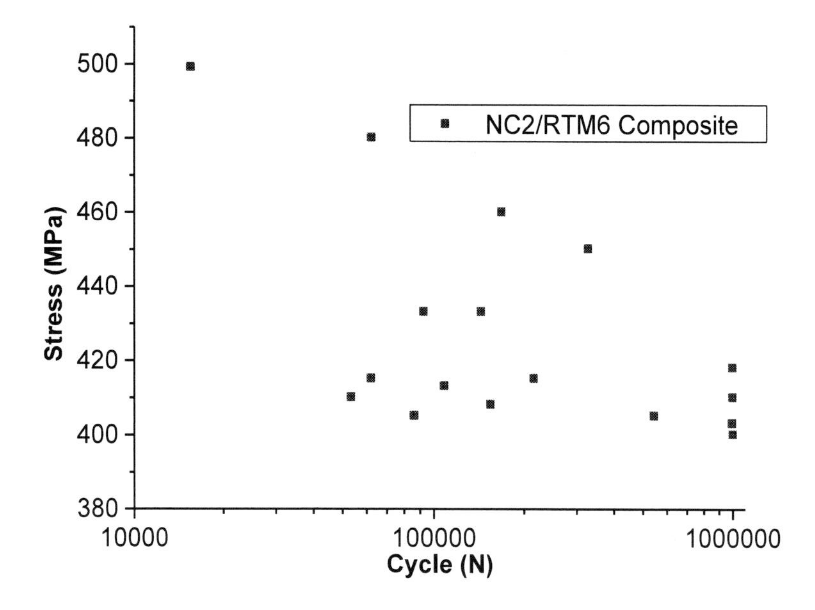

Figure 10. A [90/-45/045]S mutiaxial laminate processed by RTM [5].

Obviously the matrix plays an important hole in fatigue strength and for a multiaxial laminate the damage accumulates, in major part, in the matrix [41]. This information reinforces the aforementioned influence of voids in the matrix, which rapidly reduces the composite stiffness. Following the initial crack caused by voids, promoted by early cycles is the driving force to the delamination process and the stiffness reduction is severe for multiaxial laminates as observed in Figure 11 [5, 42].

Despite the fact that Figure 11 did not provide the information of stiffness reduction, it is easily to perceive that is necessary a few cycles until the final rupture, and between the delamination process and final rupture, considering uniaxial loading, there were rupture in angled plies in the following sequence: 90° plies, 45°/-45° plies, according to the last ply failure (LPF) criteria [43]. The intralaminar crack initiates in the off-axis plies; in this case the 90° plies, and propagates towards the adjacent plies and eventually leading to interlaminar delamination. Afterwards the rupture of the angled plies and 0° plies take place [44].

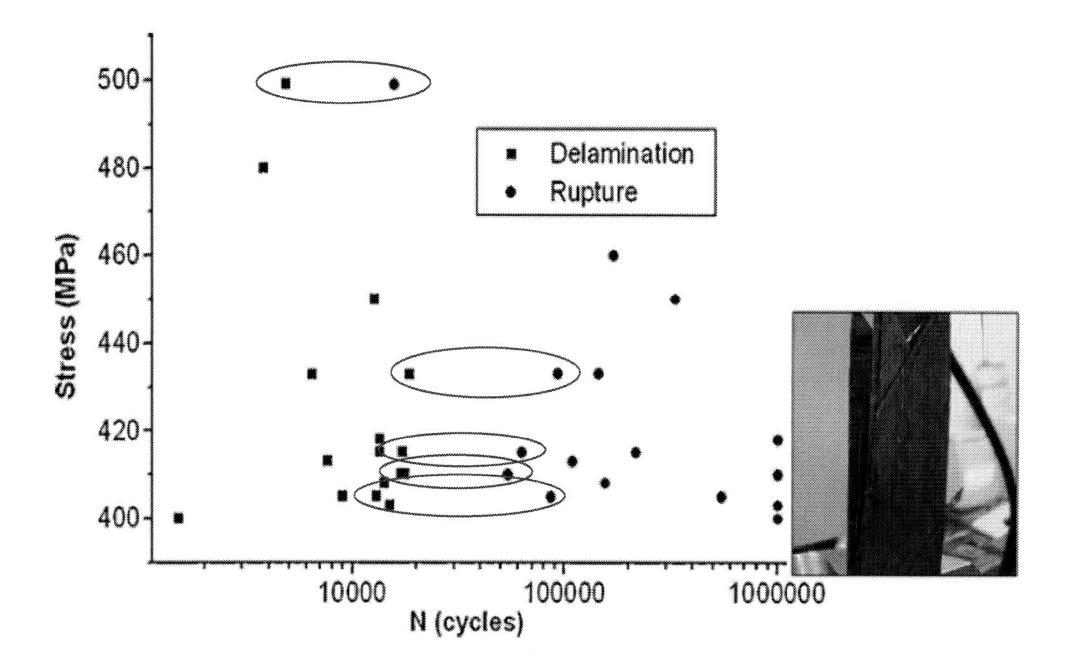

Figure 11. A delamination process in [90/-45/045]S mutiaxial laminate processed by RTM [5].

The case studied is considered homogeneous distribution of defects, as seen in Figure 12 that shows a C-scan of the laminate, which S-N curves is plotted in Figure 11 [45]. In the C-scan map 0% represents total dispersion of energy sound and 100% all the sound returned, considering the pulse-echo method. Based on this information, is exposed how the prediction derived from residual strength or stiffness, since both can evaluate fatigue life [46], fails when defects appear in the cross section and, as a consequence, the gap between the delamination and rupture becomes closed. Besides the defects matter, more variables need to be taken into account in the analysis of the decay in stiffness, mainly in regard of the matrix toughness.

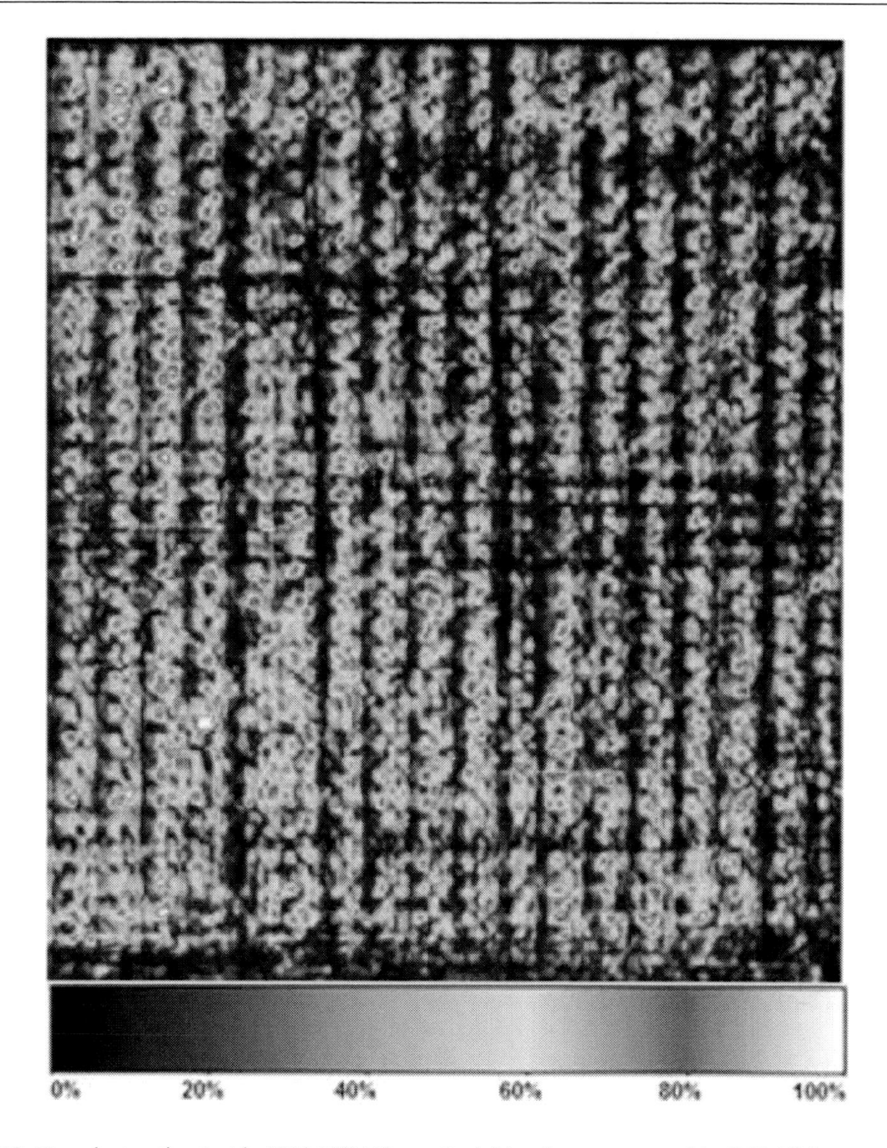

Figure 12. Non-destructive test in $[90/-45/045]_S$ mutiaxial laminate processed by RTM [7].

In an overall view the matrix is responsible to keep the integrity of the role composite. Furthermore this encloses the elasto-plastic properties of the material. The importance of this procedure is related to the toughness, the property that is responsible to absorb the vibration. For a comparison, the main material used in landing gear for aircraft, the aluminum alloy 7050-T74 has almost the same strain (~ 6%) of an epoxy matrix, however the damage of the latter spread all over the component, different in the aluminum alloy that presents a sharp nucleation which propagates until the rupture. The behavior of the thermoset matrix is, in part, connected to the fragile nature, well described by the cross linked structure [47, 48].

Along the years, authors have proposed models to predict the degradation process, but not often one model was enable to be employed in a range of laminates, besides a lack of information about the presence of void content that has influence on a typical stiffness degradation curve, as shown in Figure 13 [44,49].

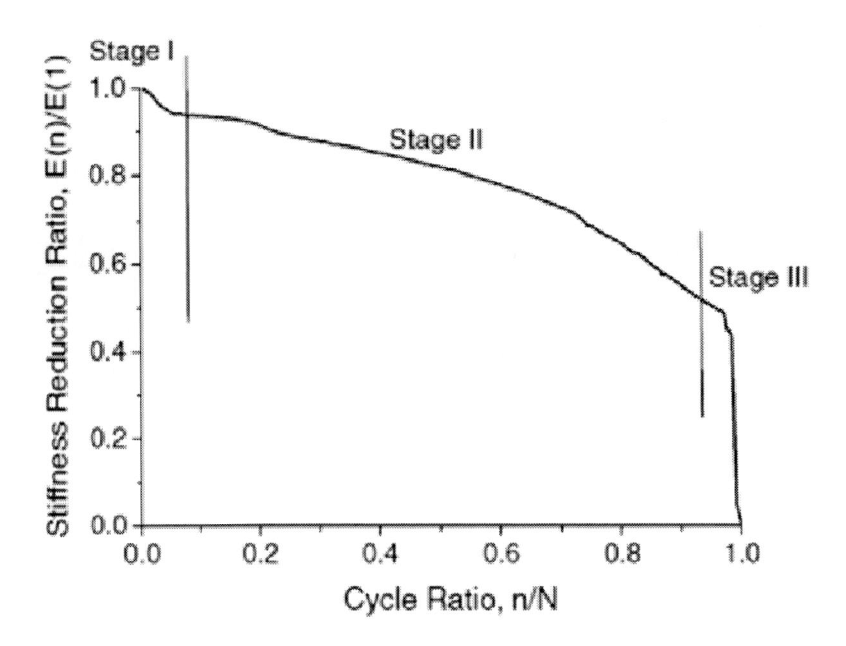

Figure 13. Stiffness degradation curve until rupture in carbon/epoxy braided composite at 75% UTS [50].

As proposed by Mao and Mahadevan 2002, the degradation curve is plotted by applying Eq. (1); where fatigue damage calculated by the Young modulus or stiffness is obtained through the stress-strain curve slope at a specific cycle [51].

$$D \equiv 1 - \frac{E}{E_0} \tag{1}$$

In which:

D is the accumulated fatigue damage;
E_0 is initial Young modulus of the undamaged material
E is the Young modulus of the damaged material.

Experimentally was seen that the stiffness before the complete failure is not zero and, at the final accumulated damage is $1\text{-}E_f = E_0$ as opposed to the unity at the material fail, been E_f the Young modulus when the fracture occurs, which can be consider as a new damage parameter

$$D \equiv \frac{E_0 - E}{E_0 - E_f} \tag{2}$$

According to Eq. (2), the accumulated damage will be in the range between 0 and 1.

As occur in metals, but in a more unpredictable way, the degradation curve in composite material also changes with fatigue parameters, e.g. stress ratio, amplitude, cycling frequency.

By combining materials variables with test parameters, models to describe the degradation process achieved a wide range [50].

Progressive damage model is based on the heterogeneous and anisotropic natures of composite materials. Damage evolution functions are constituted basically by three separated terms: damage initiation, growth and the final progressive damage evolution, as proposed by Paepegem and Degrieck [38]. The damage evolution is formulated by means of the material effort, given by the global failure according to Tsai/Wu criterion [33].

Experimentally, applied cyclic load on composites generates damages, which accumulate in early stage as matrix microcracks, interface debonding, in case of weak fiber/matrix interface and also as fibers fracture when the reinforcement presents any defect. According to these characteristics the macroscopic mechanical properties can be changed. In the following stage, a stable and slow damage growing occurs. Based on that, the loading cycle growing promotes the damages increase in a non-linear evolution form causing delamination until the final fracture of the specimen. These steps are represented in Figures 14a, 14b, 14c and 14d.

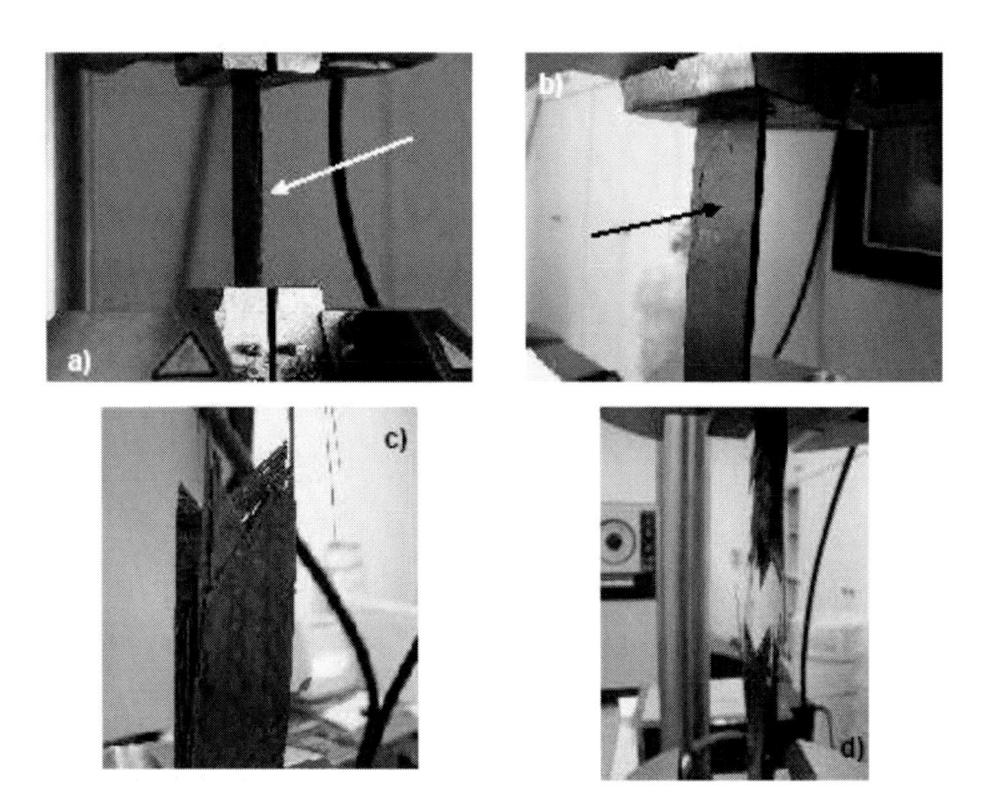

Figure 14. Delamination process: (a) matrix microcracking, (b) damage evolution, (c) delamination, (d) specimen fracture.

Figure 14 illustrates a delamination process of a carbon/epoxy composite occurred during an axial fatigue tests. As indicated in Figure 14b, the polyester stitching used in the carbon reinforcement can be a problem because it can act as a stress concentrator. Figures 14a and 14b show the delamination in layers of different reinforcement orientation until the total fail that occur in an explosive gage middle (XGM) (Figures 14c-d) [30].

A recent work [52] purposes a phenomenological fatigue damage model, which make two important considerations: damages are proportional to the fatigue life of materials and inversely proportional to the fatigue loading level. According the work, the authors describe the damage development and accumulation in composite materials subjected to variable amplitude loading.

In constant amplitude fatigue load, the materials damage development is described by eq. 3.

$$D_n \equiv \frac{E_0 - E_{(n)}}{E_0 - E_f} \equiv 1 - \left(1 - \left(\frac{n}{N}\right)^B\right)^A \tag{3}$$

Under variable amplitude loading, generated damage at initial stage can affect the one produced in the precedent stage, consequently the accumulated damage $D(n_i)$ in the composite submitted to the ith loading could be calculated by eq. 4.

$$D_{(n_i)} \equiv 1 - \left(1 - \left(\frac{n_i + n_{i,i-1}}{N_i}\right)^{B_i}\right)^{A_i} \tag{4}$$

Where $n_{i,i-1}$ is specified in the eq. 5

$$n_{i,i-1} \equiv 1 - \left(\left(1 - \left(\frac{n_{i-1} + n_{i-1,i-2}}{N_{i-1}}\right)^{B_{i-1}}\right)^{\frac{A_{i-1}}{A_i}}\right)^{1/B_i} \tag{5}$$

Where A_i, A_{i-1}, B_i, B_{i-1} are parameters at i^{th} and $(i-1)^{th}$ loadings, respectively, n_i and n_{i-1} are the cycles under the ith and $(i-1)^{th}$ cyclic loadings, N_i and N_{i-1} are the fatigue lives corresponding to the ith and $(i-1)^{th}$ applied loadings, $n_{i,i-1}$ are the equivalent cycles.

According to the same generated damage, the equivalent cycles $n_{i,i-1}$ under i^{th} loading are calculated by the sum of cycles $(n_{i-1} + n_{i-1,i-2})$ under the $(i-1)^{th}$ cyclic loading, and $i \geq 2$; $n_{1,0} = 0$.

When the fatigue failure occurs at the Mth cyclic loading, last step of loading, and the critical damage is defined by eq. 6

$$D_{n_M} \equiv 1 \tag{6}$$

To complete an investigation program of composite fatigue behavior, optical and scanning electron microscopy analyses plays an important role. Composite fracture patterns define phenomena, which occurs during the applied cyclic loading and contribute for the fail.

A confocal reconstruction of NCF/RTM6 composite fatigue fractures specimen is presented in Figure 15. The reconstruction by using an optical microscopy is built taking multiple photos from different depths, the number of photos depends on the depth in this case was approximately 3000 photos, afterwards the photos are assembled, using the public domain software Image J, to generate 3D image that provides a complete information of the fracture.

This image presents a fracture region with different depth levels, discontinues, voids, and resin-rich regions.

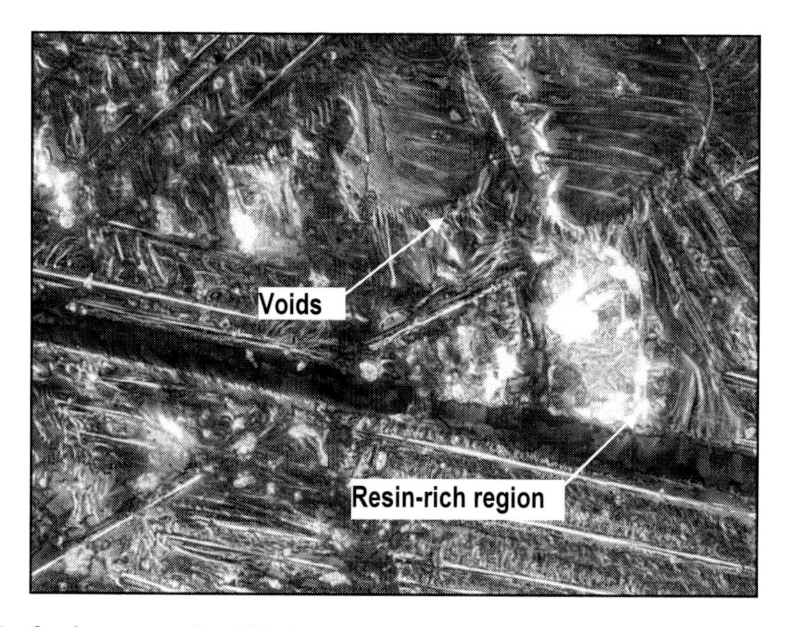

Figure 15. Confocal reconstruction (500x).

This Figure shows resin-rich regions and voids which contributed to the microcracking initiation followed by the delamination process.

Fracture surfaces obtained from fatigue tests should be also analyzed by scanning electron microscopy following, for example, a procedure where tungsten filament is the electron source, the energy in kV is established according to the thermal resistance of material. In the case presented it was used 15 kV, low vacuum technique and images was obtained by scattering electron method when the intention is the patterns visualization. The back scattering method defined the material by atomic weight, then the use is justified by the need of depth material knowing. In the case of material with low or no electrical conduction, samples are covered with a thin gold layer.

Fig. 16a shows a typical radial topography on broken fiber ends, indicating crack propagation for each individual fiber. In Fig. 16b, debonding of fiber from the polymeric matrix and hackles formation as result of deformation in the matrix, typical events present in polymeric material submitted to cyclic load, is observed.

In polymer solids, irreversible deformations are generated by brittle fractures, ductile fracture, rotation of chemical linkages or changes in the orientation of molecular linking, or even a combination of these mechanisms [53].

Figure 16. Fatigue Fracture Surfaces. a) Fiber topography. b) Deformation on the matrix by cyclic load.

Microcracks and shear bands are some common failure modes, which occur during the fatigue tests in polymeric matrices below the glass transition temperature [53]. Fig. 17a shows a delamination fracture and hackles formation, characteristic of polymer matrix submitted to cyclic load. The presence of scallops aspects or hackles associated to the shear load during the fracture process was also observed (Fig. 17b) [54].

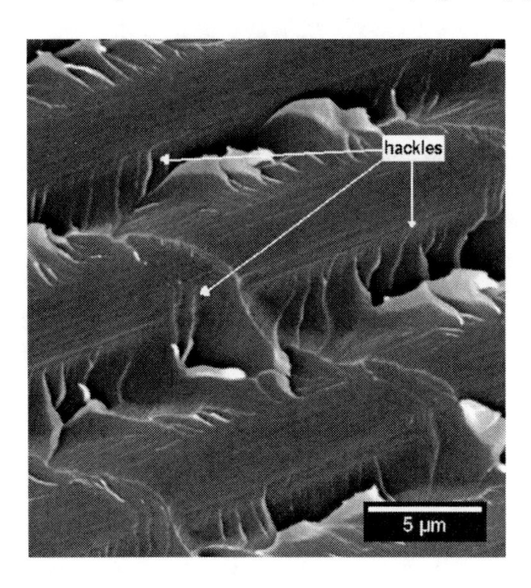

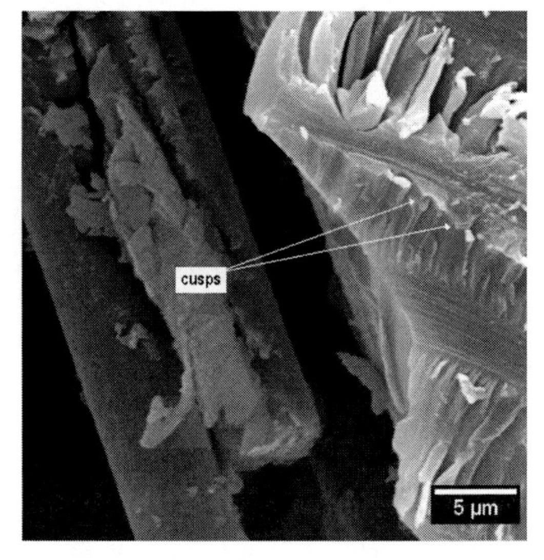

Figure 17. Fatigue Fracture Surfaces. a) Delamination fracture. b) Shear bands.

Polymer failure modes are associated to the molecular structure, specimen geometry, processing methods, temperature tests, loading ratio, stress statement, reinforcement and plastification degree [53]. With respect to the polymeric matrix composites laminates, matrix behavior shows also load dependence. Tensile load in the transversal direction produces ruptures as observed in Fig. 18a and shear load introduces bands as scallops aspects observed in Fig. 18b, similar those found in literature [54-55].

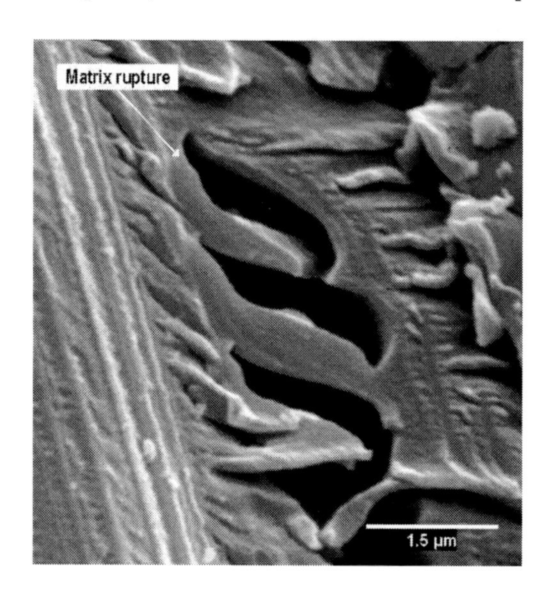

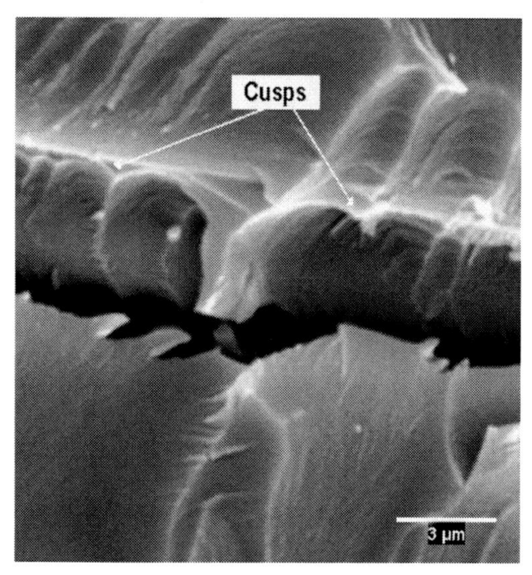

Figure 18. Fatigue Fracture Surfaces. a) Matrix Rupture. b) Scallops formation.

Fig. 10a also indicates high fiber/matrix adhesion and, at the same time, resin cohesive failure [56], associated to the low toughness of the matrix. Fig. 10b shows deformation of matrix and the presence of scallops aspects, which were atributed to the shear load acting in composites during the fatigue test [54].

References

[1] Cioffi, M. O. H., Táparo R. S., Voorwald, H. J. C., Rezende, M. C., Ambrosio, L., Ortiz, E. C., *Evaluation of dynamical-mechanical properties of NC2-RTM6 laminates.* Proceedings of the Polymer Processing Society 24th Annual Meeting ~ PPS-24 ~ June 15-19, 2008. S19-446.

[2] Msallem Y. A., Jacquemin F., Boyard N., Poitou A., Delaunay D., Chatel S. *Material characterization and residual stresses simulation during the manufacturing process of epoxy matrix composites.* Composites: Part A 41 (2010) 108–115.

[3] P. Mouritz, K. H. Leong, I. Herszberg. *A review of the effect of stitching on the in-plane mechanical properties* . Composites Part A: Applied Science and Manufacturing, 28 (12) (1997) 979-991.

[4] *Composite materials revolutionise aerospace engineering.* Ingenia issue 36 september 2008. 27.

[5] Cioffi, M. O. H., Voorwald, H. J. C., Camargo, J. A. M., Rezende, M. C., Ortiz, E. C., Ambrosio, L. *Fractography analysis and fatigue strength of carbon fiber/RTM6 laminates.* Materials Science and Engineering: A, 527 (15)(2010) 3609-3614.

[6] Voorwald, H. J. C., Volpiano, G., Cioffi, M. O. H., Rezende, M. C., Ortiz E. *Fatigue behavior of NC2/RTM6 composites: statistical analysis by Weibull distribution.* Proceedings of the Polymer Processing Society 24th Annual Meeting ~ PPS-24 ~ June 15-19, 2008. S19-1098.

[7] Vinson, J. R., Sierakowski, R. L. *in Mechanics of structural systems – the behavior of structures composed of composite materials* (kluwer Academic Publishers, 1993).

[8] Hull, D., Clyne, T. W. *in An Introduction to Composite Materials.* (Cambridge Solid State Science Series, 1996) 1-8.

[9] Antonucci, V., Giordano, M., Nicolais, L., DI Vita, G. *A simulation of the non isothermal resin transfer molding process.* Polymer Engineering and Science. 40 (12) (2000) 2471-2481

[10] Hillermeier, R. W., Seferis, J.C. *Interlayer toughening of resin transfer molding composites.* Composites: part A. 32 (2001) 721-729.

[11] Luo, J., Liang, Z., Wang, B. *Optimum tooling design for resin transfer molding with virtual manufacturing and artificial intelligence.* Composites: part A. 32c (2001) 877-888

[12] Opperer, J. G., Kim, S. K., Daniel, I. M. *"characterization of local perform defects in resin transfer molding by the gas flow method and statistical analysis".* Composites Science and Technology. 64 (2006)1921-1935.

[13] Pandita, S. D., Huysmans, G., Wevers, M., Verpoest, I. *"Tensile fatigue behaviour of glass plain-weave fabric composites in on- and off-axis directions".* Composites: Part A. 32 (2001)1533-1539.

[14] Salekeen, S., Jones D. L. *Fatigue response of thick section fiberglass/epoxy composites.* Composites Structures 79 (1) (2007) 119-124.

[15] *[http://www.io.tudelft.nl/live/pagina.jsp?id=b66d141c-d01e-4474-85db-062c93a4ea7a&lang=en.* August, 23rd 13:30 h].

[16] Potter, K.D. *The early history of the resin transfer moulding process for aerospace applications.* Composites: part A. 30 (1999) 619-621.

[17] Pearce, N. R. L., Guild, F.J., Summerscales J. *The use of automated image analysis for the investigation of fabric architecture on the processing and properties of fibre-reinforced composites produced by RTM.* Composites: part A. 29 (1998) 829-837.

[18] Cairns, D., Humbert, D. R., Mandell, J. F. *Modelling of resin transfer molding of composites materials with oriented unidirectional plies.* Composites: part A. 30 (1999) 375-383]

[19] Lawrence, J. M., Hsiao, K. T., Don, R. C., Simacek, P., Estrada, G., Sozer, E. M., Stadfeld, H. C., Advani, S. G. *Ana approach to couple mold design and on line control to manufacture complex composites parts by resin transfer molding.* Composites: part A. 33 (2002) 981-990

[20] Bang, K. G., Kwon, J. W., Lee, D. G., Lee, J. W. *Measurement of the degree of cure of glass fiber-epoxy composites using dielectrometry.* Journal of Materials processing technology. 113 (2001) 209-214.

[21] Young, W. B., Lai, C. L. *Analysis of the edge effect in resin transfer molding.* Composites: part A. 28 (1997) 817-822.

[22] Lokakou, C., Johari, M. A. K., Norman, D., Bader, M. G. *Measure techniques and effects on in-plane permeability of woven cloth in resin transfer molding. Composites: part A*. 27 (1996) 401-408.

[23] Weitzenböck, J. R., Shenoi, R. A., Wilson, P. A. *Radial flow permeability measurement.* Part A: Theory. Composites: Part A. 30 (1999) 781-796.

[24] Gorovaya, T. A., Korotkok, V. N. *Quick cure of thermosetting composites*. Composites: Part A. 27 (1996) 953-960.

[25] Abraham. D., Mattews, S., Mcllhagger, R. *A comparison of physical of glass fibre epoxy produced by wet lay-up with autoclave consolidation and resin transfer moulding. Composites: part A*. 29 (1998) 795-801.

[26] Song, Y., Chui, W. Glimm, J. Lindquist, B. Tangerman, F. *Application of front tracking to the simulation of resin transfer molding. Computers & mathematics applications*. 33 (9) (1997) 47-60.

[27] L. A. L. Franco, M. L. A. Graça, F. S. Silva. *Fractography analysis and fatigue of thermoplastic composite laminates at different environmental conditions* Materials Science and Engineering: A,488 (2008) 505-513.

[28] K. Vallons, M. Zong, S. V. Lomov, I. Verpoest. Comp.: Part A 38 (2007) 1633–1645; B. Harris. *Fatigue in composites*. 1.ed. Cambridge: Woodhead Publishing Limited, 742p.

[29] N. Himmel, C. Bach. *Cyclic fatigue behavior of carbon*.Inter. J. Fat 28 (2006) 1263–1269.

[30] ASTM D3039/06. *Standard Test Method for Tensile Properties of Polymer Matrix Composite Materials*. Fong editor.

[31] ASTM D3479/02. *Standard Test Method for Tension-Tension Fatigue of Polymer Matrix Composite Materials1*. Fong editor.

[32] Carlson, R. L., Kardomateas, G. A. *in Na introduction to Fatigue in metals and Composites*. Chapman & Hall. 1st ed. 1996.

[33] M. Gude *, W. Hufenbach, I. Koch. *Damage evolution of novel 3D textile-einforced composites under fatigueloading conditions.* Composites Science and Technology 70 (2010) 186–192.

[34] Jayantha A. Epaarachchi. *A study on estimation of damage accumulation of glass fibre reinforce plastic (GFRP) composites under a block loading situation* Composite Structures 75 (2006) 88–92 and M. Gude *, W. Hufenbach, I. Koch.

[35] Wu, F., Yao, W. *A fatigue damage model of composite materials*. Internacional Jornal of Fatigue 32 (2010) 134–138.

[36] Cioffi, M. O. H., Voorwald, Rezende, M. C., V. Ambrogi, C. Carfagna, V. Antonucci. *Thermal regulation in NCF/RTM6 composites*. Proceedings of COMATCom 2009.

[37] Baker, A., Dutton, S., Kelly, D. *in Composite materials for aircraft structures*. 2nd edition, AIAA Education Series 2004.

[38] Vanpaepegem, W, Degrieck, J. *Modelling damage and permanent strain in fibre-reinforced composites under in-plane fatigue loading*. Comp. Sci. Tech. 63 (2003) 677–694.

[39] ASM - *American Society For Metals*. Handbook Fatigue and Fracture.

[40] Adden S, Horst P. *Stiffness degradation under fatigue under multiaxially loaded non-crimp-fabrics*. International Journal of Fatigue 32 (2010) 108 – 102.

[41] Belingardi G, Cavatorta MG. *Bending fatigue stiffness and strength degradation in cabon-glass/epoxy hybrid laminates: Cross-ply vs. angled-ply specimens*. International Journal of Fatigue 28 (2006) 815 – 825.

[42] Duan X, Yao WX. *Multi-directional stiffness degradation induced by matrix cracking in composite laminates.* International Journal of Fatigue 24 (2002) 119 – 125.

[43] Kaw AK. *Mechanics of composite materials.* Second Edition, Taylor & Francis, United States, 2006.

[44] Taheri-Behrooz F, Shokrieh MM, Lessard LB. *Residual stiffness in cross-ply laminates subject to cycled loading.* Composite Structures 85 (2008) 205-212.

[45] Shiino MY, *Cioffi MOH, Voorwald HCJ. Flexural and Nondestructive Tests on NC2/RTM6 Composites Manufactured by RTM for Aerospace Applications.* Procedings COMATCOMP 2009.

[46] Lee JL, Fu Ke, Yang JN. *Prediction of fatigue damage and life for composites laminate under service loading spectra*. Composites Science and Technology 56 (1996) 635-648.

[47] Sobczyk K, Trebicki J. *Stochastic dynamics with fatigue-induced stiffness degradation.* Probabilistic Engineering Mechanics 15 (2000) 91-99.

[48] MIL-A-22771D-MILITARY SPECIFICATION. *Aluminun Alloy Forgings, Heat treated.* Superseding, 1984.

[49] Tate JS, Kelka AD. *Stiffness degradation model for biaxial braided composites under fatigue loading.* Composites: Part B 39 (2008) 548-555.

[50] Degriek J, Paepegem WV, *Fatigue damage modelling of fibre-reinforced composite material: Review.* Applied Mechanics Review 54 (4) 279-300.

[51] Mao, H., Mahadevan, S. *Fatigue damage modelling of composite materials*. Composite Structures 58 (2002) 405–410.

[52] Wu, F., Yao, W. *A fatigue damage model of composite materials*. International Journal of Fatigue 32 (2010) 134–138.

[53] S. Suresh. *Fatigue of Materials. 2.ed.* Cambridge: Cambridge University Press, 1998, 679p.

[54] Sjögren, L. E. Asp. *Effects of temperature on delamination growth in a carbon/epoxy composite under fatigue loading* International Journal of Fatigue 24 (2002) 179–184.

[55] E. Totry, C. González, J. Llorca. *Failure locus of fiber-reinforced composites under transverse compression and out-of-plane shear Composites* Science and Technology 68 (2008) 829–839.

[56] T. J. Vogler, S. Y. Hsu, S. Kyriakides. *Composite failure under combined compression and shear International* Journal of Solids and Structures 37 (2000) 1765-1791.

In: Advances in Mechanical Engineering Research, Volume 3 ISBN: 978-61209-243-0
Editor: David E. Malach

Chapter 7

DESIGN AND MODEL OF A SERIES HYBRID PROPULSION SYSTEM FOR A LIGHT URBAN VEHICLE

Felipe Jiménez** *and José María López
Universidad Politécnica de Madrid, University Institute for Automobile Research, Insia, Campus Sur UPM, Carretera de Valencia km 7, 28031, Madrid, Spain

ABSTRACT

This chapter presents the model of the components of a hybrid propulsion system and their integration of a light urban vehicle. This concept of vehicle originated from a specific requirement to design a vehicle with a concrete application: to collect used batteries from urban bus-shelters. The propulsion system is a series hybrid configuration with an internal combustion engine as the main charge source for the power batteries feeding an electric motor. The system is also fitted with a regenerative brake and solar panels to recharge the batteries.

The models of the components try to describe the internal processes that take place in them, so they are modelled in detail. More specifically, a complete internal combustion engine is developed considering the following submodels: Air intake model, Fuel injection model, Air-fuel mixture model, Combustion model, Vehicle dynamics model, Exhaust Gas Recirculation (EGR) model and Vapour Canister Purge (VCP) model. The model of the batteries gives information on the state of charge (SOC) at any time and calculates the changes in stored energy as a result of: charge and discharge cycles, self-discharge and variations in temperature. Finally, control strategies are proposed.

The model can be used with two main purposes:

- During the design phase, it can be used for components selection according to the initial specifications.
- It can also be used to evaluate the vehicle performance under different operating conditions.

Keywords: hybrid vehicle, series configuration, greenhouse effect, battery, solar panel

---

*Email: felipe.jimenez@upm.es, Telephone: +34 91 336 53 17, Fax: + 34 91 336 53 02

1. INTRODUCTION

The current trend towards increased mobility in the most advanced societies runs counter to the criteria for controlling the greenhouse effect, local pollution and the exploitation of fuel resources.

If energy policies remain as now, world energy demand will have grown by 65% and CO_2 emissions by 70% between 1995 and 2020 [1]. The IEA is investigating two possible solutions, one based on regulation to confront pollution and another based on increasing fuel prices. The first solution would achieve a reduction in CO_2 emissions of around 50%. The other half of the reduction in emissions would be achieved by replacing the fossil fuels used to generate electrical energy by non-fossil fuels. The second scenario contemplates a levy on the price of fossil fuel depending on its carbon content. This additional charge would be sufficient to reduce CO_2 emissions by half, the amount required to meet the Kyoto objectives.

However, a 12 to 15% increase is predicted in the number of world population becoming motorised by 2020 [2]. This would see an increase in the demand for primary energy, and if fossil energy is used, it will mean a major increase in local emissions (nitrogen oxides NO_X, carbon monoxide CO, unburnt hydrocarbons HC and particles matter PM) and in the corresponding greenhouse gases emissions.

The European Commission's goal is for mean carbon dioxide emissions to settle at around 120 grams per kilometre in the year 2010. The three main instruments for reaching this goal are: commitment from the car industry, vehicle labelling and tax instruments [3][4].

In recent years, many studies have been carried out to assess and compare fuel consumption reduction or fuel economy potentials for conventional and advanced technologies options [5][6][7].

The sustainability of the transport sector will depend heavily on the introduction of technologies to reduce polluting emissions and the consumption of fossil fuels. Strategies have recently been put forward that are interrelated with the intention of reducing both local emissions and the greenhouse effect. The solutions put forward [8] include:

- Reciprocating internal combustion engines
- Exhaust treatment
- Hybrid and battery electric powertrains
- Fuel cells
- Transmissions

Hybrid propulsion systems were proposed many years ago as a feasible and efficient solution for reducing consumption and emissions [9][10][11][12]. This work proposes the design of a hybrid propulsion system in a special urban scope vehicle for the city of Madrid (Spain). A simulation model has also been developed to facilitate evaluating the performance and features of the vehicle's powertrain as well as a tool to aid designing the vehicle.

2. Vehicle Specifications and Layout

The design of the vehicle presented in this chapter is aimed at a specific application. The two-seat vehicle with a load bay in the rear (figure 1), is designed to collect used batteries from urban bus-shelters in large cities. The vehicle has been devised to have a hybrid propulsion system which in order to attain an acceptable energy efficiency combines a internal combustion engine, an electric motor, storage cells and solar panels. The vehicle has a rear-wheel drive, must reach 50 km/h and have an electrical autonomy of at least 30 kilometres.

Figure 1. General view of the vehicle designed.

Of the usual hybrid configurations (series, parallel, power-split and srigear configurations [13]), the series configuration has been chosen, which unlike other applications enables an exact control of the power transfer to be had at any instant so its operation can be optimised for ultra-low petrol consumptions or even zero if circumstances require this. In addition, this configuration is optimum for urban operation with many short journeys with stops where power requirements are not excessive. Thus, the drive is provided by the battery-fed electric motor or the generator- internal combustion engine set. The ultimate objective is to replace the internal combustion engine by a fuel cell, a technology under development that has been identified as one of the most promising solutions for vehicle propulsion.

This series configuration involves the following basic energy management:

- The main storage source of the electrical energy is the power batteries (high voltage), that feed the electric induction motor through the inverter, whose mission is to transform direct current into three-phase alternating current according to the needs of the electric motor.
- The internal combustion engine operates the electrical generator. Part of this energy drives the electric motor and the surplus is stored in the batteries.

- The batteries are also recharged when the electric motor regains energy during braking (regenerative braking).

In addition, the client's specifications refer to other aspects of energy management such as:

- It must be possible to externally charge the batteries from the 220 V AC mains supply.
- Solar panels are included to supply additional energy, either to the low voltage circuit (vehicle accessories, control electronics, etc) when the vehicle is in motion, or to the power batteries when the vehicle is stationary and these batteries are not fully charged.
- The driver, as well as having access to the vehicle's standard accessory controls (lights, radio, air conditioning, etc) must be able to control features related to propulsion system management like regenerative braking or disabling the combustion engine regardless of the battery state of charge for circulation in zones requiring zero pollution levels.

3. Simulation Model

Preliminary studies on the usual working conditions of this type of vehicle allow an initial sizing to be made for the vehicle parts. On the other hand, there are numerous mathematical models of hybrid propulsion systems [14][15][16][17][18][19][20][21][22] There are also advanced techniques for choosing the propulsion system for a hybrid vehicle, defining the energy control strategy and studying the sensitivity of different solutions on variables such as consumption [23][24][25][26][27][28][29][30][31][32]. However, the particular configuration of the vehicle designed, fitted with solar panels in addition to the series hybrid structure, together with a need to set control strategies that conform to pre-set energy management specifications, and a requirement to simulate working in line with variables that would be measured when the prototype is built, justify developing a specific simulation model capable of evaluating hybrid propulsion system performance in this series configuration.

The computer tool used to design the models is Matlab/Simulink. The model comprises modules that represent the different parts making up the vehicle. Simulating the whole enables the influence that the driving cycle, the control strategy, and the subsystems forming the powertrain have on energy and fuel consumption.

Figure 2 shows a general schematic outline by blocks of the full simulation model. Starting out from a set cycle speed the torque and speed requirements demanded of the electric motor are calculated according to the vehicle data. For the electric motor model, the electric power consumed to provide the required torque and speed is calculated. The batteries model determines the state of charge in line with the power levels demanded by the engine and other loads, and the power released by the motor/generator set and the solar panels. Internal combustion engine ignition is controlled by the battery state of charge, depending on the energy management strategy.

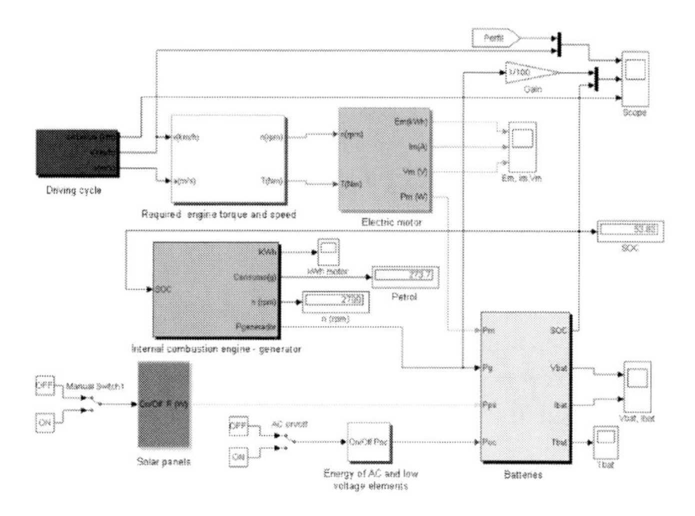

Figure 2. Matlab/Simulink model of the series hybrid configuration.

Below, we go on to describe the most relevant blocks of the developed model.

3.1. Calculating the Torque and Speed Required

The performance that will be demanded of the electric motor is estimated from the operating conditions. The input signals to the motor are speed and acceleration. The mission of this block is to determine at what point of its torque-speed curve the electric motor should be situated at every instant in order for the vehicle to follow the set cycle.

The resistances considered are expressed in the following classic form:

$$R = \frac{1}{2} \cdot \rho \cdot C_x \cdot A_f \cdot v^2 + M \cdot g \cdot \left(f_r + \sin\theta + a\right) \tag{1}$$

where ρ is air density, C_x the aerodynamic coefficient, A_f is the front area of the vehicle, v is vehicle speed, M is vehicle weight, g is gravity acceleration, f_r is the coefficient of rolling resistance, θ is the ramp and a is vehicle acceleration.

Transforming wheel variables to engine variables is done by means of the following expressions:

$$\text{Engine speed } n = \frac{30 \cdot v \cdot \xi}{\pi \cdot r_e} \tag{2}$$

$$\text{Engine torque } T = \frac{R \cdot r_c}{\xi \cdot \eta} \tag{3}$$

where r_e is effective tyre radius, r_c is the radius under load, ξ is the transmission ratio and η transmission performance.

3.2. Motor-Generator Set

A internal combustion engine running as part of a hybrid application allows for greater efficiency, as it is no longer necessary to optimise a wide range of engine speeds or be oversized in order to meet the severest operating requirements since other energy sources absorb these peaks [13].

The engine model is divided into various subgroups:

- Air intake model. This calculates the airflow mass entering the cylinders ($\dot{m}_a$) and the intake pressure (p_{adm}) from the position of the accelerator throat (α) and the engine speed (n).
- Fuel injection model. Calculates the fuel mass flow entering the combustion chamber
- ($\dot{m}_f$) from the injection time (t_i) and the engine speed (n).
- Air-fuel mixture model (parameter λ). This calculates the actual value of λ from air mass flows ($\dot{m}_a$) and fuel mass flows ($\dot{m}_f$).
- Combustion model. Given λ, mass flow ($\dot{m}_a$), intake pressure (p_{adm}) and the engine speed (n) the brake power (N) is calculated.
- Vehicle dynamics model. Given the brake power (N), the engine speed (n) is calculated.
- Exhaust Gas Recirculation (EGR) model. Given the intake pressure (p_{adm}) and the engine speed (n) the exhaust mass flow is calculated.
- Vapour Canister Purge (VCP) model. Given the intake pressure (p_{adm}) and the external pressure (p_{ext}) the fuel vapour mass flow is calculated.

The interconnection of the different blocks is shown in figure 3.

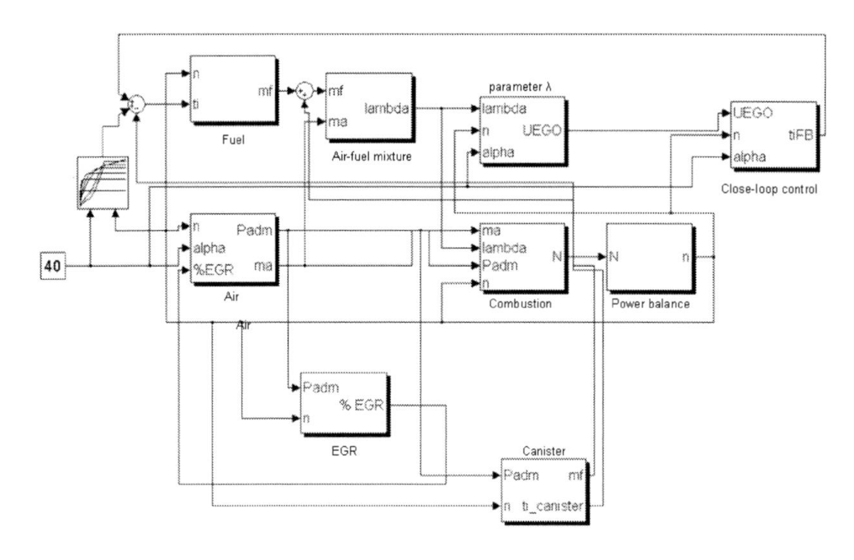

Figure 3. Engine blocks diagram.

The new systems included in this engine model are the Exhaust Gas Recirculation (EGR) subsystem and the Vapour Canister Purge (VCP) subsystem. Exhaust Gas Recirculation (EGR) is controlled by a subsystem of the engine electronic control system that regulates the amount of exhaust gases returning to the cylinders. The temperature, on the inside of the cylinders can reach over 1,650 °C under normal operating conditions. High temperatures increase nitrogen dioxide emissions (NOx). A small amount of exhaust gases (between 5 and 20% as shown in figure 4) enters the cylinder diluting with the air from the intake. The result is a drop in the mean temperature, and therefore in NOx emissions.

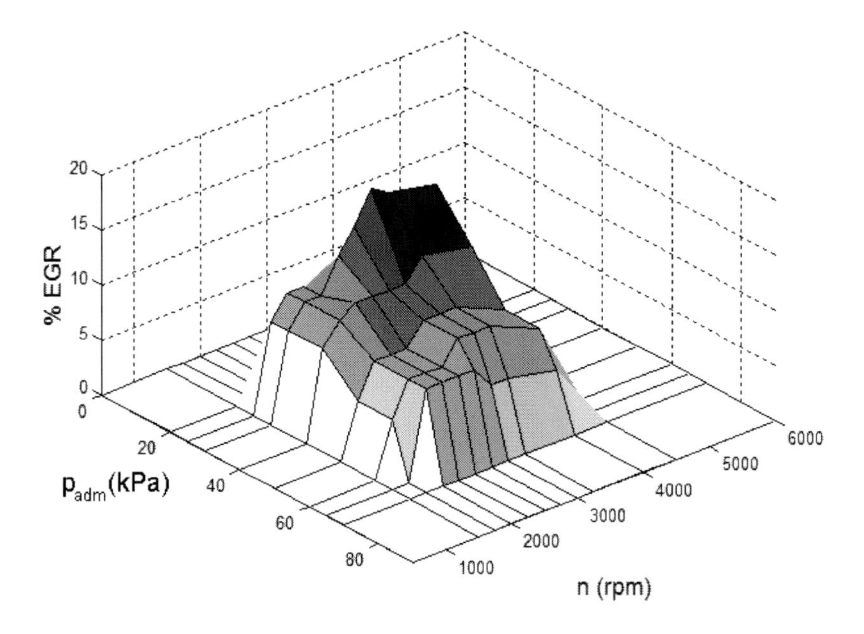

Figure 4. EGR map.

The control algorithm determines when the EGR valve should open or close. The valve must remain closed during ignition, when the engine is cold, ticking over, during rapid accelerations or in other circumstances requiring a high torque. The EGR control signal is determined taking account of the engine speed (n), the temperature of the coolant and the load level (intake pressure). The EGR signal can control the valve opening which is detected by a positional sensor or a differential pressure sensor [33]. This sensor sends an electric signal needed for the open loop injection to know at any one time the amount of exhaust gases re-circulating and take action to ensure the right air/fuel ratio. The electronic control unit (ECU) calculates the required EGR flow according to the operating conditions of the engine and uses the difference between this number and the actual EGR flow as the input to a digital filter, whose output determines the percentage amount of the duty cycle of the signal applied to the EGR valve [34], the air intake block has to be modified.

$$\dot{m}_a = \dot{m}_{a_{int\,ake\,valve}} \left(1 - \%EGR\right) \tag{4}$$

where $\dot{m}_{a_{int\,ake\,valve}}$ is the air mass flow throw intake valve.

The model also include a subsystem to regulate the fuel vapours entering the engine from the canister vent, and the adjustment needed to the injector control signal due to the additional fuel vapour.

The total mass flow purged is calculated by using Bernoulli's equation for a section:

$$\dot{m}_p = A_{eff} \frac{P_0}{\sqrt{RT_0}} f\left(\frac{P_{adm}}{P_0}\right) \tag{5}$$

where P_{adm} is the intake pressure, P_0 is the external pressure and A_{eff} is the effective area and depends exclusively on the solenoid valve duty cycle (vdc). It can be approximated to an equation by determining its parameters experimentally [35]:

$$A_{eff} = 5 \cdot 10^{-5} \cdot vdc^{0,8} - 8 \cdot 10^{-6} \cdot vdc \tag{6}$$

where vdc is the valve control signal, which is 0 when the solenoid valve is fully closed and 1 when it is fully open.

The engine control instructions should let it work at its optimal operational condition. Since the generator will only have one operational stage, its behaviour can be simulated by inserting its performance at that stage. Depending on the state of charge, the ignition control takes over a block called the "Engine control" and must deal with the additional loads that may be caused by the air-conditioning while keeping the engine speed constant. Figure 5 shows this control over the engine. The situation of the engine being switched off is also considered. Cold engine behaviour can be simulated by applying corrective coefficients to its running under standard operating conditions [36].

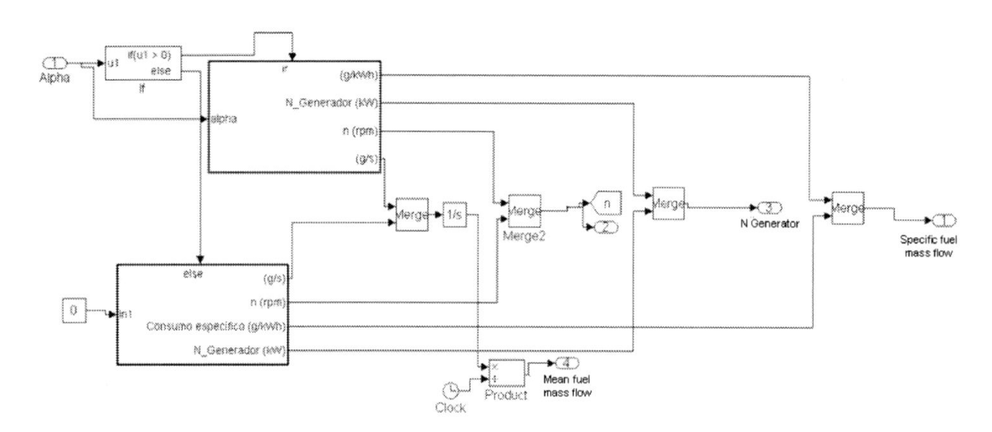

Figure 5. Internal combustion engine control.

3.3. Batteries

As electrochemical storage components, batteries present many uncertainties when evaluating their performance and a substantial variability when different parameters are changed. [37][38]. Phenomena of hysteresis, variations in capacity according to temperature, variations in useful life depending on the extent of discharge, etc. are all difficult to evaluate. Many attempts have been made to model battery performance ([36][39][40][41][42], among others). The objectives to be met by any battery model for simulating an electrical hybrid vehicle powertrain are as follows:

- To give information on the state of charge (SOC) at any time. This parameter is essential since hybrid vehicle control strategies are based on this for evaluating electrical autonomy.
- To describe battery energy, power, voltage, current and efficiency as a function of state of charge during the changing charge/discharge conditions of a driving cycle.
- To calculate the changes in stored energy as a result of:
  - Charge and discharge cycles.
  - Charge recovery during regenerative braking.
  - Self-discharge.
  - Variations in temperature.
- To determine the limits for the batteries to accept charge.
- To incorporate a thermal model to allow estimating the variations in temperature according to the different control strategies.

The simplest simulation models use an equivalent circuit comprising an internal resistance (which varies during the life of the battery [37]) in series with an ideal voltage source, even if dynamic models include condensers in the equivalent circuit [43]. The main drawback to this is estimating the characteristic parameters of the circuit elements. The state of charge is calculated as the quotient between the surplus energy and the maximum energy the batteries can supply, which, in turn, depends on the charge or discharge rate. Although the state of charge of the batteries can be evaluated from the open circuit voltage with which there is a linear relationship [37] as figure 6 shows, or from the constant discharge rate curves (figure 7), these measurements are unacceptable during in-vehicle use and other procedures have to be used [38]. Different alternatives that attempt to resolve the dependence of capacity on the rate of discharge can be found in the bibliography. So, while methods can be found to calculate discharge in respect of the capacity corresponding to the average rate of discharge in an interval of time [44], others refer the magnitudes to a reference current using corrective coefficients [45] and among the latter there are developments that take account of electrolyte dispersion during pauses [38].

To evaluate the state of charge a balance of powers is conducted to calculate the power the batteries need to supply or receive. The state of charge calculation (SOC) is performed from the battery's state of charge at the instant prior to the direct count of the amperes-hour input or output of the batteries.

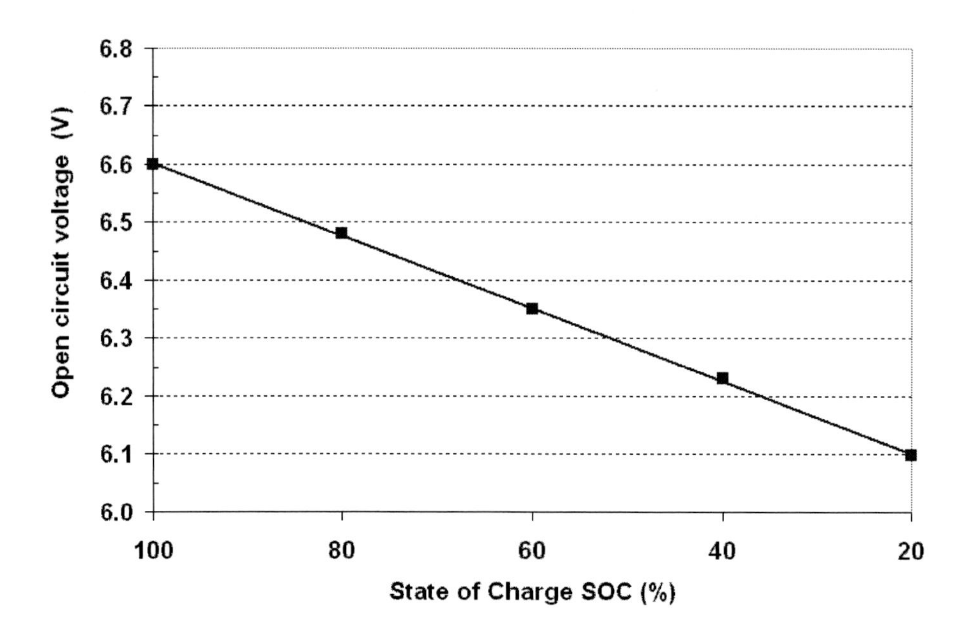

Figure 6. Relation between open circuit voltage and state of charge (6V 20 Ah Orbital HEV modules).

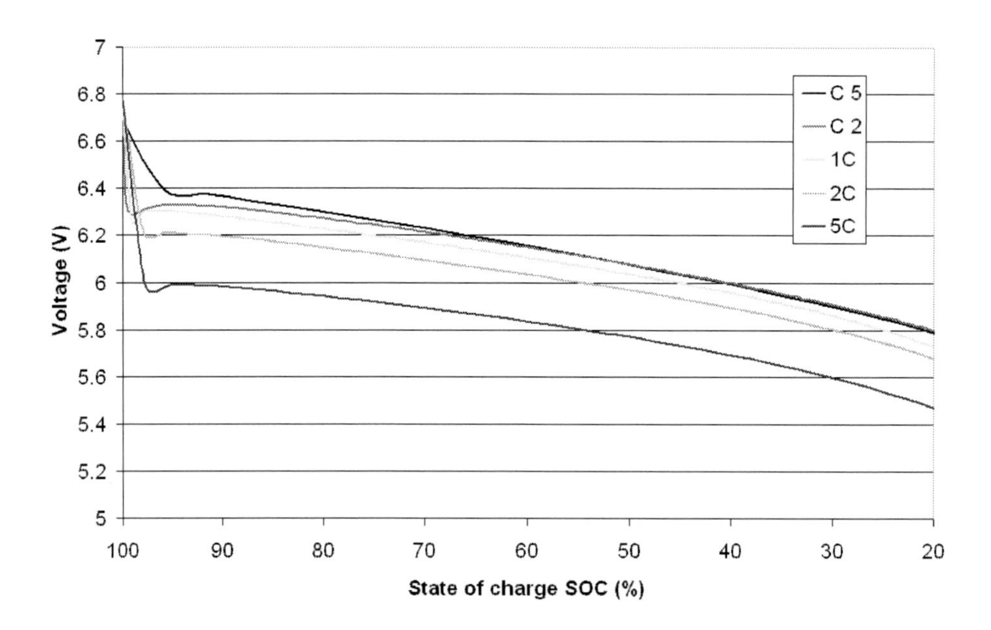

Figure 7. Discharge curves under constant current.

$$\mathrm{SOC}_2 = \mathrm{SOC}_1 - \Delta\mathrm{SOC} \tag{7}$$

$$\Delta SOC = \frac{\int_0^t I dt}{C_n} \tag{8}$$

where C_n is battery capacity according to discharge rate. For cases where the current is being extracted from the batteries, capacity is calculated using Peukert's equation, $I^n \times t = C$, where C is a coefficient of adjustment and n is Peukert's exponent, depending on battery type [37]. When current is being input to the battery, a value of $C_n = C_{charge}$, is used, a constant that is adjusted in line with test results.

Apart from calculating the ampere-hours consumed, loss power must be determined as this will be responsible for heating the electrolyte. For this purpose, the charge and discharge performance curves are applied according to the SOC, as shown in figure 8.

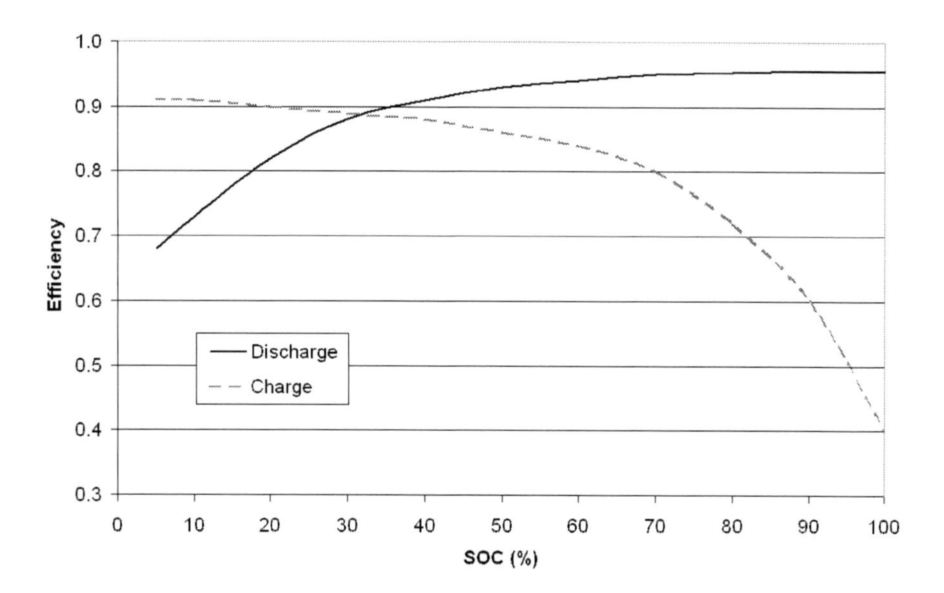

Figure 8. Charge and discharge performance curves depending on SOC.

In the thermal model included to estimate the variations in temperature of the battery pack, this temperature changes in line with the following equations:

$$P_{cool} = \frac{T - T_{out}}{R_{bat\_therm}} \tag{9}$$

$$\frac{dT}{dt} = \frac{P_{loss} - P_{cool}}{C_{bat\_therm}} \tag{10}$$

Resulting in:

$$\frac{dT}{dt} = \frac{1}{C_{bat\_therm}} \cdot \left(P_{loss} - \left(\frac{T - T_{out}}{R_{bat\_therm}} \right) \right) \tag{11}$$

where T_{out} is the outdoor temperature, P_{cool} is the cool power, P_{loss} is the power losses, C_{bat_therm} and R_{bat_therm} are respectively the capacity and the thermal resistance of the batteries.

Besides the influence of temperature, the battery model may also empirically take account of the effect of the useful life that influences the internal features and performance, affecting the characteristic charge and discharge curves [37][46].

Figure 9 shows the Simulink implementation of the full power batteries model of the hybrid vehicle. As with [36], the effects of any possible differences between individual battery modules are not taken into account due to the difficulty in quantifying and modelling them.

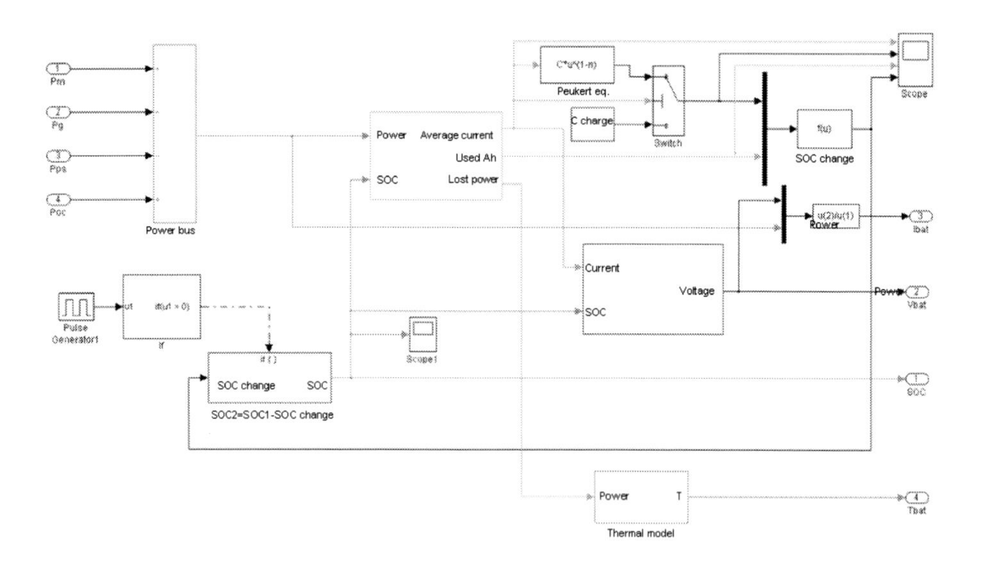

Figure 9. Battery simulation model.

3.4. Electric Motor

The electric motor is modelled using curves that give us motor performance according to torque and speed, since an equivalent circuit-based model requires more data, usually unknown, which has to be estimated. Thus, the electrical power consumed by the motor at any one time can be calculated. In order to calculate the voltages and intensities flowing round the motor stator winding and its frequencies, the motor operating zone is separated into two zones depending on whether the engine speed is less or greater than the speed of synchronism.

The performance of the motor when it acts as a generator in regenerative braking processes can also be simulated with the model. In such an instance, this means a positive yield in energy recovery and the kinetic energy recovered is stored in the batteries.

3.5. Low Voltage System and Solar Panels

The solar panels are modelled using a constant that represents the power delivered by the photovoltaic panels, which can be modified according to the conditions to be simulated.

Modelling the low voltage system is done by inserting constants to represent the different electrical charges consumed by the vehicle accessories. This block also includes the air-conditioning system load, which may be activated or not, and which has an influence on the operating point of the internal combustion engine.

4. Defining the Control Strategy

Taking account of the pre-established specifications for the vehicle, electronic management for the whole system is approached. This control strategy is configurable, but the most suitable one is presented now. In it, three possible operating states have been defined for the vehicle:

- Vehicle switched off (E0): State of the vehicle when the ignition key is removed or is in the off position.
- Electrical operation (E1): In this state the combustion engine is not running and the only power source for the electric motor is the batteries.
- Operation of the combustion engine (E2): The combustion engine ignites to operate the air-conditioning or recharge the batteries, setting itself at the constant engine speed operating point suited to the generator and providing the torque required.

Table 1 shows the characteristic input signals of the system and figure 10 the states of energy control diagram showing the transitions between pre-defined states.

Table 1. Control unit input binary signals.

Input signal		States	Control
Vehicle ignition	e1	0: Off 1: On	Driver
Combustion engine enabling/disabling	e2	0: Disabled 1: Enabled	Driver Petrol level
Air-conditioning switch	e3	0: Off 1: On	Driver
Battery charge level	eSOC1 eSOC2	00: Complete (90%-100%) 01: High (70%-90%) 10: Medium (50%-70%) 11: Low (0%-50%)	SOC evaluation module
External battery charge	eCharge	0: No external charge 1: With external charge	Driver
Solar panel charge		0: Low voltage circuit 1: Batteries	Vehicle ignition
Regenerative braking enabling/disabling		0: Disabled 1: Enabled	Driver

It should be pointed out that the combustion engine can be disabled by two signals: one operated by the driver (for circulation in zones requiring zero pollution levels) and another given by the petrol level. Moreover, it should be emphasised that the regenerative braking management acts directly on the inverter regardless of the operating state, so it is not included in the diagram, nor is solar panel management which is directly dependent on the ignition of the vehicle. Furthermore, following the tests presented in [37] and the conclusions drawn from them and from figure 8, the optimum operating range of the batteries is to be found between 50% and 70% of the state of charge. Therefore, the hybrid operation control strategy should be adapted to these values in the absence of other requirements, as this kind of strategy is particularly suited to optimising consumption [47]. This conclusion has been checked by simulation results.

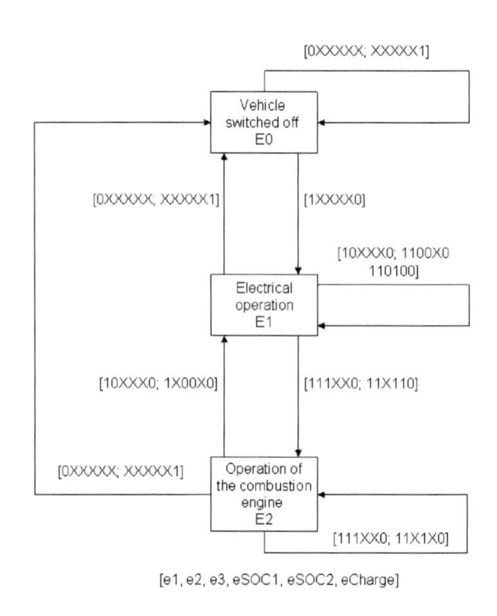

Figure 10. Energy control system states diagram.

5. Results of Simulations

The model described above was used to size the vehicle's components and to arrive at the most effective control strategy that was compatible with specification requirements. Table 2 shows the features of the components that were finally chosen and which were used for the simulations shown below.

Given the initial specifications of the vehicle, the driving cycles usually taken into account would not be representative of actual operation. Therefore, in the study, an adapted cycle was designed whose urban cycle coincided with the standard European driving cycle, while the interurban cycle was replaced by driving at a top constant speed of 50 km/h with a 120 second duration (figure 11). This cycle attempts to simulate a standard operation of the vehicle designed to include the exit from depots, journeys between bus-shelters to collect used batteries and the return to the depot, mainly during night-time journeys. Figure 12 shows how the state of charge evolves and the activation of the internal combustion engine, starting from different initial states of charge.

Table 2. Components characteristics.

BATTERIES (25 MODULES)	
Type	Acid Pb
Nominal voltage	150 V
Capacity (2h)	20 Ah
Weight	117.5 kg
ELECTRIC MOTOR	
Type	Induction
Nominal power	15 kW
Nominal torque	50 Nm
Nominal speed	2850 rpm
Maximum speed	9000 rpm
Weight	41.5 kg
Cooling	Water
INTERNAL COMBUSTION ENGINE	
Fuel	Unleaded petrol
Cylinders	2
Maximum power	15 kW (5000 rpm)
Maximum torque	37 Nm (3000 rpm)
Weight	49 kg
Cooling	Water
Emissions Regulations	EURO 4
GENERATOR	
Power	9.4 kW
Rotation speed	2600 rpm
Output voltage	135 V (130 Hz)
Voltage control	Close-loop
Cooling	Air
SOLAR PANELS	
Type	Photovoltaic
Nominal voltage	12 V
Maximum power	120 W

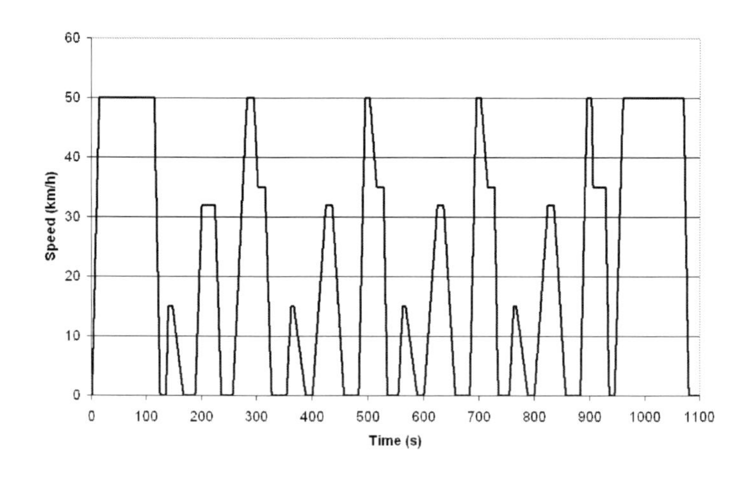

Figure 11. Driving cycle used.

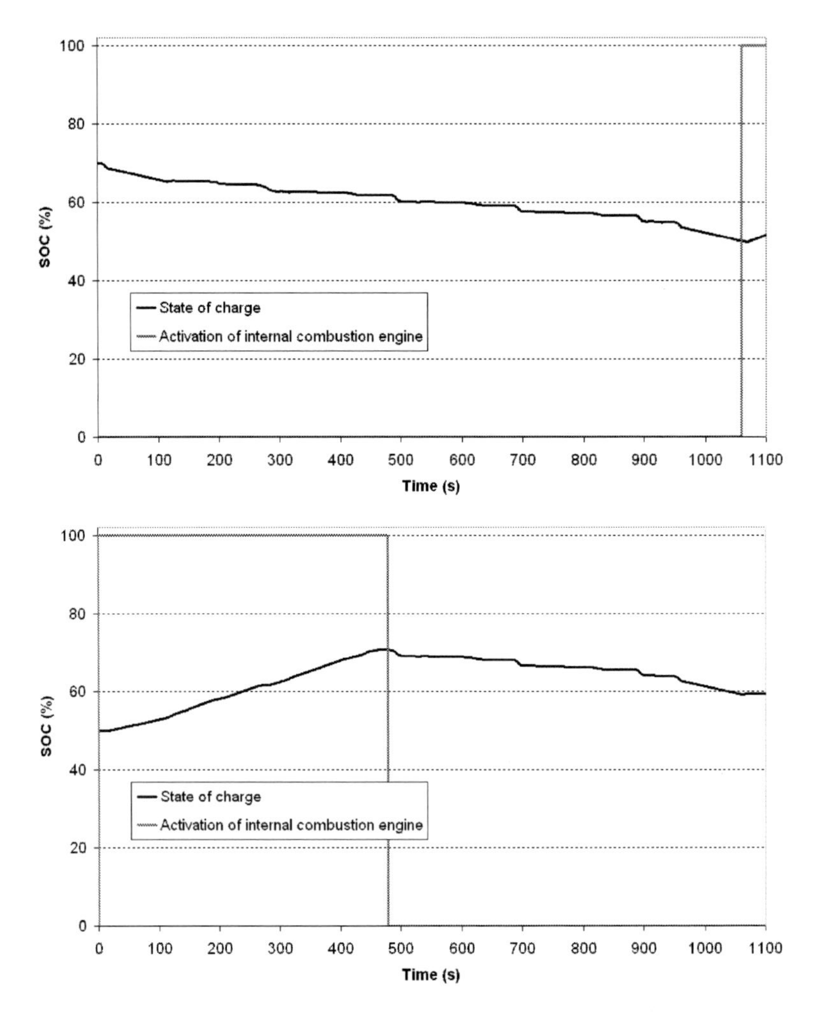

Figure 12. Evolution of the state of charge of the batteries and activation of the internal combustion engine (different initial SOC values: 70% and 50%).

As shown, the simulation model was used as a support during the design stages. Thus, it was used to make decisions, such as how to operate the air-conditioning, for instance, a system which involves a substantial but non-permanent load as it is driver operated. Two possibilities were taken as the starting point:

- To operate the compressor with the internal combustion engine, as with conventional vehicles.
- To operate the compressor using an electric motor driven by the energy from the batteries.

The first option has the advantage of being simpler and more economical. The main drawback is that if the air-conditioning is switched on, the internal combustion engine needs to be started. The second option is more complicated but makes the air-conditioning possible without having to switch on the combustion engine. However, the combustion engine has to intervene more frequently since the power batteries discharge more quickly. The simulations performed gave the results shown in table 3. It should be pointed out that since the final state of charge of the batteries, after the cycle, is not the same, the results of the simulations are not directly comparable. The solution proposed is to work with a fictitious estimate of the amount of fuel that would be needed to increase the state of charge with the vehicle stationary. As may be seen, fuel consumption is lower when the air-conditioning compressor is operated by the combustion engine, so this solution is adopted, as previously stated.

Table 3. Simulation results with the air-conditioning operating.

Initial SOC (%)	Air-conditioning operating	Final SOC (%)	Consumption (l)	Equalised consumption (l)
100	Batteries	66	1.72	2,40
	Combustion engine	59	1.37	2.19
75	Batteries	69	2,05	2,17
	Combustion engine	57	1,73	2,09
50	Batteries	62	2,78	2,54
	Combustion engine	60	2,09	1,89

An element that differentiates the designed vehicle from other hybrid vehicles is the fitting of solar panels to compensate part of the low voltage circuit charge during operation and charge the power batteries when the vehicle is stationary. The fuel saving study obtained when the initial state of charge is 100 % appears to be representative. According to the simulation, another 12.8 km could be run without any need to operate the combustion engine to recharge batteries compared to if this charge was not used.

Finally, exclusively in electrical mode, autonomy is evaluated from an initial 100% state of charge (either through external or solar panel charging). The simulation shows that it is possible to cover 35.1 km without needing to switch on the combustion engine. The figures are higher than the initial specifications, which makes this design suitable for working in electrical mode in an urban zone where a reduction in emissions is sought.

CONCLUSION

This chapter has presented the process for modelling and designing the energy part of a hybrid vehicle (combustion engine- generator, electric motor, batteries and solar panels) configured in series for a specific application with defined operating specifications.

Due to the particularities of the propulsion system and the need to optimise control strategies that observe vehicle design specifications, a specific model has been developed in Matlab/Simulink that enables components to be selected and studied and the full system to be analysed.

Simulations were performed on a standard European driving cycle with its interurban part modified to the characteristics of the vehicle. The results led to the selection of the components and decision making regarding design, an example of which was operating the air-conditioning with the combustion engine instead of the electric motor. The utility of the results was demonstrated in the preliminary design stages as well as their high flexibility for optimising control strategies.

Based on these results, the prototype is currently in the construction and testing stage. As already stated, as a future development it is hoped to replace the combustion engine by a fuel cell.

REFERENCES

[1] Argiri, M. and Birol, F. *World Energy to 2020: prospects and challenges.* International Energy Agency (IEA). 1999

[2] General Motors. *Hydrogen and Fuel Cells*. Future of Automobility. General Motors, 2004

[3] European Commission. Green Paper: *Towards a European strategy for the security of energy supply.* Luxembourg, 2001 (http://europa.eu.int/comm/energy_transport/doc-principal/pubfinal_en.pdf)

[4] European Commissions. *White Paper: European transport policy for 2010:* time to decide.Luxembourg.2001. (http://europa.eu.int/comm/energy_transport/library/lb_texte_complet_en.pdf)

[5] An, F. and Santini, D.J. *Mass Impact on Fuel Economics of Conventional Vs. Hybrid Electric Vehicle.* SAE paper nº 2004-01-0572, 2004

[6] Graham, R. et al. *Comparing the Benefits and Impacts of Hybrid Electric Vehicle Options.* Final Report, Electric Power Research Institute, Palo Alto, California, USA. 2001

[7] An, F., Decicco, J. and Ross, M. *Assessing the Fuel Economy Potential of Light Duty Vehicles.* SAE paper nº 2001-01FTT-31, 2001

[8] Gott, P., Linna, J-R- and Mello, J. P. *The evolution of powertrain technology 2008 and beyond: engines, hybrids, battery electric, fuel cells and transmissions*. Proceedings of FISITA 2004 Congress. Barcelona, May 2004.

[9] Unnerwehr, L. E., Auiler, J. E., Foote, L. R., Moyer, D. F. and Stadler, H. L. *Hybrid vehicle for fuel economy.* SAE paper nº 760121, 1976

[10] Weiss, et al. On Road in 2020: *A Life Cycle Analysis of New Technologies*. MIT Energy Laboratory Report nº MIT-EL00-003, 2000

[11] General Motors Corp. *Well to Wheel Energy Use and Greenhouse Gas Emissions of Advanced Fuel/Vehicle System*. North America Analysis. Executive Summary Report, 2001

[12] Papasavra, S., Weber, T.R. and Cadle, S.H. *Tank to Wheels Preliminary Assessment of Advanced Powertrain and Alternative Fuel Vehicles for China*. SAE paper nº 2007-01-1609, 2007

[13] Inman, S., El-Gindy, M., Haworth, D.C. *Hybrid electric vehicles technology and simulation: Literature review.* Heavy Vehicle Systems, 2003, 10 (3), pp 167-187

[14] Powell, B.K. Pilutti, T.E. *A range extender hybrid electric vehicle dynamic model,* Proceedings of the 33rd IEEE Conference on Decision and Control, 1994, 12, pp 2736–2750.

[15] Bailey, K. E. and Powell, B.K. *A hybrid electric vehicle powertrain dynamic model.* Proceedings of the American Control Conference, 1995, pp 1677-1682

[16] Hubbard, G.A., Youcef-Toumi, K. *Modelling and simulation of a hybrid-electric vehicle drivetrain,* Proceedings of the 1997 American Control Conference, 1997, pp 636–640.

[17] Harb, J.N., Johnson, V.H., Rausen, D. *Use of a fundamentally based lead-acid battery model in hybrid vehicle simulations,* Tutorials in Electrochemical Engineering – Mathematical Modeling, 1999, 14, pp 163–177.

[18] Campbell, A., Rengan, A., Steffey, J., and Ormiston, J. *Simulation of 42-volt hybrid electric vehicles*, www.mth.msu.edu, 2002.

[19] Rousseau, A. *Simulation and Validation of Hybrid Electric Vehicles,* ADVISOR Users Conference, Costa Mesa, California, August 24, 2000.

[20] Butler, K. L., Stevens, K. M. Ehsani, M. *A versatile computer simulation tool for design and analysis of electric and hybrid drive trains*. SAE Technical paper nº 970199, 1997

[21] Evans, C. and Stone, R. *Adaption of ADVISOR 2.0 for dual hybrid vehicle modelling, C575/001/99:* International Conference on Integrated Powertrain Systems for a Better Environment, Birmingham, UK, November 10, 1999.

[22] Braun, C. G., Busse, D. T. *A modular simulink model for hybrid electric vehicles.* SAE Technical paper nº 961659, 1996

[23] Wei, X., Utkin, V. and Rizzoni, G. *Sliding optimal control in hybrid electric vehicle energy management optimizations*. Proceedings of FISITA 2004 Congress. Barcelona, May 2004.

[24] Filipi, Z., Louca, L., Daran, B., Lin, C.-C., Yildir, U., Wu, B., Kokkolaras, M., Assanis, D, Peng, H., Papalambros, P., Stein, J., Szkubiel, D. and Chapp, R. *Combined optimisation of design and power management of the hydraulic hybrid propulsion system for the 6 × 6 medium truck*. Int. J. of Heavy Vehicle Systems, 2004, Vol. 11, Nos 3/4

[25] Hung, Y.-H, Lin, P.-H. and Hong, C.-W. *A rule-based control algorithm for hybrid electric motorcycle powertrain systems,* Proceedings of the 6th International Symposium on Advanced Vehicle Control, Hiroshima, Japan, 2002

[26] Won, J.-S. and Langari, R. *Fuzzy torque distribution control for a parallel hybrid vehicle,* Expert Systems, February, 2002, 19 (1), pp. 4-10.

[27] Wu B., Lin, C.-C., Filipi, Z., Peng, H. and Assanis, D. *Optimization of power management strategies for a hydraulic hybrid medium truck*. 6th International Symposium on Advanced Vehicle Control, Hiroshima, Japan, 2002

[28] Lin C., Filipi Z., Wang Y., Louca L., Peng H., Assanis D. and Stein J. *Integrated, feed-forward hybrid electric vehicle simulation in SIMULINK and its use for power management studies,* SAE Paper 2001-01-1334, 2001

[29] Wei, X., Guzzella, L., Utkin, V. I. and Rizzoni, G. *Model-based fuel optimal control of hybrid electric vehicle using variable structure control systems*. Journal of dynamic systems measurement and control, 2007, 129 (1), pp 13-19

[30] Han, Z., Yuan, Z., Guangyu, T., Quanshi, C. and Yaobin, C. *Optimal energy management strategy for hybrid electric vehicles*. SAE paper nº 2004-01-0576, 2004

[31] Yoon, H.-J. and Lee, S.-J. *An optimized control strategy for parallel hybrid electric vehicle*. SAE paper nº 2003-01-1329, 2003

[32] Hellgren, J., Jacobson, B. *A systematic way of choosing driveline configuration and sizing components in hybrid vehicles.* SAE Technical paper nº 2000-01-3098, 2000
[33] Olbrot, A. W. et al. *Parameter Scheduling Controller for Exhaust Gas Recirculation (EGR) System.* Society of Automotive Engineers, Inc. 1997.
[34] Azzoni, P.M. et al. *A Model for EGR Mass Flow Rate Estimation.* Society of Automotive Engineers, Inc. 1997
[35] Sultan, M. C. et al. *Closed Loop Canister Purge Control System.* Society of Automotive Engineers, Inc, 1998
[36] Van Mierlo, J., Van Den Bossche, P. and Maggetto, G. *Models of energy sources for EV and HEV: fuel cells, batteries, ultracapacitors, flywheels and engine-generators.* Journal of Power Sources 2004, 128, pp 76-89
[37] Trinidad, F., Gimeno, C., Gutiérrez, J., Ruiz, R., Sainz, J., Valenciano, J. *The VRLA modular wound design for 42 V mild hybrid systems.* Journal of Power Sources, 2003, 116, pp 128–140
[38] Caumont, P. Le Moigne, C. Rombaut, X. Muneret, and P. Lenain. *Energy Gauge for Lead-Acid Batteries in Electric Vehicles.* IEEE Transactions on Energy Conversion, 2000, 15 (3), pp 354-360.
[39] Ceraolo, M. *New dynamical model of lead-acid batteries,* IEEE Trans. Power Systems, 2000, 15, pp 1184–1190.
[40] Esperilla, J. J., Félez, J., Romero, G.,Carretero, A. *A model for simulating a lead-acid battery using bond graphs.* Simulation Modelling Practice and Theory, 2007, 15, pp 82–97
[41] Salameh, M. S., Casacca, M. A. and Lynch, W. A. *A mathematical model for lead-acid batteries.* IEEE Trans. on Energy Conversion, 1992,. 7 (1), pp. 93–97.
[42] Lee, J., Kim, C., NamGoong, E. *Dynamic State Battery Modeling of Energy and Power as States Variables for EVs Application.* SAE Technical Paper nº 97241, 1997
[43] Zoelch, U., Schroeder, D. *Comparison of electric drives for a hybrid vehicle*, in: Proceedings of the European Power Electronics and Drives Association, Sevilla, Spain, 1995.
[44] Karden, E., Mauracher, P. and Lohner, A. *Battery management system for energy-efficient battery operation: Strategy and practical experience.* EVS 13, 1996, 2, p. 91
[45] Guogang, Q., Jianming, L. and Hang, J. *A new battery state of charge indicator for electric vehicles.* EVS 13, 1996, 2, p. 631
[46] Trinidad, F., Sáez, F., Valenciano. J. *High power valve regulated lead-acid batteries for new vehicle requirements.* Journal of Power Sources, 2001, 95, pp 24-37
[47] Burke, A. F. *On-off engine operation for hybrid/electric vehicles.* SAE Technical paper nº 930042, 1993

In: Advances in Mechanical Engineering Research, Volume 3 ISBN: 978-61209-243-0
Editor: David E. Malach

Chapter 8

STRATEGIES TO PROMOTE THE PREMIXED COMBUSTION PHASE IN DIESEL ENGINES FOR THE SIMULTANEOUS REDUCTION OF NOX AND PM

J.M. Riesco-Avila[1], A. Gallegos-Muñoz[1], J.M. Belman-Flores[1], V.H. Hernández-Rangel[1], S. Martínez-Martínez[2] and F.A. Sánchez-Cruz[2]

[1]División de Ingenierías, Universidad de Guanajuato, Mexico
[2]Facultad de Ingeniería Mecánica y Eléctrica, Universidad Autónoma de Nuevo León, Mexico

ABSTRACT

The combustion by compression of a lean and homogeneous (or premixed) air-fuel mixture has emerged in the last few years as an effective alternative to achieve simultaneous reductions in nitrogen oxides (NOx) and particulate matter (PM) in internal combustion engines. The main subject of this chapter focuses its attention on the analysis of different strategies that promote the premixed burn phase in a heavy duty diesel engine, typical of those used in on-road transportation, with the aim of reducing NOx and PM emissions simultaneously.

The methodology used in this study is based on a parametric experimental study of these strategies. Using both, cylinder pressure measurements and operating conditions of the engine, the combustion diagnosis is carried out. This makes it possible to obtain relevant parameters for the combustion analysis such as pressure evolution, mean gas temperatures, adiabatic flame temperature, premixed burnt mass fraction, fifty percent of heat release, etc. When all these parameters are obtained systematically, the way in which the operating conditions affect the combustion process and the relationship between these parameters and the performance of the engine (measured in terms of efficiency and emissions) can be properly analysed.

Even though a completely premixed burn could not be achieved, the studied strategies have proven to be effective ways of achieving a simultaneous reduction in NOx and PM. Nevertheless, there was also an increment in the HC and CO emissions and higher fuel consumption.

NOTATION

ABF50: Angle at which 50% of fuel has been burnt
$ABF50_{TDC}$: Angle at which 50% of fuel has been burnt referred to TDC
EGR: Exhaust gas recirculation
HRR: Heat release rate
PBF: Premixed burnt fraction
PMC: Premixed combustion
SoI: Start of injection
SoC: Start of combustion
η_i: Gross indicated efficiency
τ_{id}: Ignition delay time
τ_{fi}: Fuel injection time

1. INTRODUCTION

In recent years, heavy-duty Diesel engines have been widely employed because of their high efficiency and reliability. Unfortunately, the characteristics of the diffusive combustion of Diesel fuel sprays promote the formation of nitrogen oxides (NOx) and particulate matter (PM). It is important to highlight that world emission standards are becoming increasingly and require simultaneous reductions in these pollutants (Moser et al. 2001). The past, present and future emission regulations in Europe and USA are shown in Figure 1 together with the requirements with respect to fuel consumption. It seems clear that, despite the introduction of new technologies and future research, it will be not possible to fulfil the limits imposed by the upcoming legislation operating only with the combustion process. Consequently, after-treatment of the exhaust gases with external systems will be essential to attain the target values of emissions in a near future but they are expensive and, in most cases, they need maintenance and affect negatively the brake specific fuel consumption.

However, an alternative combustion process is being investigated with the objective of enhancing the premixed combustion (PMC) phase, which is able to reduce both NOx and PM without affecting fuel efficiency in particular operating conditions. This combustion can be considered as a combination between the spark ignition (SI) and the compression ignition (CI) combustions. To attain it, generating a homogeneous (or highly premixed) mixture of fuel and air (such as SI combustion) is necessary, but this mixture is compressed up to its autoignition limit (such as DI combustion). In this way, the combustion starts almost simultaneously at different and well-distributed points inside the combustion chamber, without flame propagation and free of zones with high local temperatures. This reduces or even inhibits NOx formation. Moreover, the diffusive combustion and the fuel-rich zones during combustion are eliminated, avoiding the formation of PM.

Homogeneous Charge Compression Ignition (HCCI) is the most widely accepted name to identify this combustion (Thring 1989), but there are a great variety of systems and denominations based on the enhancement of the PMC concept with slight differences. Some of them are contained in Table 1.

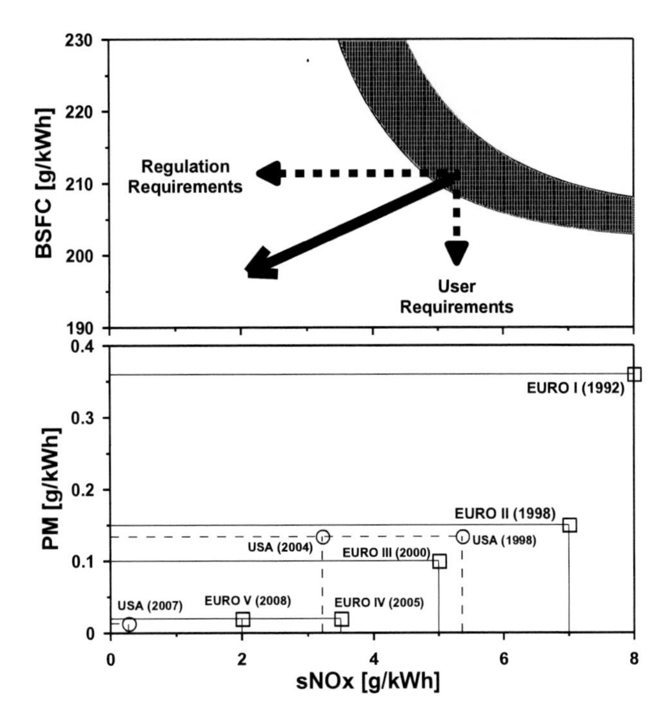

Figure 1. Evolution of the pollutant emission regulations in Europe and USA and the market requirements.

Table 1. Some combustion processes based on the PMC enhancement.

ACRONYM	NAME	REFERENCE
ATAC	Active-Thermo Atmosphere Combustion	Onishi et al 1979
AR	Active Radical	Ishibasi and Asai 1996
CIHC	Compression-Ignited Homogeneous Charge	Najt and Foster 1983
PCCI	Pre-mixed Charge Compression Ignition	Aoyama et al. 1996
CAI	Controlled Auto-Ignition	Lavy et al. 2000
HiMICS	Homogeneus Charge Inteligent Multiple Injection Combustion System	Yokota et al. 1997
UNIBUS	Uniform Bulky Combustion System	Yanagihara 1996
PCI	Pre-mixed Compression-Ignited Combustion	Iwabuchi et al. 1999
PREDIC	Pre-mixed Lean Diesel Combustion	Takeda et al. 1996
MULDIC	Multiple Stage Diesel Combustion	Hashizume et al. 1998
MK	Modulated Kinetics	Kimura et al. 1999
HCDC	Homogeneous Charge Diesel Combustion	Suzuki et al. 1997

One of the newest approaches to this alternative combustion system is the MK concept (see Table 1). In this concept, fuel is injected directly into the combustion chamber near or after the TDC. Therefore, it is essential to retard the start of combustion (SoC) beyond the end of injection (EoI). If combustion starts after the end of injection, there is enough time to provide an adequate air/fuel mixture to achieve the PMC emission characteristics. Large amounts of cooled EGR and a reduction in the engine compression ratio are necessary to extend the ignition delay whereas vigorous swirl enhances the air/fuel mixing process.

The purpose of the study reported in this chapter was to investigate the potential of the PMC in HD diesel engines for simultaneously reducing NOx and soot emissions. To attain the PMC conditions, an approach to the MK concept was selected due to its easy implementation without important hardware changes in the engine and particularly in the injection system.

Special attention has been paid to effects produced by retarded start of fuel injection (SoI) and fuel injection pressure (IP) at low engine load.

2. Objective and Methodology

The main objective of the research project is to investigate the potential of the PMC in HD Diesel engines for simultaneously reducing NOx and soot emission. Contrary to the classical HCCI concept, injection is produced after TDC, in order to avoid the problems involved with the control of the auto-ignition delay. However, the basic goal remains unchanged, namely obtaining a shorter injection and mixing time than the auto-ignition delay. In order to achieve this condition, the auto-ignition delay time (τ_{id}) is extended by retarding the SoI past TDC. The injection time can be shortened and the air/fuel mixing improved by increasing the injection pressure. The ideal situation would be when injection and mixing occurs completely before the start of combustion. This is very difficult to achieve even at light engine load, and in this chapter, intermediate conditions will be also considered as PMC.

In order to ensure compatibility with normal injection and combustion conditions, which may be required at other engine operation modes, standard injection hardware has been used in this study, as well as a commercial fuel formulation. Hence, the remaining strategy has been to increase the ignition delay by modifying the in-cylinder gas temperature and pressure during injection by retarding the SoI. When the SoI is retarded beyond the TDC, gas pressure and temperature during the injection event decreases, and τ_{id} increases. An additional free parameter is the injection pressure, which was increased up to the maximum value, even though it is known that this measure would reduce the auto-ignition delay (Rosseel and Sierens 1996).

Since it is known that NO_x values attained with the PMC concept are still too large to meet upcoming restrictions, it made sense to explore the sensitivity of PMC to high cooled EGR rates.

The focus of the study has been placed on a light-load engine operation mode, since the long injection events at medium and high load make the condition of τ_{id} longer than fuel injection time (τ_{fi}) practically impossible to achieve.

From the engine tests, direct information on performance behaviour can be obtained in terms of brake specific fuel consumption (BSFC) and pollutant emissions.

The recorded values of in-cylinder pressure were processed by means of a combustion diagnosis code CALMEC (Armas 1998; Lapuerta et al. 1999, 2000). This model is mainly based on the solution of first law of thermodynamics in the cylinder, with the assumption of uniform pressure and temperature over the volume. Valuable information can be extracted, such as the heat release (HR) and several associated parameters, like combustion duration, ignition delay, etc.

To define parameters like τ_{id} or combustion duration, the start of combustion was defined as the point where the heat release rate (HRR) start to rise from zero and the combustion duration is defined by the point at which the cumulative heat release (HR) curve reaches 90% of the total value.

3. Experimental Facility

The engine used in this work is a single-cylinder, four-stroke, direct-injection diesel research engine with 1.85 litres displacement, well representative of large truck engines. Detailed basic specifications of the engine are given in Table 2.

Table 2. Engine characteristics.

Engine Type	Direct-injection Diesel, 4 Stroke-cycle single cylinder
Bore x Stroke	123 x 156 mm
Displacement	1.85 dm^3
Compression Ratio	16.3
Fuel Injection System	Common rail
Injection Nozzle	8 holes x 0.178 mm in two crowns

The engine was installed in a fully instrumented test cell, with all the auxiliary facilities required for the operation and control of the engine. Figure 2 shows a scheme of the installation.

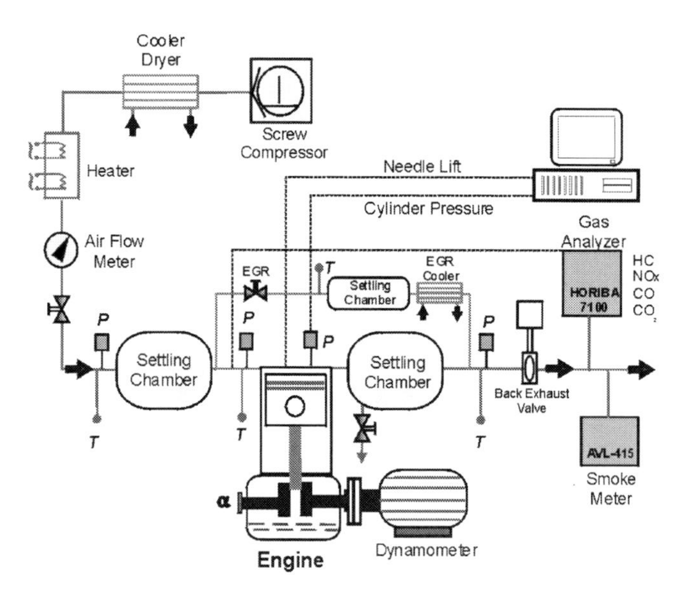

Figure 2. Scheme of the experimental test cell.

To achieve intake air conditions like those found in a real production engine, a screw compressor supplied the required boost pressure, while air temperature in the engine intake was controlled by an external cooler, keeping the desired inlet conditions in a settling chamber upstream of the intake pipe. Intake air flow rate was measured by a volumetric flow meter. The exhaust backpressure produced by the turbine in the real engine was replicated by a valve placed in the exhaust system, controlling the pressure in the exhaust settling chamber.

When EGR needs to be produced, exhaust gas is taken from the exhaust settling chamber, cooled by a gas-water intercooler to the desired temperature, and then introduced into the intake pipe, as shown in Figure 2. The temperature at the EGR settling chamber is controlled electronically and in this study it was kept at a constant value of 80ºC in all the tests. In order to ensure a large enough EGR mass flow rate, exhaust pressure was maintained 0.01 MPa above the intake pressure in all the operating conditions. The exact EGR rate was controlled by means of a valve between the EGR settling chamber and the intake pipe.

At every engine operating point, the in-cylinder pressure traces from a piezo-electric transducer were recorded during 25 engine cycles in order to compensate for dispersion in engine operation. The filtered average pressure signal was used as input for the combustion diagnostic model.

The determination of the EGR rate was carried out using the experimental measurement of intake and exhaust CO_2 concentration. The concentration of NOx, CO, HC, intake and exhaust CO_2, and O_2 was measured with specific state-of-the-art analysers. Smoke emission was measured with a variable sampling smoke meter, providing results directly in FSN (Filter Smoke Number) units that were transformed into dry soot mass emissions by means of a correlation (Christian et al. 1993).

The fuel injection rate was measured in a commercial test rig following the Bosch system (Bosch 1966). Besides the shape of the fuel injection event, other interesting parameters obtained in these tests are the start of injection (SoI) and the end of injection (EoI).

4. Results and Discussion

4.1 Effect of Late Injection on Combustion and Exhaust Emissions

To analyse the effect of retarding the injection event, the SoI was moved from 8 cad before the TDC to 6 cad after the TDC (aTDC). It was not possible to retard beyond this point due to combustion degradation. Table 3 contains the basic engine operation conditions in the tests.

Table 3. Test conditions to analyse the effect of retarding the SoI.

n [rpm]	IP [bar]	BP [bar]	EP [bar]	T_{INTAKE} [ºC]	F_r [-]	SoI [cad aTDC]
1500	1000	1.50	1.60	35	0.32	From -8 to +6

The first part of the analysis is focused on the evolution of the combustion process. As the SoI is retarded, the maximum value of the in-cylinder pressure decreases monotonically, as shown in Figure 3 (a). The main reason is the retard of the combustion farther along the expansion stroke.

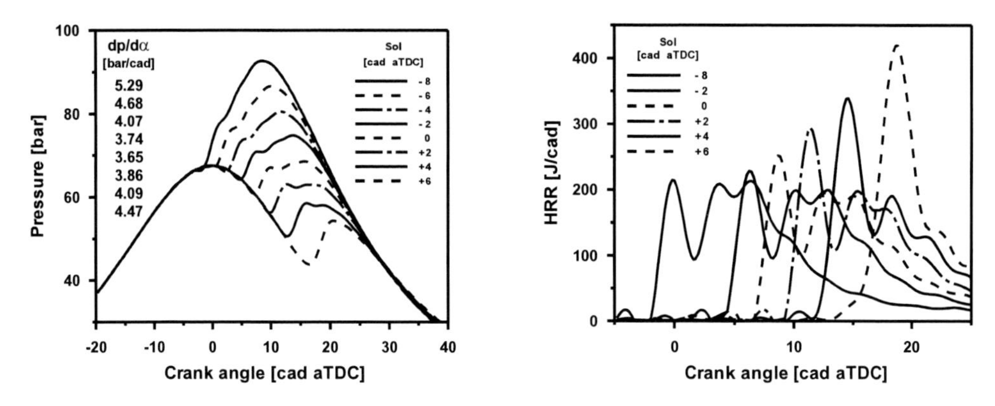

Figure 3. (a) In-cylinder pressure evolution for different SoI and (b) HRR evolution for different SoI.

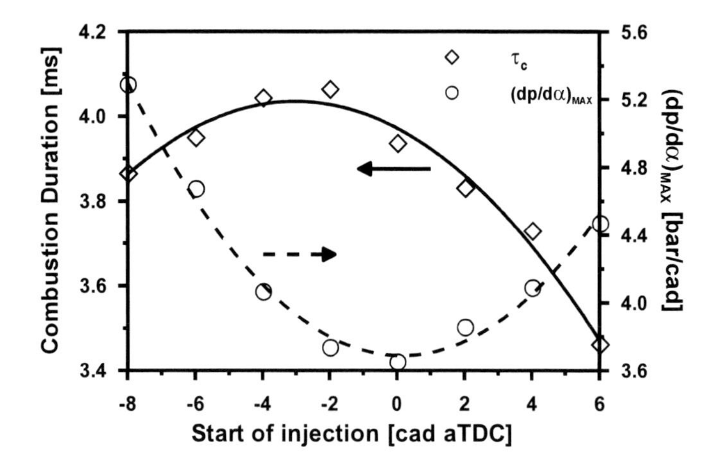

Figure 4. Combustion duration and maximum pressure gradient at different SoI.

However, the maximum pressure gradient initially decreases, but it starts to rise when the SoI is placed beyond the TDC. This means that, at these points, the premixed combustion phase is more intense resulting in a sharp increment in the gas pressure, and the balance between the premixed combustion and the diffusive combustion changes in favour of the former. This effect is clearly presented in Figure 3 (b), from SoI at -8 cad aTDC to TDC the shape of the HRR hardly changes, but when the SoI is placed progressively retarded from TDC, the premixed combustion phase is strongly enhanced.

Figure 4 shows the variation in the maximum pressure gradient and the combustion duration as a function of SoI. Initially, as SoI is retarded, combustion duration increases due to the degradation of the late diffusive combustion. However, as with pressure gradient, combustion duration presents a change in its trend and it starts to reduce when the SoI is placed after the TDC. This reduction in combustion duration reflects the promotion of the premixed combustion phase, which is considerably faster than the diffusive combustion.

Ignition delay is one of the most important factors affecting the combustion process, noise and exhaust emissions among others (Rosseel and Sierens 1996). Perhaps the most important effect pursued retarding the SoI is an increment in τ_{id}, as can be seen in Figure 5. In this figure, the percentage of total fuel burnt in premixed conditions (PBF) is also represented. Obviously, this parameter is closely related to τ_{id}. Again, the behaviour of both parameters is steady with SoI before the TDC, but from TDC to 6 cad aTDC, a sharp increment in τ_{id} and therefore, in PBF takes place.

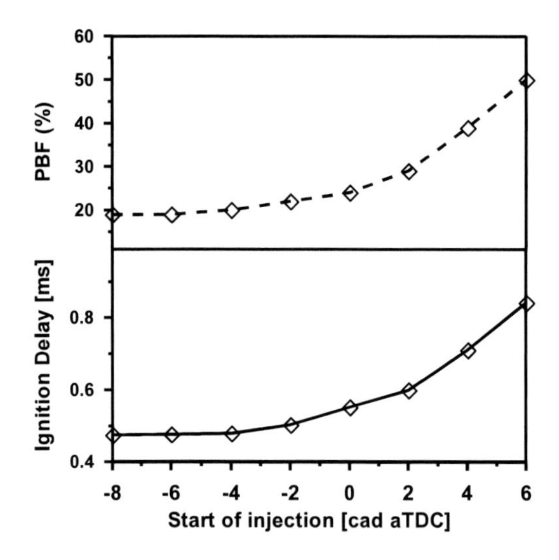

Figure 5. Ignition delay and premixed burnt fraction at different SoI.

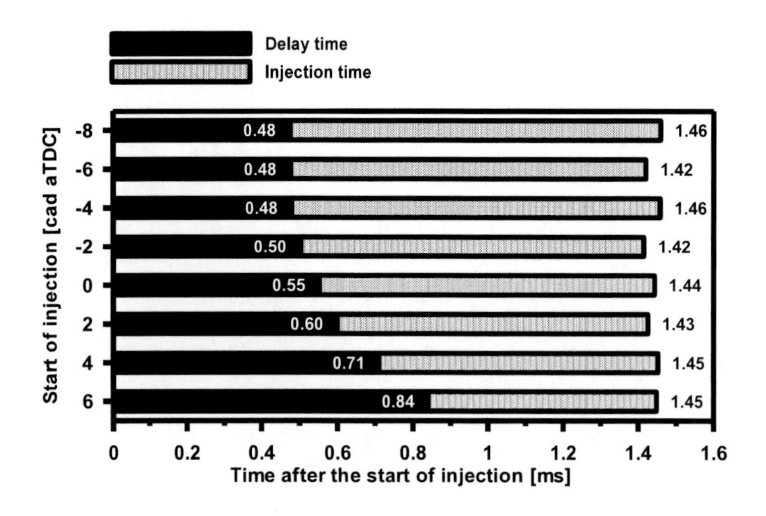

Figure 6. Relation between ignition delay time and fuel injection time at different SoI.

Figure 6 presents the relationship between τ_{id} and τ_{fi}. At -8 cad aTDC, τ_{id} is only 32.9% of the total injection time, whereas at 6 cad aTDC this percentage rises to 57.9%. An important conclusion extracted from Figure 5 and Figure 6 is the necessity to extend greatly τ_{id} with these injection conditions to attain a complete PMC, but the SoI cannot be retarded any more due to combustion failure, so other actions focused on reducing τ_{fi} will be needed.

Another important parameter related to the potential of retarding SoI is the gross indicated efficiency (η_i) of the engine. This is the ratio between the work carried out during the compression and expansion strokes and the total energy released by the injected fuel.

Figure 7 shows how retarding SoI affects η_i and the angular position when 50% of the fuel mass has been burnt, both referred to TDC, $ABF50_{TDC}$, and to the start of combustion, ABF50. These two parameters can shed light on the position and evolution of the combustion process.

Retarding SoI shifts the combustion process later in the expansion stroke, as indicates its $ABF50_{TDC}$ and this causes the thermal efficiency to steadily decrease. The ABF50 confirms and highlights the difference between traditional combustion with SoI before TDC, and highly premixed combustion obtained retarding the SoI after TDC. Due to an increment in combustion speed, the combustion process is shortened.

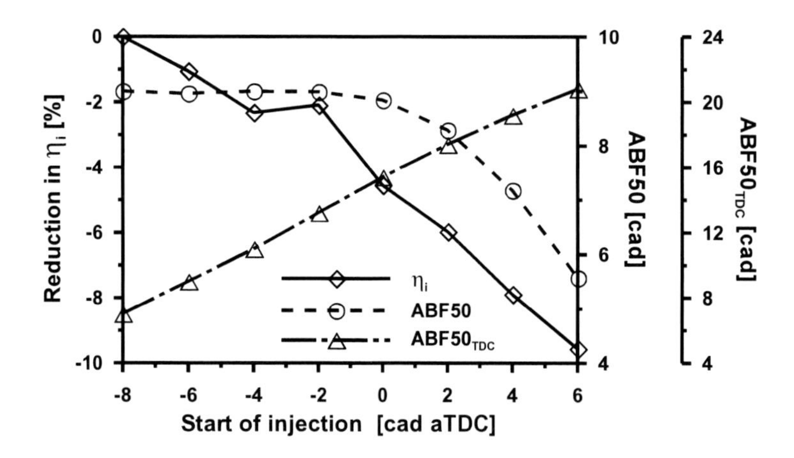

Figure 7. Gross indicated efficiency and 50% burn fraction angle at different SoI.

The most important effect of the PMC on NO_x and soot emissions is its ability to reduce simultaneously both pollutants, as demonstrated in Figure 8.

As expected, retarding SoI results in an important and steady reduction in NOx emissions due to a decrease in the combustion temperature. According to the model proposed by Dec, NO_x are mainly formed at the fuel-poor flame region, when both the oxygen and high temperature necessary for the nitrogen oxidation are available (Dec 1997). In addition, it is accepted that the NO formation starts just at the end of the premixed combustion. In this phase, the development of the reaction of NO formation is not possible because the mixture is very rich and the temperatures are too low. This model was confirmed later by the same author (Dec 1998) and, in general, it coincides with the results of other authors (Voiculescu and Borman, 1978; Donahue et al. 1994; Nakagawa et al. 1997).

This conceptual model clearly explains why the NOx are dramatically reduced when the PMC is enhanced and agrees with the results shown in Figure 8 (a).

In comparison with NOx, the mechanisms of formation and oxidation of soot are more difficult to understand, even in simple combustion processes. The plots in Figure 8 (a) show that initially, retarding SoI causes the usual increase in soot emissions, but from TDC to 6 cad aTDC the smoke opacity declines monotonically.

The reason for the latter behaviour could be the increment in τ_{id}, which promotes the PMC and consequently reduces the fuel mass burnt in the last phase by diffusion, where soot is mostly formed. In the premixed combustion phase, local air/fuel ratio is high and soot formation is avoided.

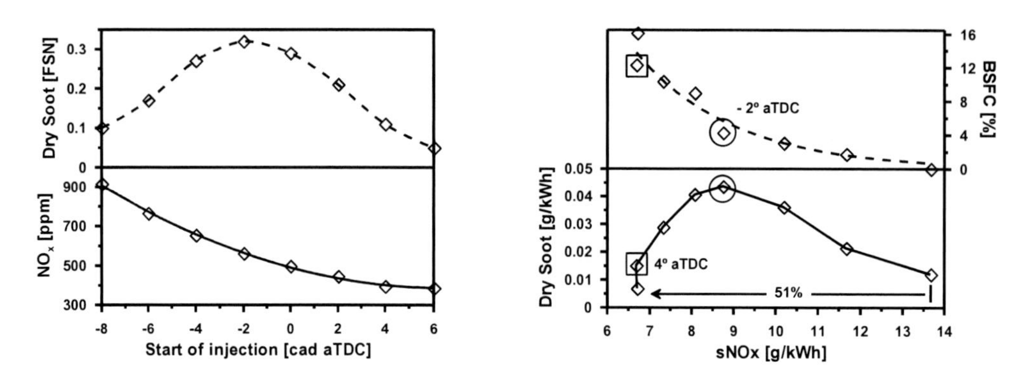

Figure 8. (a) NOx and soot emissions measured at different SoI and (b) Trade-off showing the global results obtained.

The mentioned behaviour of NOx and soot can be corroborated in the trade-off presented in Figure 8 (b). The outcome is a curious hook-like curve, with an important reduction in NOx emissions, but with absolute values very far from the objectives of emission regulations. Dry soot levels remain at very low values using this strategy. This figure also shows the BSFC-NOx trade-off, where a monotonic increase in BSFC is detected. Maximum BSFC values range about 15% higher than with the most advanced SoI. This outcome is, perhaps the worst aftermath of the strategy of retarding SoI.

Finally, as it was expected, HC and CO are both increased due to the deterioration of the last diffusive stage, which is developed at low in-cylinder pressure and temperature. Obviously, this trend is more pronounced at most retarded SoI (4 and 6 cad aTDC).

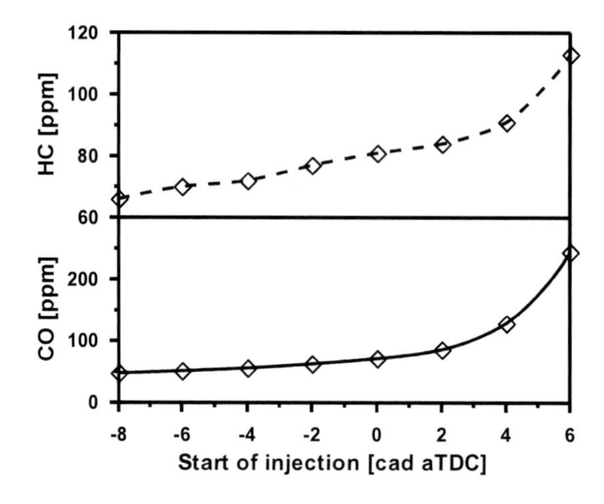

Figure 9. CO and HC emissions measured at different SoI.

4.2 Effect of Increasing Injection Pressure at Retarded Soi

Increasing injection pressure, aside from reducing the auto-ignition delay time, has a direct effect on the fuel injection rate, as it can be derived from equation (1), below. Consequently, if the injected fuel mass remains constant, the injection time will be reduced. For the purposes of enhancing PMC, the question is whether the difference between injection time and auto-ignition delay is reduced or not.

Theoretically, if the flow through the nozzle holes is quasi-steady, incompressible and one-dimensional, the mass flow rate is given by equation (1).

$$\dot{m}_f = C_d \cdot A_t \cdot \sqrt{2 \cdot \rho_f \cdot \Delta p} \tag{1}$$

In this equation, A_t is the total area of the holes, ρ_f is the fuel density, Δp is the pressure drop between the inlet and the outlet of holes and C_d is the discharge coefficient of the nozzle (Heywood 1988). Hence, the measure of raising IP produces an increase in injection rate which is fairly proportional to $(IP)^{1/2}$ and, consequently τ_{fi} would be reduced in the same proportion. Both effects are shown below in Figure 10, where it can be corroborated that the values of maximum injection rate and total injection time roughly follow this trend.

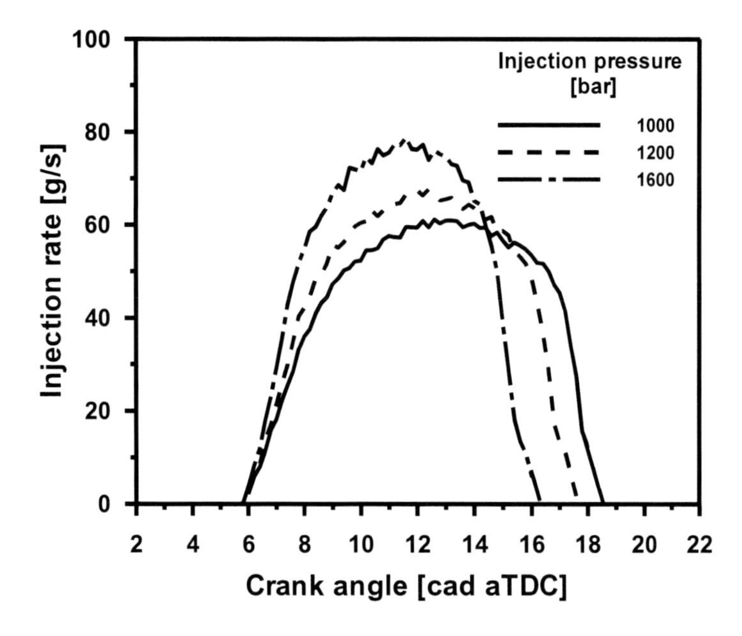

Figure 10. Injection rate obtained with different injection pressures at SoI 4 cad aTDC.

For the purpose of promoting PMC the most important effect produced by increasing IP is the enhancement of the fuel/air mixing process. Wakuri et al. proposed equation (2) to relate IP with air entrainment rate in the spray along the injection event. In this equation, ρ_f and ρ_a are fuel and air density respectively and t is the time after the SoI (Wakuri et al. 1960). According to this equation, for a given time, the air entrainment rate is also proportional to $IP^{1/4}$.

$$\frac{\dot{m}_a}{\dot{m}_f} \propto \frac{IP^{1/4}}{d_0^{1/2}} \cdot \left(\frac{\rho_a}{\rho_f}\right)^{1/4} \cdot t^{1/2} \tag{2}$$

In order to analyse the effect of increasing injection pressure at retarded SoI point, the SoI was fixed at 4 cad aTDC. At this point, the engine operates with PMC mechanism and the combustion process has sufficient stability to allow this parametrical analysis without misfires or other failures. Table 4 contains the basic engine operating conditions for these tests.

Table 4. Test conditions to analyse the effect of increasing IP on PMC.

n [rpm]	IP [bar]	BP [bar]	EP [bar]	T_{INTAKE} [°C]	F_r [-]	SoI [cad aTDC]
1500	1000	1.50	1.60	35	0.32	+4
	1200					
	1600					

When IP is increased, the maximum value of in-cylinder pressure also increases monotonically, as Figure 11 (a) shows. This is the result of enhancing of the HRR during the combustion process, but especially during the premixed phase. The HRR is presented in Figure 11 (b) and it is clear how the premixed stage is promoted by IP.

Two direct consequences of accelerating the combustion process are an increment in the maximum pressure gradient and a significant reduction in combustion duration (see Figure 12). It is important to point out that the diffusive phase is also improved due to better air entrainment promoted by high IP values, as discussed before. The enhancement of the diffusive phase and the increment in the quantity of fuel burnt in the premixed phase explains the reduction in combustion duration.

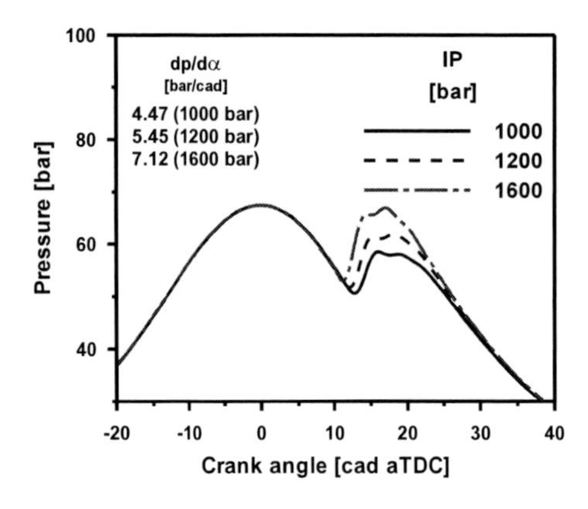

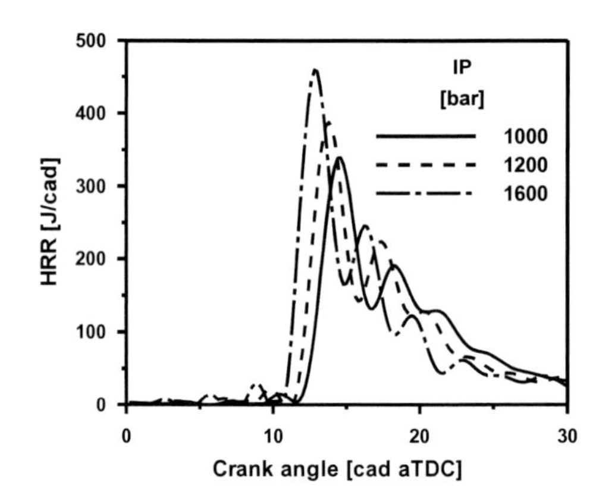

Figure 11. (a) In-cylinder pressure evolution for different IP and (b) HRR evolution for different IP at SoI, both at 4 cad aTDC.

Figure 13 shows the effect of IP on τ_{id}. Increasing IP from 1000 to 1600 bar, τ_{id} is reduced by 16.9%. Although it is generally accepted that the chemical part of ignition delay is more important than the physical part, Rosseel and Sierens suggested that in low temperature combustions, the physical processes are more important than the chemical (Rosseel and Sierens, 1996). IP mainly affects fuel atomisation and air/fuel mixing processes and this explains the strong influence on τ_{id}.

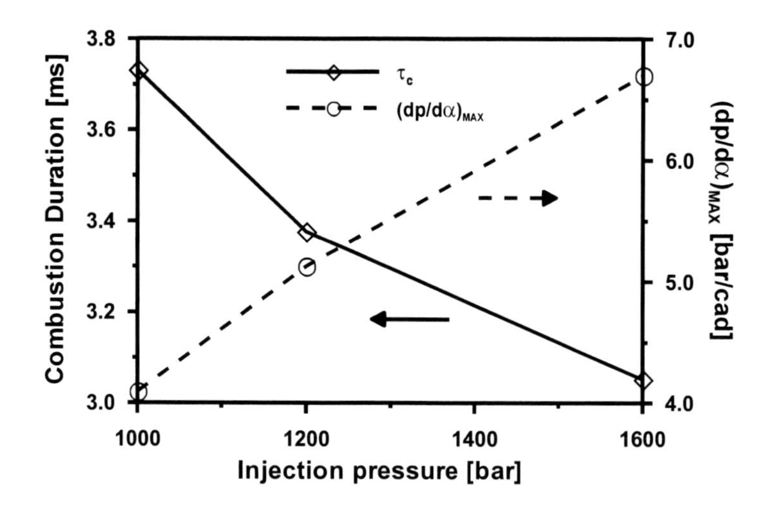

Figure 12. Combustion duration and maximum pressure gradient at SoI 4 cad aTDC.

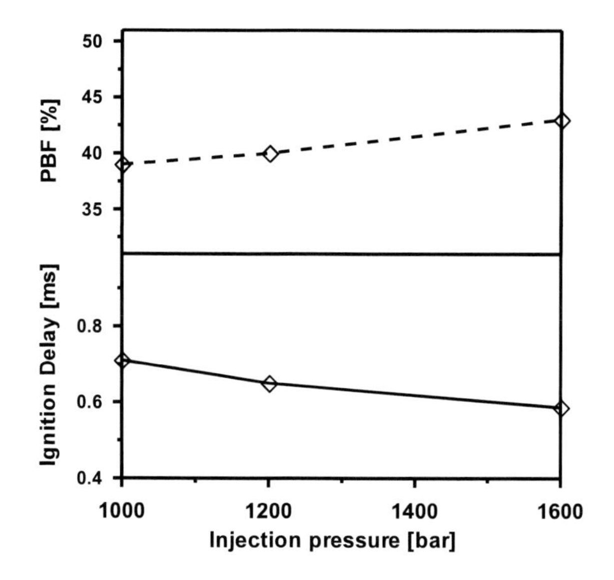

Figure 13. Ignition delay and premixed burnt fraction for different IP at SoI 4 cad aTDC.

However, despite the important reduction in τ_{id}, the PBF increases by 10.2%. To explain this opposite trend, it is interesting to analyse the results shown in Figure 14. Figure 14 (a) shows how fuel injection time (τ_{fi}) is reduced by 18.6%. This reduction is slightly greater than the reduction in τ_{id}, but the amount of fuel injected before the start of combustion is nearly the same, as shown in Figure 14 (b). This confirms that the cause for the increase in PBF is the better air/fuel mixing promoted by IP during τ_{id}, which increases the amount of fuel in conditions to burn when the combustion starts. As a final consequence, the ratio between fuel burnt in the premixed phase and fuel injected after the start of combustion increases.

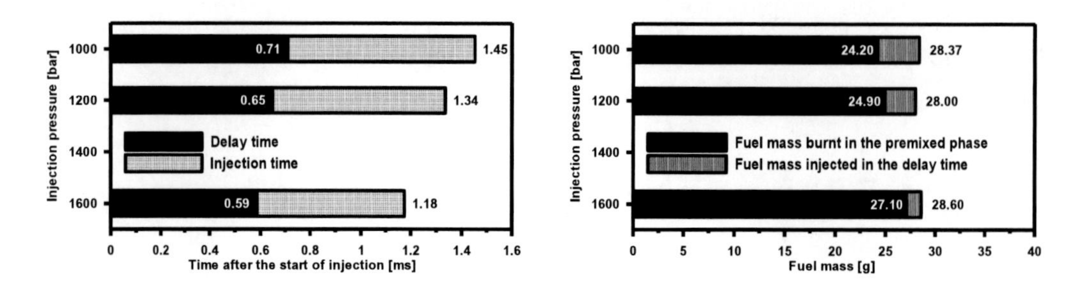

Figure 14. (a) Relation between ignition delay time and fuel injection time for different IP at SoI 4 cad aTDC and (b) Fuel mass injected during τid and fuel mass burnt in the premixed phase.

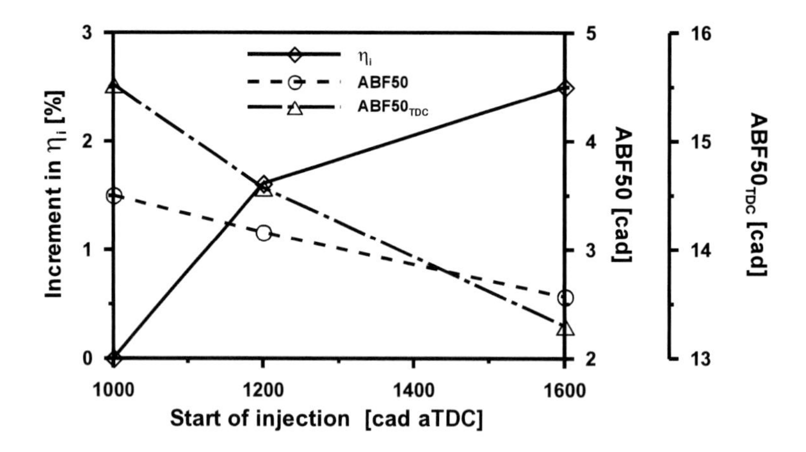

Figure 15. Gross indicated efficiency and 50% burn fraction angle for different IP at SoI 4 cad aTDC.

Regarding the efficiency of the combustion process, it is clear that an increment in IP increases the indicated efficiency. The reason is the position of the combustion process with respect to the TDC. As can be seen in Figure 15, as IP increases, the combustion is placed near TDC and $ABF50_{TDC}$ decreases. This improves the degree of constant volume of the process, which results in improved thermal efficiency. Basically, two reasons explain the displacement of combustion process towards TDC as IP increases: the reduction in τ_{id} and the reduction in combustion duration.

The effect of IP on exhaust emissions is analysed using the trade-off graph shown in Figure 16. As expected, NOx emissions increase with IP due to higher combustion temperatures. Placing combustion close to TDC increases overall combustion temperatures. In spite of the slight promotion of premixed combustion phase, in which NO_x formation is avoided (Dec 1997), diffusive phase cannot be sufficiently inhibited. Local flame temperatures during the remaining diffusive phase and oxygen availability around the flame are also higher because of the better air/fuel mixing, resulting in high amounts of NOx formation in this phase.

This better air entrainment is the main reason for which soot emissions are also reduced with an increment in IP. The light promotion of premixed combustion phase play an important role, but the improvement in late soot oxidation during the last diffusive stage due to the improvement in the air entrainment seems to be the most important factor in reducing soot emissions.

Finally, the fuel consumption will be presented in terms of indicated specific fuel consumption (ISFC). The main reason for this is to discard errors in BSFC as a consequence of losses in the fuel pump. At 1600 bar, the fuel pump installed in the test rig operates far from the nominal pressure and its efficiency drops sharply, distorting the mechanical losses. However, ISFC is improved when IP increases because the combustion process is developed closer to TDC, as commented before (see Figure 15).

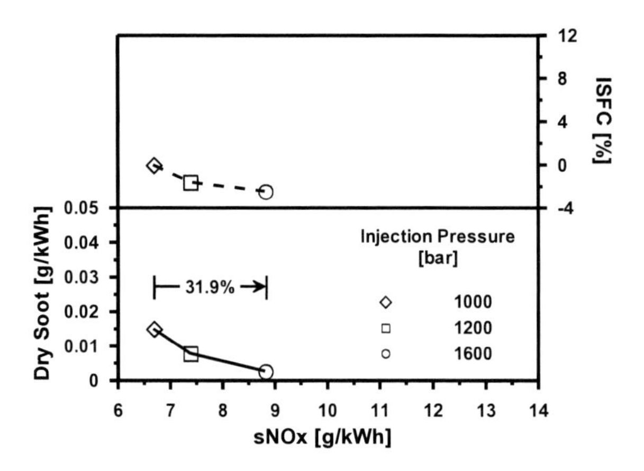

Figure 16. Trade-off showing the global results obtained with different IP at SoI 4 cad aTDC.

4.3 Effect of EGR

Looking at the results, it is interesting to analyse the potential of EGR on exhaust emissions when the engine operates under PMC conditions, especially the effect on soot levels. Two SoI values were tested to compare the effect of EGR with and without PMC strategy. Table 5 contains the most relevant settings selected in this study.

Table 5. Test conditions to analyse the effect of EGR on PMC.

n [rpm]	IP [bar]	BP [bar]	EP [bar]	T_{INTAKE} [°C]	SoI [cad aTDC]	EGR [%]	T_{EGR} [°C]	F_r [-]
1500	1200	1.50	1.60	35	-6	0	80	0.32
						30		0.42
						40		0.47
					+4	0		0.32
						30		0.42
						40		0.47

Obviously, NOx emissions drastically drop with EGR in both cases as Figure 17 shows. As it is well known, reducing oxygen concentration gives a proportional reduction in local flame temperatures, avoiding chemical reactions of NOx formation (Ladommatos et al. 2000). In the case of the retarded SoI case, EGR increases τ_{id} and enlarges the premixed combustion phase. This also reduces NOx formation.

But one of the strongest differences between traditional Diesel combustion and PMC mechanism is the effect of EGR on soot emissions. It is believed that the EGR only affects the soot oxidation process, whereas the formation hardly changes (Dec and Kelly-Zion, 2000). In the first case, when the engine operates at SoI -8 cad aTDC with traditional Diesel combustion, the introduction of EGR produces an exponential increment in soot emissions because the formation process remains unchanged, but the oxidation rate drops strongly. When operating under PMC conditions, soot emissions are not modified with EGR, they can even be reduced due to the shortening of the diffusive combustion phase. This particular behaviour confirms the most important characteristic of PMC: when the fuel is premixed with air before the start of combustion, soot formation is avoided as a consequence of high air/fuel ratio. Therefore, despite the soot oxidation process being worse, soot emissions do not increase with EGR.

Finally, introducing EGR strongly worsens BSFC at retarded point because of the displacement of the combustion process to the expansion stroke, far from TDC. The overall combustion process is deteriorated by EGR, and this also contributes to the BSFC worsening.

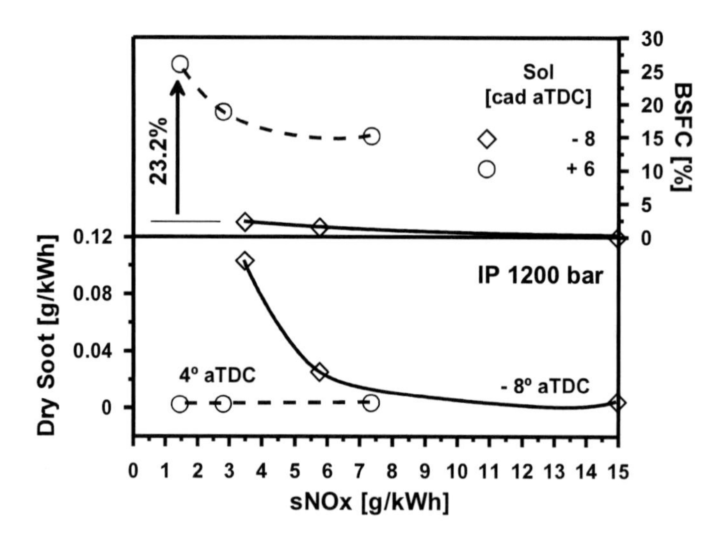

Figure 17. Trade-off showing the global results obtained with EGR at IP 1200 bar and SoI -8 and 4 cad aTDC.

CONCLUSION

The possibility of promoting the premixed combustion phase (PMC) in a HD Diesel engine operating at light load has been investigated by adjusting the injection parameters in a standard engine configuration. The potential of EGR in reducing NOx emissions in this combustion condition has been also studied.

The most interesting results obtained by retarding the SoI from -8 cad aTDC to 6 cad aTDC are summarised below:

- Ignition delay was extended by 57%, but even in these conditions, it was not possible to attain a totally premixed combustion.
- Premixed burnt fraction was increased from 19% to 50%. With this important enhancement of PMC, it was possible to avoid the usual trend of increasing soot emissions as the SoI is retarded due to the great quantity of fuel burnt in premixed conditions. The NO_x emissions were also reduced by 51% due to low combustion temperatures.
- In spite of reducing combustion time, retarding SoI shifts the combustion to the expansion stroke. This reduces the indicated efficiency of the process and increases HC and CO emissions.
- BSFC is worsened due to the commented reduction in indicated efficiency.

With the aim of improving the air/fuel mixing during the ignition delay time and reducing the injection time, IP was increased from 1000 to 1200 and 1600 bar. The most important effects identified were:

- Injection time was reduced by 18.6% and ignition delay time was shortened by 16.9%. As a result, the quantity of fuel injected before the start of combustion was almost the same.
- In spite of the previous conclusion, the fuel mass burnt in premixed conditions was clearly increased, plausibly due to an improvement in the fuel/air mixing process.
- The better fuel/air mixing improves the overall combustion process (including the diffusive phase) and greatly reduces the combustion duration, thus improving the indicated efficiency and the BSFC.
- Soot emissions were reduced to inappreciable values and an important increment in NOx emissions of 31.9% was observed, probably due to a more intense air entrainment, especially during the last diffusive stage.

The effect of EGR combined with PMC on NO_x and soot emissions and fuel consumption gives the following results:

- NOx are drastically reduced as occurs at advanced SoI points. In the case of retarded SoI, EGR increases the ignition delay time and enlarges the premixed combustion phase, which should contribute to the common causes for the NOx reduction.
- Soot emissions are not increased, even with high EGR rates of 40%. Since a large portion of fuel is mixed with air before the start of combustion, soot formation is limited because of high air/fuel ratio. This result confirms that the most important characteristic of PMC is attained.
- BSFC slightly increases with high values of EGR rate.

References

Aoyama T, Hattori Y, Mizuta J and Sato Y (1996) *An experimental study on premixed-charge compression ignition gasoline engine.* SAE Paper 960081.

Armas O (1998) *Experimental diagnostic of combustion process in direct injection diesel engines. Doctoral Thesis (in Spanish),* Universidad Politécnica de Valencia, Spain.

Bosch W (1966) *The fuel rate indicator: A new instrument for display of the characteristics of individual injection.* SAE Paper 660749.

Christian R, Knopf F, Jasmek A and Schindler WA (1993) *New method for the filter smoke number measurement with improved sensitivity (in German).* MTZ 54, pp.16-22.

Dec JE (1997) *A conceptual model of Diesel combustion based on laser-sheet imaging.* SAE Paper 970873.

Dec JE and Canaan RE (1998) *PLIF imaging of NO formation in a D.I. diesel engine.* SAE Paper 980147.

Dec JE and Kelly-Zion PL (2000) *The effects of injection timming and diluent addition on late-combustion soot burnout in a DI Diesel engine based on simultaneous 2-D imaging of OH and soot.* SAE Paper 2000-01-0238.

Donahue RJ, Bormann GL and Bower GR (1994) *Cylinder-averaged histories of nitrogen oxide in a D.I. Diesel with simulated turbocharging.* SAE paper 942046.

Hashizume T, Miyamoto T, Akagawa H and Tsujimura K (1998*) Combustion and emissions characteristics of multiple stage Diesel combustion.* SAE Paper 980505.

Heywood J (1988) *Internal combustion engine fundamentals.* McGraw-Hill.

Ishibashi Y and Asai M (1996) *Improving the exhaust emissions of two-stroke engines by applying the activated radical combustion.* SAE Paper 960742.

Iwabuchi Y, Kawai K, Shoji T, and Takeda Y (1999) *Trial of new concept diesel combustion system –premixed compression-ignited combustion–.* SAE Paper 1999-01-0185.

Kimura S, Aoki O, Ogawa H, Muranaka S and Enomoto Y (1999) *New combustion concept for ultra-clean and high-efficiency small DI diesel engines.* SAE Paper 1999-01-3681.

Ladommatos N, Abdelhaim SM and Zhao H (2000) *The effects of exhaust gas recirculation on diesel combustion and emissions.* Int. J. Engine Research, Vol.1. No.1, pp.107-126.

Lapuerta M, Armas O and Bermúdez V (2000) *Sensitivity of diesel engine thermodynamic cycle calculation to measurement errors and estimated parameters.* Applied Thermal Engineering, Vol.20 (9), pp.843-861.

Lapuerta M, Armas O and Hernández JJ (1999) *Diagnostic of D.I. diesel combustion from in-cylinder pressure signal by estimation of mean thermodynamic properties of the gas.* Applied Thermal Engineering, Vol.5, pp.513-519.

Lavy J, Dabadie J, Angelberger C, Duret P, Juretzka A, Schaflein J, Ma T, Lendresse Y, Satre A, Schulz C, Kramer C, Zhao H and Damiano L (2000) *Innovate untra-low NOx controlled auto-ignition combustion process for gasoline engines: the 4-SPACE project.* SAE Paper 2000-01-1837.

Moser FX, Sams T and Catellieri W (2001) *Impact of future exhaust gas emission legislation on the heavy duty truck engine.* SAE Paper 2001-01-0186.

Najt PM and Foster DE (1983) *Compression-ignited homogeneous charge combustion.* SAE Paper 830264.

Nakagawa H, Endo H, Deguchi Y, Noda M, Oikawa H and Shimada T (1997) *NO measurement in Diesel spray flame using laser induced fluorescence.* SAE Paper 970874.

Onishi S, Jo SH, Shoda K, Jo DP and Kato S (1979) *Active termo-atmosphere combustion (ATAC) – A new combustion process for internal combustion engines.* SAE Paper 790501.

Rosseel E and Sierens R (1996) *The physical and the chemical part of the ignition delay in Diesel engines.* SAE Paper 961123.

Suzuki H, Koike N and Okada M (1997) *Exhaust purification of Diesel engines by homogeneous charge with compression ignition, Part 1: Experimental investigation of combustion and exhaust emission behaviour under pre-mixed homogeneous charge compression ignition method.* SAE Paper 970313.

Suzuki H, Koike N and Okada M (1997) *Exhaust purification of Diesel engines by homogeneous charge with compression ignition, Part 2: Analysis of combustion phenomena and NOx formation by numerical simulation with experiment.* SAE Paper 970315.

Takeda Y, Keiichi Na and Keiichi Ni (1996) *Emission characteristics of premixed lean diesel combustion with extremely early staged fuel injection.* SAE Paper 961163.

Thring RH (1989) *Homogeneous-charge compression-ignition (HCCI) engine.* SAE Paper 892068.

Voiculescu IA and Borman GL (1978) *An experimental study of Diesel engine cylinder-averaged NOx histories.* SAE Transactions, Vol.87, Sec.1, pp.1001-1014, paper no. 780228.

Wakuri Y, Fujii M, Amitani T and Tsuneya R (1960) *Studies on the penetration of fuel sprays in a Diesel engine.* Bull. Of the JSME, Vol.13, No.9.

Yanagihara H.A (1996) *A simultaneous reduction of NOx and soot in diesel engines under a new combustion system (uniform bulky combustion system UNIBUS).* 17th. Int. Vienna Motor Symposium.

Yokota H, Kudo Y, Nakajima H, Kakegawa T and Susuki T (1997*) A new concept for low emission diesel combustion.* SAE Paper 970891.

In: Advances in Mechanical Engineering Research, Volume 3 ISBN: 978-61209-243-0
Editor: David E. Malach

Chapter 9

ENHANCING THE WRITING QUALITY OF AIRCRAFT MAINTENANCE TECHNICAL ORDERS AND ESTABLISHING A MANAGEMENT MECHANISM FOR MAINTENANCE TECHNICIANS USING THE SIX SIGMA PROCESS

***Tai-Chang Hsia*[1]*, Su-Chen Huang*[*2] *and Hsi-Tien Chen*[3]**

[1]Department of Industrial Engineering and Management, Chienkuo Technology University, Changhua, Taiwan, R.O.C.

[2]Department of International Trade, Overseas Chinese University, Taichung, Taiwan, R.O.C

[3]Department of Leisure Industry Management, National Chin-Yi University of Technology, Taichung, Taiwan, R.O.C

ABSTRACT

Aircraft maintenance technical orders are guidelines for aircraft system maintenance and repair. Maintenance technicians must follow instructions of technical procedures step by step. For aviation safety the writing quality of technical orders must be enhanced to provide aircraft maintenance technicians with a reliable management system. This study adopts the Six Sigma Process improvement procedure known as DMAIC. The readability index is defined as readability of statements written in maintenance technical orders, while the importance index is defined as the significance of aviation safety obtained through composing maintenance technical orders. These indices evaluate quality and vigilance in writing maintenance technical orders. The DMAIC simultaneously rates and analyzes maintenance of technical orders with a high satisfactory value on the importance index, but low on readability. It proposes a way to enhance writing quality while devising

[*]Correspondence: Dr. Su-Chen Huang, Full postal address: Department of International trade, Overseas Chinese University, 100, Chiao Kwang Rd., Taichung 40721, Taiwan, R.O.C., Telephone: 886-4-27016855 Ext.2102, Fax: 886-4-27075420, E-mail address: tb033215@gmail.com

a management mechanism for maintenance technicians in order to ensure quality technical procedures to enhance aviation safety.

Keywords: Aircraft Maintenance Technical Order, Aviation Safety, Six Sigma Process, Management Mechanism, Maintenance Technician

1. INTRODUCTION

In the current global economy, the expansion of airline safety has received the most attention in terms of human safety issues, followed by mechanical issues (Krause, 2003). There are two main reasons for mechanical failure: the quality of the materials, and the maintenance skill. No aircraft manufacturing company can afford to neglect the quality of construction and high levels of maintenance, lest they run the risk of air safety incidents. In order to assure the quality of maintenance and repair, mechanics must pay close attention to the maintenance protocols and instructions of manufacturers, in particular the steps involved in technical manuals. The key in aircraft manufacturing and safety is to elevate the quality of technical orders used by mechanics.

The protocols for technical orders pertain to the maintenance, repair, testing, and verification and validation of aircraft. Therefore, technical orders must be composed not only with a clear understanding of all facets of aircraft design, materials, and systems, but also in a way that is sure to be (verbally) clear to repair workers. Maintenance technicians lacking relevant technical knowledge and a sophisticated background of professional experience in technical design make poor candidates for composing technical orders. Providing technical authors with a set of satisfaction criteria and efficient technical orders will go a long way toward solving a pressing difficulty in the field.

However, for an aircraft maker to produce satisfactory, high-level instruction manuals for all repair workers is no simple task. The ISO Manual 9241, part 11, “Guidance on usability specification and measure,” states, "The usability of a product or software application depends on whether it can be effectively and satisfactorily used in a specialized setting to achieve a specialized goal." This shows the great importance of the quality of composition of technical orders and of a good management (quality control) system.

The Six Sigma Process stems from a statistical concept referring to how regulations with a margin of error of 3.4/1,000,000 achieve a nearly perfect limit-value. The five steps that were developed in the DMAIC procedure (viz., Define, Measure, Analyze, Improve, and Control) seek key cause-effect factors that raise organizational effectiveness. As early as 1987 Motorola implemented the procedure, and in the 1990s G.E., Allied Signal, and Seagate also adopted the procedure with great effect. DMAIC has thus become an important tool by which modern corporations transform problems into successes (Brue and Launsby, 2003).

This study continues the work in Hsia (2007) Evaluating the Writing Quality of Aircraft Maintenance Technical Orders Using a Quality Performance Matrix, and Hsia et al. (2008) Measuring the Readability Performance (RP) of Aircraft Maintenance Technical Orders by Fuzzy MCDM Method and RP Index. These research programs use technical orders published by a Taiwan aircraft manufacturing company for the maintenance of a single type of fighter aircraft, selecting the most important and appropriate examples from them for research in order to assess the quality of executive editing and to judge the quality of writing in each manual's technical maintenance orders.

This research employs the DMAIC Six Sigma Process proposed by Thomas (2003), focusing on inadequate technical maintenance orders and suggesting ways to improve the quality of composition. It also provides concrete actions to be taken with respect to realizing a sound management system for technical maintenance staff so as to ensure quality of repair and maintenance of aircraft safety. First of all, the Define section looks for and tabulates the number of unsatisfactory descriptive sentences in technical orders. Further, it isolates and quantifies the number of sentences that were labeled “warning” or “caution” in the text of technical maintenance orders. In the Measure section, the research focuses on the level of satisfaction readers have while reading the technical orders. It also examines the safety and (perceived) importance of technical orders, and then assesses how aberrations or faults in the composition affect the quality of the technical orders. The Analyze section employs a special cause-effect diagram to focus on technical orders with a high level of safety and importance but a low level of (reader) satisfaction. Analyzing the readability elicits the main reasons for this cause-effect complex. In the Improve section, Quality Function Development technology is used to once more indicate key factors which can improve technical orders. Finally, the Control section deploys the TRIZ creative method for solving problems to attain gradual improvement of composition quality of aircraft maintenance technical orders and in order to establish protocols for repair workers management for a better final outcome.

2. Defining Satisfaction Index and Safety Importance Index

With technical orders, every step and visual illustration must be extremely precise so that maintenance technicians can accurately adhere to them and thereby attain the high levels of quality required for safely maintaining aircraft. From the perspective of human factor engineering, a sound technical instruction manual must include the following efficacy criteria: visibility, legibility, and readability (Sanders and McCormick, 1993).

Documents in hard copy can satisfy the requirements for visibility and legibility, so only readability must be assessed in detail. In any manual, the fewer unreadable sentences, the higher the readability rating. An opposite rating creates a detrimental impact on the execution of technical orders by maintenance technicians. Defining the readability of technical orders, Hsia (2007) points out optimal instructions must have six features which enable a maintenance technician to quickly grasp the meaning of a text and completely understand the procedure. These six features are:

1) Accessible Sentence Structure: An affirmative and active style of sentence that is easily understood and assimilated by technicians.
2) Apt Sentence Order: Sentences must be written in a way that reflects the actual order of the procedure in question.
3) Primarily Semantic and Secondarily Syntactic Sentences: Sentences that primarily appear in a semantic rather than a syntactical form are the easiest for technicians to follow in construction, diagnostics, disassembly, assembly, and even in forming a "mental model" of the repair. Semantic cognitive reference

knowledge systems are operated in the choice of diction and of how to link words amidst different lexical concepts. Syntactic commands map the arrangement and structure of words in the sentence. As the choice of vocabulary is affected by grammatical structure, a writer must first consider the words and then method (or order) of composition when writing technical orders. The appearance of technical orders must allow technicians to first follow semantic cognition, then semantic structure, so that they may easily understand the sentences while reading.

4) Simple Sentences vs. Complex Structure: The more complex a grammatical structure is, the more cognitive resources it requires to process, and the fewer resources it leaves for real comprehension. Hence, readable grammatical structure must be succinct and vivid (in other words, short on words and long on content).
5) Additional Explanatory Illustrations: Instructions with model examples are superior to instructions without them, and in a degree proportionate to the number of explanatory scenarios.
6) Practical Visual Illustrations "Manuals with only text limit readers to the lower two-thirds of their comprehension ability. If manuals include visual illustrations and schematic diagrams, reader comprehension will rise to the top third.

This study identifies sentences in a volume of aircraft technical orders which violate any of the (above) criteria for readability. Such sentences are labeled "unreadable". Additionally, any sentences in technical orders which contain typos, faulty meanings, or unclear and grammatically incoherent sentences will be labeled as unreadable. Such sentences are recorded in Table 1 as "7. Others".

A statistical analysis is then performed using Formula (1) below:

$$L_i = \sum_{j=1}^{N_i} X_{ij} \quad , \qquad i = 1,\ldots,7 \tag{1}$$

Where X_{ij} represents aberrations from the seven readability criteria, i represents a criterion, j represents any detriment caused by unreadable sentences, N_i represents the total number of unreadable sentences in aircraft technical orders, and L_i represents the total amount of lost readability in a technical order.

There are 102 volumes of intermediate level technical orders for maintenance of the fighter aircraft. Twelve of these relate to engine maintenance. This study examines the most critical and important eight volumes of the 12 aircraft engine intermediate level maintenance technical orders used to service the fighter and noted any sentences that violated the readability criteria, labeling them "unreadable". The total number of unreadable sentences was calculated, taking 100 pages as a discrete unit for measurement. The amount of unreadability in the technical orders was compared with the level of satisfaction with the descriptions. The total number of unreadable sentences was then transformed by a linear function to produce a ratio, ranging from 0.1 to 0.9, for quantifying the customer satisfaction index (i.e., satisfaction with the description of tasks), indicated as I_S (Table 1).

Table 1. Statistics on Unreadable Sentences and Descriptive Satisfaction for eight volumes of Maintenance Technical Orders.

Volume	Series number	Pages	Six writing features						7. Others	Number of unreadable sentences	Number of unreadable sentences per hundred pages	Satisfaction index (I_s) [0.1, 0.9]
			1.Sentence structure	2. Sentence order	3. Order of semantic and syntactic forms	4. Simple and complex sentences	5. Examples	6. Realistic graphics				
1	UXX 2J-FX XXX-6-3	365	32	1	0	2	0	0	47	84	23.01	0.1
2	UXX 2J-FX XXX-6-4	182	1	0	2	2	0	0	9	14	7.70	0.9
3	UXX 2J-FX XXX-6-5	193	0	0	0	1	0	7	22	30	15.50	0.49
4	UXX 2J-FX XXX-6-6	238	11	0	1	1	0	0	8	21	8.80	0.84
5	UXX 2J-FX XXX-6-9	407	0	0	1	0	0	7	32	40	9.80	0.79
6	UXX 2J-FX XXX-6-10	217	1	2	2	10	0	2	17	34	15.70	0.48
7	UXX 2J-FX XXX-6-11	288	8	4	3	0	1	1	18	35	12.20	0.65
8	UXX 2J-FX XXX-6-12	510	11	3	4	12	0	0	44	74	14.50	0.54
Sum		2400	64	10	13	28	1	17	197	332		

After calculation of the value of the customer satisfaction index (I_S), the value of the safety importance index (I_I) was then calculated. Again, over 100 pages per unit, the number of instances of "Warning" and "Caution", and related sentences, were tabulated from the contents of eight maintenance technical orders. The higher the number of such instances, the more technicians need to show vigilance and caution while executing aviation maintenance and repair. Lapses directly affect the safety level of aircraft and associated apparatus. Failure to heed Warnings can cause serious system damage and/or serious harm to the technician. Failing to heed Cautions leads to system damage and/or inefficient work. As such, the hierarchical distinction between heeding and ignoring Warnings and Cautions is ranked from 1.0 to 0.5, respectively. In the formula below, for V_i, i represents the number of instances of Warning clauses in a technical orders, for L_{ij}, j represents the number of related sentences attached to a warning statement, for v_i, i represents the number of instances of Caution statements in a technical orders, for l_{ij}, j represents the number of additional sentences attached to a Caution sentence. Accordingly, the weighted value of i with respect to safety awareness and importance is W_i, as formula (2) shows.

$$W_i = \sum_{j=1}^{V_i} \square L_{ij} \square\square \sum_{j=1}^{v_i} \square l_{ij} \square/2 \qquad i = 1, 2, ..., 8 \tag{2}$$

Once the weighted value of safety consideration and importance is calculated for every manual, averaged over 100-page units, a linear transformation situates the value from [0.1, 0.9] which is called the safety importance index (I_I) (Table 2).

Table 2. Consideration of the Safety Importance Index in eight volumes of Maintenance Technical Orders.

Volume	Series number	Pages	Number of sentences after "warning" and "caution" labels				Weight of safety importance W_i	Average safety importance of every hundred pages y_i	Safety Importance Index (I_I) [0.1, 0.9]
			number labeled caution	Number of statements after the caution label	number labeled warning	Number of statements after the l warning label			
1	UXX 2J-FX XXX-6-3	365	15	119	90	682	741.5	203	0.9
2	UXX 2J-FX XXX-6-4	182	7	26	28	73	86	47	0.12
3	UXX 2J-FX XXX-6-5	193	38	169	13	49	133.5	69	0.23
4	UXX 2J-FX XXX-6-6	238	44	63	51	176	207.5	87	0.32
5	UXX 2J-FX XXX-6-9	407	64	196	107	307	405	100	0.38
6	UXX 2J-FX XXX-6-10	217	13	51	42	71	96.5	44	0.1
7	UXX 2J-FX XXX-6-11	288	28	192	41	348	444	154	0.65
8	UXX 2J-FX XXX-6-12	510	33	204	49	159	261	51	0.14
Sum		2400	242	875	421	1810			

3. Measuring the Quality of the Composition of Technical Orders

In the measuring section, a satisfaction index (IS) is gauged based on the descriptions of technical orders when technicians reading, as well as a safety importance index (II), based on the safety and importance of the technical orders when technicians implement them. These two criteria, displayed in Figure 1, are used to assess the quality of composition of technical orders. In Figure 1, the first volume (T1) falls in the B13 block for writing quality. The third, sixth, and eighth volumes (T3, T6, T8) fall in block B21 of the writing effectiveness grid. The seventh volume (T7) falls in block B22 for writing quality, the second and fourth volumes (T2, T4) fall in block B31, and the fifth volume (T5) falls in block B32. Looking at the eight volumes, we find that the first volume has a very high safety and importance rating, but a very low description-satisfaction rating. This means that the writing quality is poor and insufficiently exact, which in turn means that the technical order writers must focus on revising and improving the quality of volume 1 so that the responsible technicians can establish a management system which brings the quality of writing towards the level of the safety importance rating. Insofar as the other seven volumes have very good safety and importance ratings, matched by a high level of descriptive satisfaction, they need not be revised much beyond raising the quality of the writing in general.

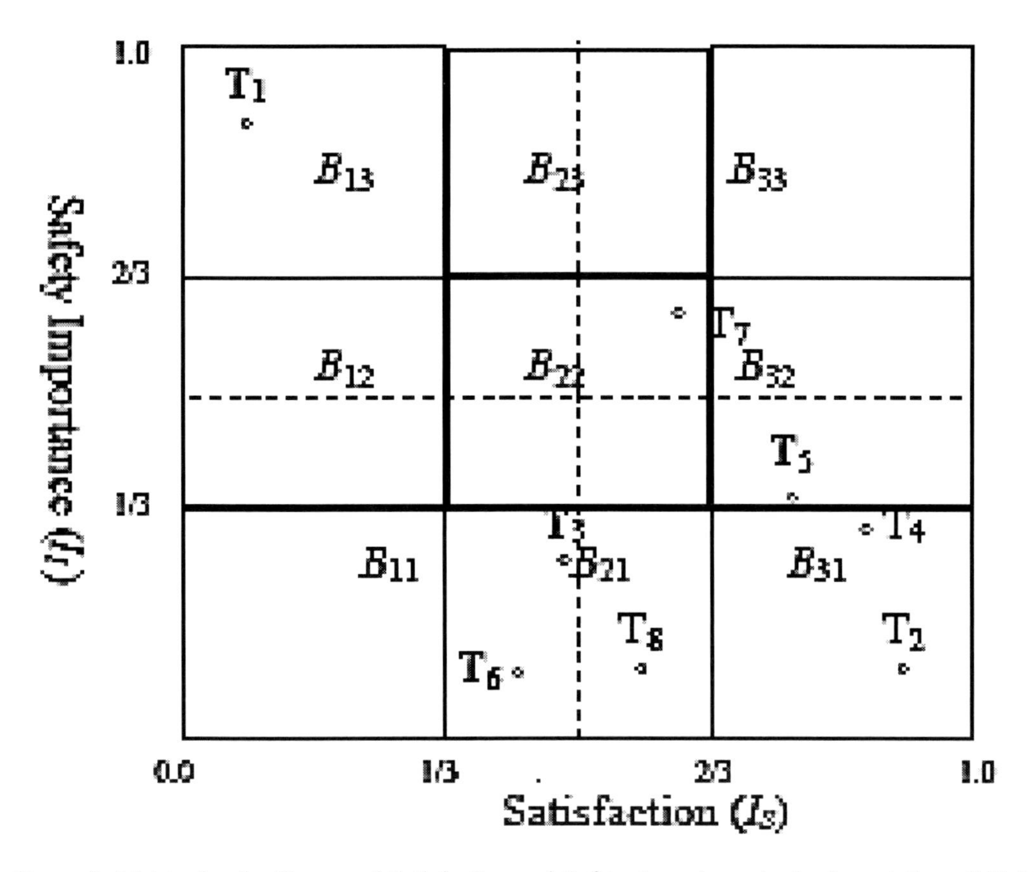

Figure 1. Matrix for the Degree of Satisfaction and Safety Importance in Implementation of Eight Volumes of Maintenance Technical Orders.

4. Analysis the Readability of Maintenance Technical Order

The quality of writing in technical instruction orders has a close connection to the effectiveness of technicians' repairs. Here the focus is on raising the quality of technical instruction in volume 1 (T_1). Among the 84 unreadable sentences, 32 are inadequate because of problems with violations of the first criteria, Sentence Structure. One sentence is faulty based on the second criteria, Sentence Order, while two sentences have problems based on the third criteria, the Order of semantic and syntactic Forms, and forty-seven other sentences violate the seventh criteria, miscellaneous.

This study uses a specialized cause-effect diagram to analyze the causes of unreadable sentences. In collaboration with the supervisor of technical instruction at the Aeronautical Research Laboratory, this research focused on the 84 unreadable sentences and discovered four main factors that increase the readability of technical orders: 1. The quality of maintenance technicians; 2. The ability of technical order writers; 3. Printing/layout; 4. Managerial system.

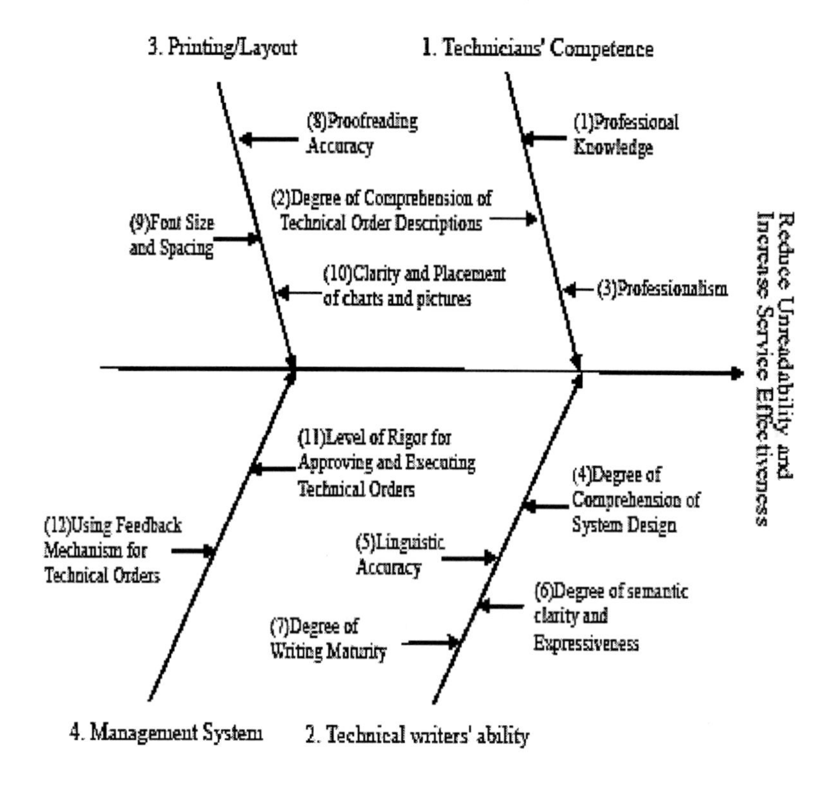

Figure 2. Using a Cause-Effect diagram to analyze the causes of unreadable sentences.

From among these four main factors, twelve other headings, termed secondary factors, were developed for better analysis of the causes of unreadability, as shown in Figure 2. The secondary factors included in the first main factor, the quality of maintenance technicians (technicians' competence) are (1) how much professional knowledge technicians have, (2) how technicians process and comprehend technical orders, and (3) the professional attitude of technicians. The second main factor, the ability of technical order writers (technical writers' ability), includes four secondary factors (4) the level of comprehension writers have about the system design, (5) the syntactic (linguistic) precision writers possess and use, (6) the level of semantic clarity writers display, and (7) the level of technical maturity writers have. In the third main factor, the heading of printing/layout, the three secondary factors are (8) proofreading accuracy, (9) the font size and spacing of letters, and (10) the appropriateness and placement of charts and pictures in the text. The fourth main factor, the heading of the management system, contains two secondary factors: (11) the level of conscientiousness and exactness in the approval of technical orders and (12) managerial feedback after executing technical orders.

5. FROM SECONDARY FACTORS TO SEEKING KEY FACTORS FOR IMPROVEMENT

In the improvement section, this study uses Quality Function Development (QFD) to find the key factors for improvement. While processing the Quality Function Development technology, it was discovered that there were related secondary factors in all four main factors for lowering unreadability and improving the efficacy of technical maintenance. Thus, we carry out a Quality Function Development and use the relationship between the 12 secondary factors and the four main factors to determine key factors. We measure TW_i as the sum of the values derived from multiplying each of the headings of the causes of unreadability, which corresponds to the weight w_i of each correlation coefficients of main factors (δ_i):

$$\text{TWi} = \sum_{i=1}^{4} (\delta i \times w_i) \tag{3}$$

In the formula, w_i represents the weight function of each secondary factors corresponding to the main factor and δ_i represents the correlation coefficients of main factors.

During the quality function development, the correlation coefficients of the main factors were determined for discussion at a joint meeting at the Aerospace Industrial Development Corporation in Taiwan attended by the head of the technical composition department, maintenance technicians, managers, and the head managers of each department. These correlation coefficients ranged from [0.0 – 1.0]. The quality of maintenance technicians was set at 0.8; the ability of technical order writers was set at 1.0; printing/layout was set at 0.4. The Managerial system was set at 0.6 as show in Table 3. The larger a coefficient, the more important the factor it represents. Formula (3) calculates the weight function of these factors, such that the larger a coefficient is, the more it can lower unreadability and improve technician effectiveness, as shown in Table 3. The top five headings of key factors are divided in terms of (4) the level of comprehension writers have about the system design, (3) the professional attitude of technicians, (1) how much professional knowledge technicians have, (8) the proofreading accuracy of printing/layout, and (12) managerial feedback after executing technical orders. These five key factors were selected gradually to improve the technical writing quality in the first volume of aircraft maintenance technical orders and the establishment of a management system as a control mechanism for improving technicians.

Table 3. Using Quality Function Development to Discover the Key Factors.

Four main factors for increasing readability in technical orders / Twelve headings of the causes of unreadability / Weight w_i: Very strong 4.0, Strong 3.0, Fair 2.0, Weak 1.0, Not relevant 0.0	Correlation Coefficients of Main Factors δ_i	(1) Professional Knowledge	(2) Degree of Comprehension of Technical Order Descriptions	(3) Professionalism	(4) Degree of Comprehension of System Design	(5) Linguistic Accuracy	(6) Degree of semantic clarity and Expressiveness	(7) Degree of Writing Maturity	(8) Proofreading Accuracy	(9) Font Size and Spacing	(10) Clarity and Placement of charts and pictures	(11) Level of Rigor for Approving and Executing Technical Ordes	(12) Using Feedback Mechanism for Technical Orders
1.Technicians' Competence	0.8	4.0	3.0	4.0	3.0	2.0	2.0	2.0	4.0	1.0	1.0	2.0	3.0
2.Technical writers' ability	1.0	3.0	3.0	3.0	4.0	2.0	2.0	3.0	1.0	0.0	0.0	2.0	2.0
3.Printing/Layout	0.4	1.0	1.0	1.0	2.0	0.0	0.0	0.0	4.0	3.0	4.0	3.0	3.0
4. Management System	0.6	3.0	1.0	4.0	4.0	4.0	2.0	1.0	4.0	2.0	2.0	4.0	4.0
Total Weight (TW_i)		8.4	6.4	9.0	9.6	3.0	3.0	2.5	8.2	3.2	3.6	7.2	8.0
Key Factor		3		2	1				4				5

6. Using the TRIZ Method for Inventive Problem Solving to Attain Technical Writing Quality and to Establish Control Mechanisms through Service Technician Management

After determining the five key factors of improvement, this study used the TRIZ method for creative problem-solving to improve the writing quality of technical orders and to set up a management system, thereby increasing the effectiveness of maintenance in general. TRIZ is from the Russian "Теория решения изобретательских задач", which translates into English as the Theory of Inventive Problem Solving (Altshuller, 1996).

According to this theory, when someone is looking for a solution to engineering problems, he or she frequently encounters contradictions in the system, and efforts to improve one feature of a system frequently result in defects in another part of the system. When faced with this kind of contradictory engineering problem, TRIZ can provide effective solutions. The first step is to pinpoint contradictory features in the system. The next step is to consult the 39 parameters that should be in a system, and then to indicate the features one hopes to improve and the features one hopes not to worsen. Using a contradiction matrix, the analyst can then choose creative principles for finding a solution, as shown in Figure 3.

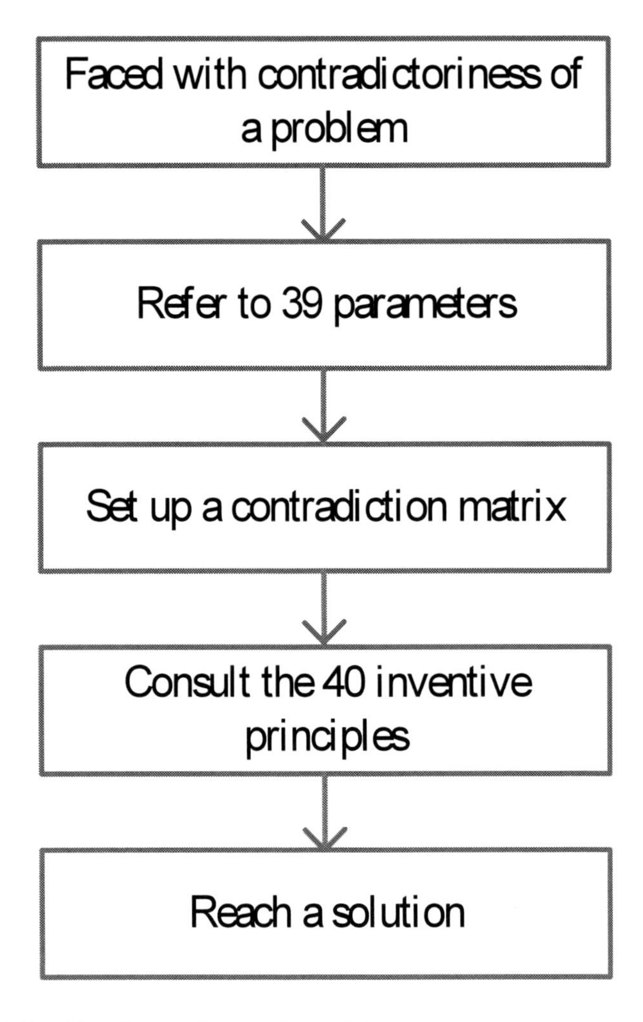

Figure 3. The TRIZ algorithm for creative problem solving.

The contradiction matrix and relevant principles for inventive problem solving enabled research to focus on the five key factors in maintenance improvement and to isolate a method for aggressively improving the writing quality in the first volume of aircraft maintenance technical orders, so that those responsible for technical maintenance can establish a sound management mechanism for issuing technical orders, as Table 4 shows.

Table 4. Using TRIZ in conjunction with five key cause-effect factors for composition quality and to establish a mechanism for system management as a control mechanism.

Item	Key Factors	TRIZ Inventive Principle	Explanation	Improve Composition Quality and Establish Management Control Mechanism
(4)	The level of comprehension authors have about the system design	1	Segmentation	Applied Training for Writers about Different Systems
		2	Extraction	Full Comprehension of the detailed function of Components in a System
		19	Periodic Action	Periodic Discussion and Training for Writers
		21	Rushing Through	Required Assessment of Writers
		27	Disposal	Dismissal of Inadequate Writers
(3)	The professional attitude of technicians	1	Segmentation	Work Segmentation/Allocation
		7	Nesting	Periodic Reward of Outstanding Technicians
		11	Cushioning	Instruct Technicians about Job Field Ethics and Workplace Principles Beforehand
		13	Do It In Reverse	Investigate and Inspection of Maintenance Technicians
		34	Rejecting and Regenerating Parts	Dismissal of Non-Qualifying Technicians
		35	Transformation of Properties	Create a Better Team through Sharing and Discussion
(1)	How much professional knowledge technicians have	1	Segmentation	Work Segmentation/Allocation
		10	Prior Action	Must Have Experience Prior
		17	Transition Into a New Dimension	Promote Experienced and Skilled Technicians to Foreman or Cadre
		19	Periodic Effects	Maintenance Technicians Must Receive Periodic Professional Evaluation
		33	Homogeneity	Maintenance Technicians Need to Obtain Professional License
		40	Composite Materials	Technicians Need to Undergo Professional Training in Various Systems
(8)	The proofreading accuracy of printing/layout	1	Segmentation	Use Different Professional Staff to Proofread His Professional Systems' Instructional Writing
		22	Convert Harm Into Benefit	Use Error-Detection and -Removal Software for Printing Layout
		32	Change Color	Use Color Printing for Important Marks and Signs
(12)	Managerial feedback after executing technical orders	13	Do It In Reverse	Authors Schedule Visits with Maintenance Technicians
		15	Dynamicity	Establish Quick-Response Feedback Mechanisms
		23	Feedback	Technician Feedback Pipeline Must Be Fluid/Smooth
		24	Mediator	Invite Customer Satisfaction Survey Company to Gauge Satisfaction with Technical Instruction and Maintenance

Based on Table 4, we use the first key factor, namely item (4), as an example.

1) Face "The level of comprehension writers have about the system design", the 39 reference headings can detect features that will produce contradictions.
2) The 31st Parameter "Harmful Side Effects" needs to be improved by generating readable sentences. For the 36th and 37th Parameters "Complexity of the System" and "Complexity of Control", controlling the writing quality of aircraft maintenance technical orders is necessary to preserve writing quality.
3) Establish a matrix of the first key factors, namely item (4) factors, as shown in Table 5.
4) In Table 5, the contradiction matrix shows viable inventive principles as numbers 1, 2, 19, 21, 27, and 31.
5) Among the six kinds of inventive principles solutions noted above, the following measures for striving for composition quality and establishing management control were identified:

    1) Heading 1 is Segmentation principle: Virtually or actually segment the system into ordered parts in order to remove, or incorporate, elements that have a harmful, or useful, character. This principle provides measures for "applied training of maintenance technical orders writers in dividing different systems."
    2) Heading 2 is Extraction. Principle: Separate harmful or beneficial elements and characteristics from an entire system. The extraction method can be imaginary or actual (i.e., virtual or applied). This principle provides measures to ensure that "technical order writers possess full comprehension of the detailed function of all components in a system."
    3) Heading 19 is Periodic Action. Principle: Alternate the method of application to achieve the desired result. This principle provides measures for scheduling "periodic lectures and training about aviation systems for technical authors."
    4) Heading 21 is the Rushing Through principle: If some matter yields errors at a certain speed, speed it up. This principle provides measures for "assessing the level of comprehension that technical authors have or should have about a system."
    5) Heading 27 is Disposal principles: Use less expensive, simpler, or easier objects to lower losses and increase convenience and durability. This principle provides measures for "dismissing authors with inferior assessments."
    6) Heading 31 is Porous Material principles: Alternate the state of gases, fluids, and solids which will alter objects with holes. This principle has no special connection to the item (4) factor the goals for improvement and so has no use in this context.

Table 5. The contradiction matrix for the(4) key factors.

Optimizable Parameters \ Stable Parameters		Complexity of the System	Complexity of Control
		Heading 36	Heading 37
Has Adverse Side Effects	Heading 31	1 19 31	12 21 27

CONCLUSION

Aircraft maintenance technical orders contain instructions for technicians to follow during maintenance procedures. Editing of written orders thus directly affects quality and reliability. This study used DMAIC Six Sigma Process to upgrade procedures in gradually enhancing quality of order compilation and establishing systematic operational management. The authors began with definitions, first ascertaining the number of unreadable sentences in technical orders as a Satisfaction Index, then reviewing content of technical orders to count sentences that followed "Warning" and "Caution" as Safety Importance Indices. Our measurement phase depended on operators' Satisfaction Index, as well as the Safety Importance Index, to determine the location of a Performance Evaluation Matrix to gauge editing quality. Analytic methodology involved a Cause-Effect diagram and pinpointing technical orders' high Safety Importance versus low Satisfaction Indices. This was used to evaluate readability by identifying its main factors and secondary factors. The improvement stage employed quality function display technology to derive key factors from secondary factors. Finally, the control phase used TRIZ creativity as a problem-solving procedure in searching for ways to augment editing quality, as well as establish better control systems for maintenance operators to ensure better results. Handling problems of the editing quality of technical orders by systematizing procedures will mean step-by-step improvement in execution, improving the quality of aircraft maintenance, a key method in raising aviation safety.

ACKNOWLEDGMENTS

The authors would like to thank Aeronautical Research Laboratory of the Chung Shan Institute of Science and Technology and Aerospace Industrial Development Corporation for the support from both institutes for this research and thanks National Science Council under Grant NSC-96-2221-E- 270-004.

REFERENCES

Brue, G. and Launsby, R. G., 2003. *Design for Six SIGMA*, McGraw-Hill Companies, Inc.

Genrich Altshuller, 1996. *And Suddenly the Inventor Appeared: TRIZ, the Theory of Inventive Problem Solving,* Natalie Dronova Uri Urmanchev.

Hsia, T. C., Chen, H. T., and Chen, W. H., 2008. *"Measuring the Readability Performance (RP) of Aircraft Maintenance Technical Orders by Fuzzy MCDM Method and RP Index,"* Quality & Quantity, Quality & Quantity, Vol. 42, pp. 795-807.

Hsia, T. C., 2007. *"Evaluating the Writing Quality of Aircraft Maintenance Technical Orders Using a Quality Performance Matrix,"* International Journal of Industrial Ergonomics, Vol. 37, Issue 7, pp. 605-613.

Krause, S. S., 2003. *Aircraft safety: accident investigations, analyses, and applications*, 2nd ed., McGraw Hill, New York, NY.

Sanders, M. S., and McCormick, E. J., *Human Factors in Engineering and Design*, 7th Ed., New York: McGraw-Hill, Inc., 1993.

Thomas, P., *The Six Sigma Project Planner*, New York: McGraw-Hill, 2003.

VDT Ergonomic Requirements Set: ISO 9241 / ISO9241, 2001 The ISO Standards Bookshop.

In: Advances in Mechanical Engineering Research, Volume 3 ISBN: 978-61209-243-0
Editor: David E. Malach

Chapter 10

MEMS FABRICATION AT CMP: PAST AND PRESENT

Bernard Courtois**, Gregory Di Pendina and Kholdoun Torki
CMP, 46 Avenue Felix Viallet, 38031 Grenoble Cedex – France

ABSTRACT

Infrastructures to provide access to custom integrated hardware manufacturing facilities are important because they allow Students and Researchers to access professional facilities at a reasonable cost, and they allow Companies to access small volume production, otherwise difficult to obtain directly from manufacturers. This paper is reviewing the developments at CMP to offer various types of MEMS manufacturing to Students, Researchers and Companies since 1994. CMP has been the first service of its type to introduce MEMS fabrication. Today CMP is offering bulk micromachining on CMOS and GaAs, and various MEMS processes. CAD tools provided by CMP are also reviewed.

I. FIRST MEMS PROCESS OFFERED BY CMP FOR PROTOTYPING

CMP has been involved for a long time in MEMS (Micro Electro Mechanical Systems) prototype fabrication. CMP started to offer MEMS manufacturing in 1994 with bulk micro machining, which is a volume etching process. This etching is a post process done at die level, based on a wet etching using a TMAH recipe that removes the silicon. This means that all the dielectrics on all areas around the suspended structures need to be removed. This MEMS fabrication was based on a 0.8μm CMOS process from austriamicrosystems foundry, using 2 metal layers and 2 polysilicon layers. An example of structure is shown on the fig. 1. At this moment, the biggest advantage was to merge electronics and mechanics on the same die, in order to have the most powerful sensor or actuator management. All parameters extracted from the MEMS was analysed and treated directly and locally on the integrated circuit. CMP started in the same time to improve packaging methods in order to test these

* Bernard.Courtois@imag.fr

MEMS easily. Lots of projects have been made and structures like matrix micro mirror, oscillators, bridges and membranes have been realised using this process and post process.

Figure 1. 0.8µm CMOS / bulk micromachining.

II. Specific MEMS Processes Available at CMP

In 1999, CMP introduced in his portfolio all the MUMPs® family (Multi User MEMS Processes) processes, offered by the foundry MEMSCAP. These processes, which are still available at CMP for prototyping, are specific MEMS processes offering the possibility to realise complex, wide and thick structures, on 1 cm^2 dies. They are 3 of them:

PolyMUMPS process, which is a Polysilicon / Gold surface micromachining, using sacrificial layers to suspend structures for which the fig. 2 shows the process cross section. This process is composed of 7 physical layers, 3 polysilicon layers and 1 metal layer. These polysicon layers are grown during the process and removed at the post process step. Most of the time, processes using sacrificial layers are used to make capacitive devices. Prototypes like microphone, sensors, accelerometers, micro fluidics and display technology devices for example have been also fabricated.

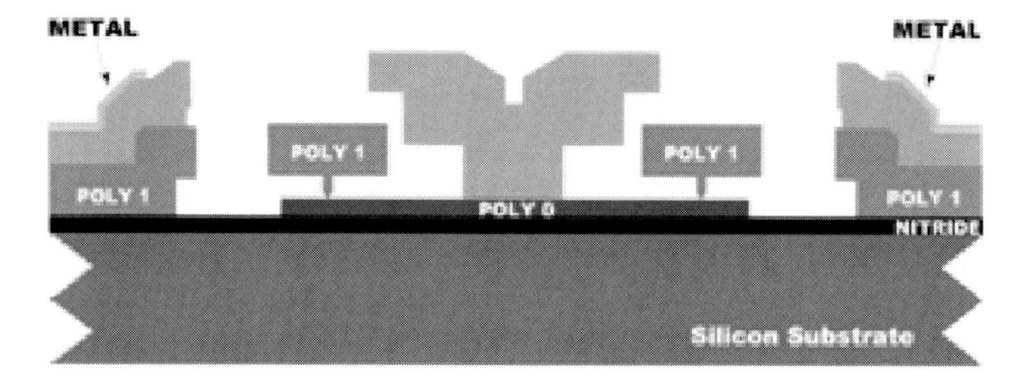

Figure 2. PolyMUMPS cross section.

SOIMUMPs process uses the DRIE etching (Deep Reactive Ion Etch) on SOI (Silicon On Insulator). This process enables to etch both front side and also back side of the wafer, to completely suspend the structures. The fig. 3 shows the cross section of this process that enables structures from 10µ to 25µm thick, offering 2 metal layers. Applications are gyros, optical devices and display technology devices.

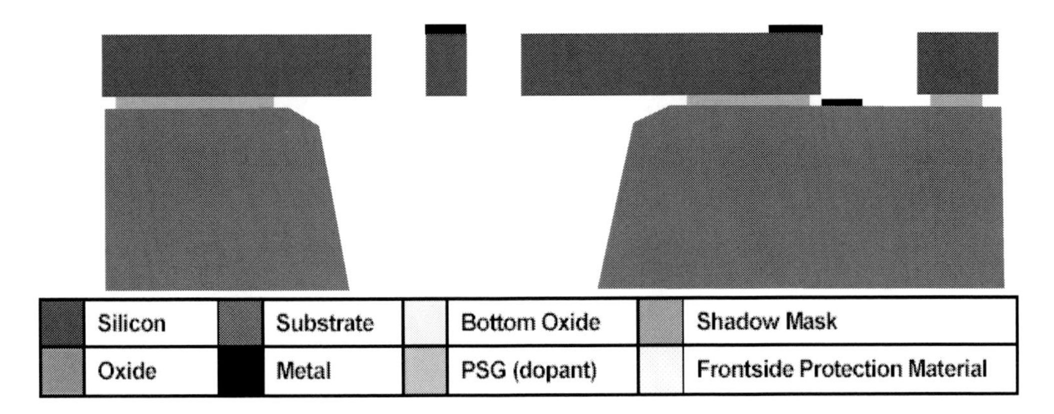

Figure 3. SOIMUMPS cross section.

MetalMUMPs process, which uses thick nickel electroplated layer, from 18 to 22µm, as illustrated on the fig. 4. On this process the substrate is etched front side releasing thick structures. Devices in RF (Radio Frequency), micro fluidics have been done, as well as relays and magnetic switches.

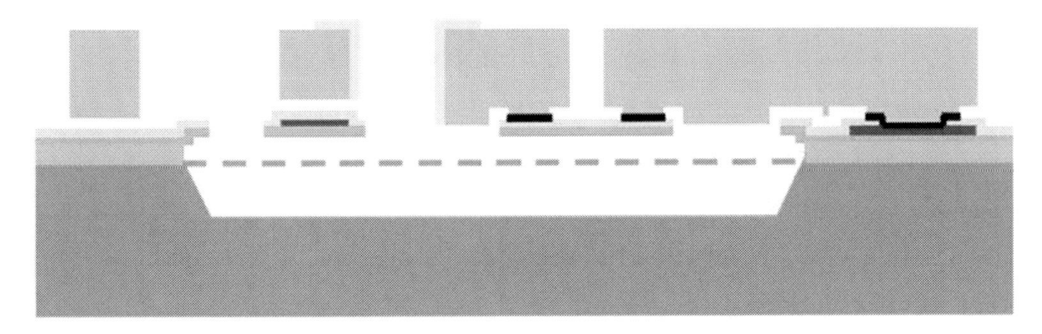

Figure 4. Metal *MUMPS* cross section.

III. Low Cost Bulk Micromachining Process

In 2006 CMP offered the possibility to fabricate MEMS on a low-cost 0.6µ CMOS process, from the foundry CSMC, which has 2 poly and 2 metal layers. This went in replacement of the 0.8µm process that was available only in mono project wafers and no more on multi project wafers, due to the fab migration to 200 mm wafers. Structures are also released after the circuit fabrication with a wet etching using a TMAH recipe to etch the silicon as well. Systems like comb drives or sensors have been made. The fig. 5 shows test structures giving the minimum and maximum structures width and also a piezoresistive sensors fabricated in 2006. Since this process stopped, this micromachining solution is no more available.

Figure 5. 0.6μm CMOS / bulk micromachining.

IV. Bulk Micromaching Post Process on BiCMOS / RF Process

In 2006, CMP adds to his portofolio a bulk micromachining possibility, based on the 0.25μm BiCMOS process from STMicroelectronics. This BiCMOS7RF process has 5 metal layers with top thick metal, vertical NPN with Ft = 55 GHz and is convenient for RF designs. It includes MIM capacitors, inductors and bipolar components. The associated post process is called ASIMPS (Application-Specific Integrated MEMS Process Service) and is made at the Carnegie Mellon University (CMU). Mechanical structures are released by RIE (Reactive Ion Etch) and then by DRIE (Deep Reactive Ion Etch as illustrated in fig. 6). Potential devices to be designed and fabricated in the process include accelerometers, gyroscopes, radio frequency (RF) MEMS communication systems (with resonator oscillators, RF filter and High-Q inductors), infrared sensors and imagers, electro thermal converters and force sensors. Some researches are currently made to develop a “bioacoustic membrane” gravimetric biosensor. This is a chip-based biosensor that is aimed at macromolecular targets.

The technology enables integration of multiple devices on the same chip. For example, high-Q inductors and micromechanical resonators can be combined for CMOS RF application. In another example, multiple accelerometers are integrated on chip to create a 3-axis inertial measurement system. Furthermore, both the communications and accelerometer systems can be combined to form a wireless micro sensor system. Some application in the biomed have been done, such as a bone implantable stress sensor, with the aim to measure with safe the bone strength. As shown on fig 7, high resolution structures can be realised.

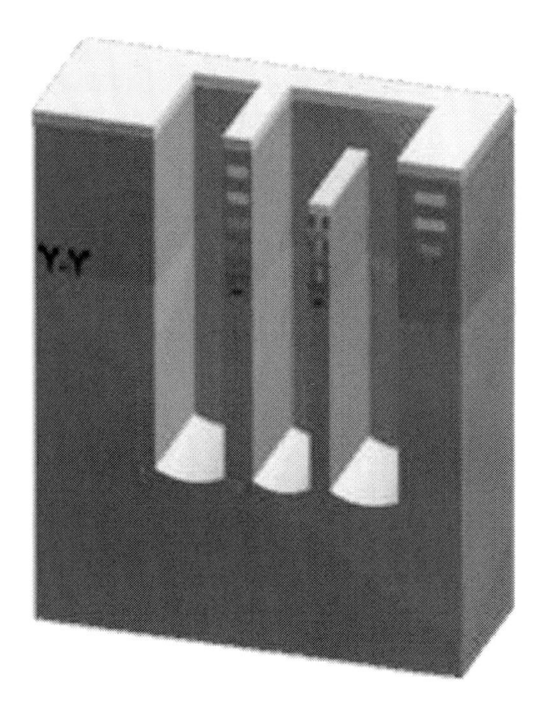

Figure 6. ASIMPS overview.

Figure 7. High resolution structure.

V. Ultra Planar MEMS Process

In 2009, CMP introduces the SUMMiT V^{TM} (Sandia Ultra planar Multi level MEMS Technology V) from Sandia. This advanced process uses 5 polysilicon layers, all planarized, as illustrated on the cross section on fig. 8. This process offers flexibility and gives a mechanical robustness in the devices. Systems like comb actuators, meshing gears and transmissions dynamometers, laminated support springs, steam engines, micro engines and micro machines, motors, mirrors and optical encoders, micro sensors, RF MEMS and linear racks can be fabricated. In the BioMed application, some works are focused on a system that may allow large samples to be handled and processed in micro channel devices that are made in sheet and rolled or stacked. Also, a component that separates nucleic acids by size has already been made.

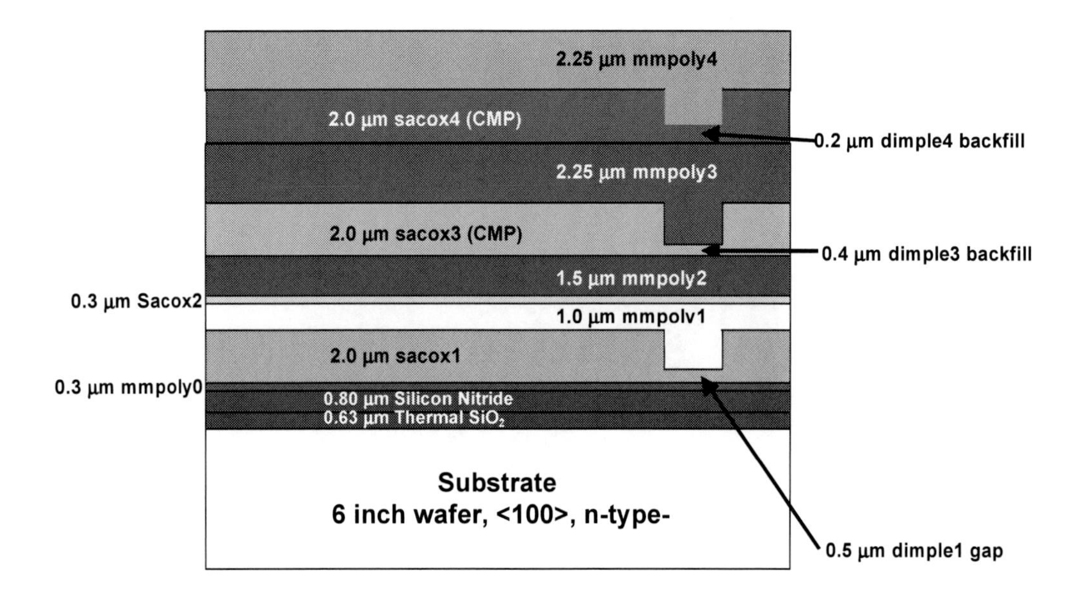

Figure 8. SUMMiT V cross section.

VI. Advanced Bulk Micromachining Process

In 2009, CMP developed a combination process / post process to replace the last bulk micromachining no more available. This one is based on the 0.35µm CMOS process from austriamicrosystems. The post process is still a wet etching with a TMAH recipe. This IC process offers 4 metal layers, 2 polysilicon layers, high resistive poly, 3.3V and 5V power supply and opto option. A first validation prototype has been manufactured in Q4 2008. Using these results, a second prototype has been fabricated Q2 2009, using an additional mask to etch stacked oxide butting MEMS structures. Results showed that both oxide and silicon have been fully etched, so the structures are well suspended and movable. This post process is available for all MPW 0.35µm CMOS MPW scheduled by CMP.

Figure 9. 0.35μm CMOS / bulk micromachining.

CONCLUSION

CMP recognized a long time ago the importance of MEMS. Since the introduction of MEMS manufacturing, many projects have been successfully manufactured for many various applications. The portfolio has been continuously expanding. It is planned to introduce soon NanoMUMPs.

INDEX

C

D

E

F

J

K

L

M

N

O

P

Q

R

S

T

U

V

W

Y